Innovations in Agricultural & Biological Engineering

TECHNOLOGICAL PROCESSES FOR MARINE FOODS, FROM WATER TO FORK

Bioactive Compounds, Industrial Applications and Genomics

Edited by

Megh R. Goyal, PhD, P.E.
Hafiz Ansar Rasul Suleria, PhD
Shanmugam Kirubanandan, MTech

APPLE
ACADEMIC
PRESS

Apple Academic Press Inc.	Apple Academic Press Inc.
3333 Mistwell Crescent	1265 Goldenrod Circle NE
Oakville, ON L6L 0A2	Palm Bay, Florida 32905
Canada USA	USA

© 2020 by Apple Academic Press, Inc.

First issued in paperback 2021

Exclusive worldwide distribution by CRC Press, a member of Taylor & Francis Group

No claim to original U.S. Government works

ISBN 13: 978-1-77463-436-3 (pbk)
ISBN 13: 978-1-77188-758-8 (hbk)

CIP data on file with Canada Library and Archives

Library of Congress Cataloging-in-Publication Data

Names: Goyal, Megh Raj, editor. | Suleria, Hafiz, editor. | Kirubanandan, Shanmugam, editor.

Title: Technological processes for marine foods-from water to fork : bioactive compounds, industrial applications and genomics / editors, Megh R. Goyal, Hafiz Ansar Rasul Suleria, Shanmugam Kirubanandan.

Description: Toronto ; New Jersey : Apple Academic Press, 2019. | Series: Innovations in agricultural and biological engineering | Includes bibliographical references and index. |

Identifiers: LCCN 2019017440 (print) | LCCN 2019018489 (ebook) | ISBN 9780429425271 () | ISBN 9781771887588 (hardcover : alk. paper) | ISBN 9780429425271 (ebook)

Subjects: | MESH: Seafood | Food-Processing Industry | Biological Products | Aquatic Organisms | Nutritive Value

Classification: LCC TP248.65.F66 (ebook) | LCC TP248.65.F66 (print) | NLM TP 248.65.F66 | DDC 664--dc23

LC record available at https://lccn.loc.gov/2019017440

Apple Academic Press also publishes its books in a variety of electronic formats. Some content that appears in print may not be available in electronic format. For information about Apple Academic Press products, visit our website at **www.appleacademicpress.com** and the CRC Press website at **www.crcpress.com**

TECHNOLOGICAL PROCESSES FOR MARINE FOODS, FROM WATER TO FORK

Bioactive Compounds, Industrial Applications and Genomics

EDITORIAL

Under the book series titled *Innovations in Agricultural and Biological Engineering*, Apple Academic Press Inc., (AAP) is publishing volumes in the specialty areas defined by the American Society of Agricultural and Biological Engineers (asabe.org) over a span of several years. AAP wants to be the principal source of books in agricultural and biological engineering.

The mission of this series is to provide knowledge and techniques for agricultural and biological engineers (ABEs). The series offers high-quality reference and academic content in agricultural and biological engineering (ABE) that is accessible to academicians, researchers, scientists, university faculty, university-level students, and professionals around the world.

ABEs ensure that the world has the necessities of life including safe and plentiful food, clean air and water, renewable fuel and energy, safe working conditions, and a healthy environment by employing knowledge and expertise of sciences, both pure and applied, and engineering principles. Biological engineering applies engineering practices to problems and opportunities presented by living things and the natural environment in agriculture.

We welcome book proposals from the readers in the area of their expertise.

For this book series, we welcome chapters on the following specialty areas (but not limited to):

1. Academia to industry to end-user loop in agricultural engineering
2. Agricultural mechanization
3. Aquaculture engineering
4. Biological engineering in agriculture
5. Biotechnology applications in agricultural engineering
6. Energy source engineering
7. Food and bioprocess engineering
8. Forest engineering
9. Hill land agriculture
10. Human factors in engineering
11. Information and electrical technologies
12. Irrigation and drainage engineering
13. Nanotechnology applications in agricultural engineering
14. Natural resources engineering

15. Nursery and greenhouse engineering
16. Potential of phytochemicals from agricultural and wild plants for human health
17. Power systems and machinery design
18. GPS and remote sensing potential in agricultural engineering
19. Robot engineering in agriculture
20. Simulation and computer modeling
21. Smart engineering applications in agriculture
22. Soil and water engineering
23. Structures and environment engineering
24. Waste management and recycling
25. Any other focus area

For more information on this series, readers may contact:

Ashish Kumar, Publisher and President
Sandra Jones Sickels, Vice President
Apple Academic Press, Inc.,
Fax: 866-222-9549
Email: ashish@appleacademicpress.com
http://www.appleacademicpress.com

Megh R. Goyal, PhD, PE
Book Series Senior Editor-in-Chief
Innovations in Agricultural and Biological Engineering
E-mail: goyalmegh@gmail.com

OTHER BOOKS ON AGRICULTURAL & BIOLOGICAL ENGINEERING BY APPLE ACADEMIC PRESS, INC.

Management of Drip/Trickle or Micro Irrigation
Megh R. Goyal, PhD, PE, Senior Editor-in-Chief

Evapotranspiration: Principles and Applications for Water Management
Megh R. Goyal, PhD, PE, and Eric W. Harmsen, Editors

Book Series: Research Advances in Sustainable Micro Irrigation
Senior Editor-in-Chief: Megh R. Goyal, PhD, PE

Volume 1: Sustainable Micro Irrigation: Principles and Practices
Volume 2: Sustainable Practices in Surface and Subsurface Micro Irrigation
Volume 3: Sustainable Micro Irrigation Management for Trees and Vines
Volume 4: Management, Performance, and Applications of Micro Irrigation Systems
Volume 5: Applications of Furrow and Micro Irrigation in Arid and Semi-Arid Regions
Volume 6: Best Management Practices for Drip Irrigated Crops
Volume 7: Closed Circuit Micro Irrigation Design: Theory and Applications
Volume 8: Wastewater Management for Irrigation: Principles and Practices
Volume 9: Water and Fertigation Management in Micro Irrigation
Volume 10: Innovation in Micro Irrigation Technology

Book Series: Innovations and Challenges in Micro Irrigation
Senior Editor-in-Chief: Megh R. Goyal, PhD, PE

Volume 1: Principles and Management of Clogging in Micro Irrigation
Volume 2: Sustainable Micro Irrigation Design Systems for Agricultural Crops: Methods and Practices
Volume 3: Performance Evaluation of Micro Irrigation Management: Principles and Practices

Volume 3: Potential of Solar Energy and Emerging Technologies in
Sustainable Micro Irrigation

Volume 4: Potential of Solar Energy and Emerging Technologies in
Sustainable Micro Irrigation

Volume 5: Micro Irrigation Management: Technological Advances and
Their Applications

Volume 6: Micro Irrigation Engineering for Horticultural Crops

Volume 7: Micro Irrigation Scheduling and Practices

Volume 8: Engineering Interventions in Sustainable Trickle Irrigation:
Water Requirements, Uniformity, Fertigation, and Crop
Performance

Volume 9: Management Strategies for Water Use Efficiency and Micro
Irrigated Crops: Principles, Practices, and Performance

Book Series: Innovations in Agricultural & Biological Engineering
Senior Editor-in-Chief: Megh R. Goyal, PhD, PE

- Dairy Engineering: Advanced Technologies and Their Applications
- Developing Technologies in Food Science: Status, Applications, and
Challenges
- Engineering Interventions in Agricultural Processing
- Engineering Practices for Agricultural Production and Water
Conservation: An Inter disciplinary Approach
- Emerging Technologies in Agricultural Engineering
- Flood Assessment: Modeling and Parameterization
- Food Engineering: Emerging Issues, Modeling, and Applications
- Food Process Engineering: Emerging Trends in Research and Their
Applications
- Food Technology: Applied Research and Production Techniques
- Modeling Methods and Practices in Soil and Water Engineering
- Processing Technologies for Milk and Dairy Products: Methods
Application and Energy Usage
- Soil and Water Engineering: Principles and Applications of Modeling
- Soil Salinity Management in Agriculture: Technological Advances and
Applications
- Technological Interventions in the Processing of Fruits and Vegetables
- Technological Interventions in Management of Irrigated Agriculture
- Engineering Interventions in Foods and Plants
- Technological Interventions in Dairy Science: Innovative Approaches in
Processing, Preservation, and Analysis of Milk Products

- Novel Dairy Processing Technologies: Techniques, Management, and Energy Conservation
- Sustainable Biological Systems for Agriculture: Emerging Issues in Nanotechnology, Biofertilizers, Wastewater, and Farm Machines
- State-of-the-Art Technologies in Food Science: Human Health, Emerging Issues and Specialty Topics
- Scientific and Technical Terms in Bioengineering and Biological Engineering
- Engineering Practices for Management of Soil Salinity: Agricultural, Physiological, and Adaptive Approaches
- Processing of Fruits and Vegetables: From Farm to Fork
- Technological Processes for Marine Foods, From Water to Fork: Bioactive Compounds, Industrial Applications, and Genomics
- Engineering Practices for Milk Products: Dairyceuticals, Novel Technologies, and Quality
- Nanotechnology and Nanomaterial Applications in Food, Health, and Biomedical Sciences
- Nanotechnology Applications in Dairy Science: Packaging, Processing, and Preservation

ABOUT THE SENIOR EDITOR-IN-CHIEF

Megh R. Goyal, PhD, P.E.

Retired Professor in Agricultural and Biomedical Engineering, University of Puerto Rico, Mayaguez Campus; Senior Acquisitions Editor, Biomedical Engineering and Agricultural Science, Apple Academic Press, Inc.

Megh R. Goyal, PhD, PE, is a Retired Professor in Agricultural and Biomedical Engineering from the General Engineering Department in the College of Engineering at the University of Puerto Rico–Mayaguez Campus; and Senior Acquisitions Editor and Senior Technical Editor-in-Chief in Agriculture and Biomedical Engineering for Apple Academic Press, Inc. He has worked as a Soil Conservation Inspector and as a Research Assistant at Haryana Agricultural University and Ohio State University.

During his professional career of 45 years, Dr. Goyal has received many prestigious awards and honors. He was the first agricultural engineer to receive the professional license in Agricultural Engineering in 1986 from the College of Engineers and Surveyors of Puerto Rico. In 2005, he was proclaimed as "Father of Irrigation Engineering in Puerto Rico for the Twentieth Century" by the American Society of Agricultural and Biological Engineers (ASABE), Puerto Rico Section, for his pioneering work on micro irrigation, evapotranspiration, agroclimatology, and soil and water engineering. The Water Technology Centre of Tamil Nadu Agricultural University in Coimbatore, India, recognized Dr. Goyal as one of the experts "who rendered meritorious service for the development of micro irrigation sector in India" by bestowing the Award of Outstanding Contribution in Micro Irrigation. This award was presented to Dr. Goyal during the inaugural session of the National Congress on "New Challenges and Advances in Sustainable Micro Irrigation" on March 1, 2017, held at Tamil Nadu Agricultural University. Dr. Goyal is slated to receive the Netafim Award for Advancements in Microirrigation: 2018 from the American Society of Agricultural Engineers at the ASABE International Meeting in August 2018.

A prolific author and editor, he has written more than 200 journal articles and textbooks and has edited over 59 books. He is the editor of three book series published by Apple Academic Press: Innovations in Agricultural & Biological Engineering, Innovations and Challenges in Micro Irrigation, and Research Advances in Sustainable Micro Irrigation. He is also instrumental in the development of the new book series Innovations in Plant Science for Better Health: From Soil to Fork.

Dr. Goyal received his BSc degree in engineering from Punjab Agricultural University, Ludhiana, India; his MSc and PhD degrees from Ohio State University, Columbus; and his Master of Divinity degree from Puerto Rico Evangelical Seminary, Hato Rey, Puerto Rico, USA.

ABOUT THE CO-EDITOR

Hafiz Ansar Rasul Suleria, PhD

Hafiz Anasr Rasul Suleria, PhD, is currently working as the Alfred Deakin Research Fellow at Deakin University, Melbourne, Australia. He is also an Honorary Fellow in the Diamantina Institute, Faculty of Medicine, The University of Queensland, Australia.

Recently he worked as Postdoc Research Fellow in the Department of Food, Nutrition, Dietetic and Health at Kansas State University, USA.

Previously, he has been awarded an International Postgraduate Research Scholarship (IPRS) and Australian Postgraduate Award (APA) for his PhD research at UQ School of Medicine, the Translational Research Institute (TRI) in collaboration with Commonwealth and Scientific and Industrial Research Organization (CSIRO, Australia).

Before joining the UQ, he worked as a Lecturer in the Department of Food Sciences, Government College University Faisalabad, Pakistan. He also worked as a Research Associate in the PAK-US Joint Project funded by the Higher Education Commission, Pakistan, and Department of State, USA, with the collaboration of the University of Massachusetts, USA, and National Institute of Food Science and Technology, University of Agriculture Faisalabad, Pakistan.

He has a significant research focus on food nutrition, particularly in the screening of bioactive molecules—isolation, purification, and characterization using various cutting-edge techniques from different plant, marine, and animal source, and *in vitro*, *in vivo* bioactivities, cell culture, and animal modeling. He has also done a reasonable amount of work on functional foods and nutraceutical, food and function, and alternative medicine.

Dr. Suleria has published more than 50 peer-reviewed scientific papers in different reputed/impacted journals. He is also in collaboration with more than ten universities where he is working as a co-supervisor/special member for PhD and postgraduate students and is also involved in joint publications, projects, and grants. He is Editor-in-Chief for the book series on *Innovations in Plant Science for Better Health: From Soil to Fork*, published by AAP. Readers may contact him at: hafiz.suleria@uqconnect.edu.au.

ABOUT THE CO-EDITOR

Shanmugam Kirubanandan, MTech

Shanmugam Kirubanandan, MTech, is a young engineer and researcher in the field of chemical engineering and biotechnology. He obtained his BTech (Chemical Engineering) from the University of Madras in 2001 and his MTech in Biopharmaceutical Technology from the Centre for Biotechnology, Anna University, Chennai, India, in 2006 and his MASc (Chemical Engineering) from Dalhousie University, Halifax, NS, Canada, in 2015.

His MASc thesis is titled "Collagen biomaterials and extraction of Omega 3 PUFA in mini channels." He has published numerous research articles in biomaterials and extraction.

He was working as an Assistant Professor at Sri Venkateswara College of Engineering, Sriperumbdur, India, and Visiting Faculty at Department of Chemical Engineering at A.C. Tech, Anna University, Chennai, India.

With expertise in the field of chemical engineering, he has worked as a Process Engineer in a well-established private company and performed design of electrochemical reactors for production of various electro-chemicals and operation of anodizing plant and a wastewater treatment plant. In addition to that, he played a significant role in quality control and assurance in the anodizing plant.

He started his research career at the Central Leather Research Institute, India, and has contributed extensively on process development of bio-diesel from non-edible vegetable oils and fats; assisted in the commercialization of biodiesel from rice bran oil; and worked on the application of chemical engineering and biotechnology in leather processing. While working on his MTech research, he developed collagen and keratin biomaterial for wound care management in the Biomaterial Division of the Central Leather Research Institute, Chennai, India. He also worked on the research on recovery of value-added products from slaughter waste to develop various biomaterials for construction of drug delivery vehicles and scaffolds for tissue engineering.

He has applied the concepts of nanotechnology in the construction of a tissue engineering scaffold for drug delivery to the cartilage tissue repair at the Indian Institute of Technology Kanpur (India). During this tenure, he has also contributed in the area of scaffold development and biosurface modification for the immolation of bioactive proteins delivery to the damaged tissue site.

Teaching is an integral part of his career, and he has taken a challenging teaching assignment at the Department of Biotechnology, Sri Venkateswara College of Engineering and Department of Chemical Engineering; and A.C. Tech, Anna University, India. He has mentored many undergraduate and postgraduate student projects in the field of biochemical and bioprocess engineering. He primarily worked on the collagen with other natural polymers for developing a scaffold for dendritic culture and supported many consultancy projects in the field of wastewater treatment in metal finishing industries.

During his tenure at Dalhousie University, Halifax, Canada, he has performed research on multiphase flows and its applications in chemical engineering and food processing. He developed mini-fluidic flow reactor for extraction of omega-3 PUFA from fish oil and an oscillatory flow reactor in mini scale. In addition to that, he was involved in the extraction of chitin/ chitosan from green crab shells at Saint Mary's University, Halifax, Canada. Presently, he is a PhD Research Scholar at Monash University, Australia, and is working on the cellulose nanofibers for high-performance applications in the field of chemical engineering.

CONTENTS

CONTRIBUTORS

Munawar Abbas
Research Associate, Institute of Home and Food Sciences, Government College University, Faisalabad, Pakistan, Mobile: +92-332-0677170, E-mail: foodian2007@gmail.com

Rai Muhammad Amir
Assistant Professor, Institute of Food and Nutritional Sciences, PMAS-Arid Agriculture University, Rawalpindi – 46000, Pakistan, Tel: +92-(0)-333-6972246, E-mail: raiamir87@yahoo.com

D. Priscilla Mercy Anitha
PhD Research Scholar, Department of Food Process and Engineering, SRM Institute of Science and Technology, Kattankulathur – 603203, Tamil Nadu, India, Mobile: +91-9940560215, E-mail: priscillaanitha18@gmail.com

P. Anand Babu
PhD Research Fellow, Department of Food Process Engineering, School of Bioengineering, SRM University, Kattankulathur – 603203, Tamil Nadu, India, Mobile: +91 8682957434, E-mail: anmicrobiology@gmail.com

Huma Bader-Ul-Ain
Research Associate, Institute of Home and Food Sciences, Government College University, Faisalabad, Pakistan, Mobile: +92-333-1672734, E-mail: humahums@yahoo.com

Megh R. Goyal
Retired Faculty in Agricultural and Biomedical Engineering from the College of Engineering at University of Puerto Rico – Mayaguez Campus; and Senior Acquisitions Editor and Senior Technical Editor-in-Chief in Agricultural and Biomedical Engineering for Apple Academic Press Inc., PO Box 86, Rincon – PR – 006770086, USA, E-mail: goyalmegh@gmail.com

Hafiz Anasr Rasul Suleria
PhD, McKenzie Fellow, Food Nutrition Department of Agriculture and Food Systems, The University of Melbourne, Level 1, 142 Royal Parade, Parkville Victoria, 3010 Australia; Mobile: +61- 470439670; E-mail: hafiz.suleria@unimelb.edu.au

Vidushi Kansal
Student, Department Food Technology Management, National Institute of Food Technology Entrepreneurship and Management (NIFTEM), Kundli, Sonepat – 131028, Haryana, India, Mobile: +91-8683033513, E-mail: vidushiknsl@gmail.com

Sana Khalid
Research Associate, Department of Pharmaceutical Sciences, Government College University Faisalabad, Pakistan, Mobile: +92-334-4709423, E-mail: sanakhalid436@gmail.com

Shanmugam Kirubanandan
PhD Research Scholar, BioPRIA, Department of Chemical Engineering, Monash University, Australia, Mailing Address: Unit 7, 9, Kelvinside Road, Noble Park 3174, Victoria, Australia, Telephone: +61 4 04570457, Mobile: +61-4-2080-4215, E-mail: skirubanandan80@gmail.com

Kshitiz Kumar
Assistant Professor, Department of Food Engineering, National Institute of Food Technology Entrepreneurship and Management (NIFTEM), Kundli, Sonepat – 131028, Haryana, India, E-mail: kshitizyp@gmail.com

M. Mahesh Kumar
Assistant Professor, Department of Food Process Engineering, School of Bioengineering, SRM
University, Kattankulathur – 603203, Tamil Nadu, India, Mobile: +91-9884278027,
E-mail: mk1591983@gmail.com

Glindya Bhagya Lakshmi
PhD Research Scholar, School of Biosciences, Mahatma Gandhi University, Priyadarsini Hills P. O.,
Kottayam 686560, Kerala, India, Mobile: +91-9745534258, E-mail: bhagyabalan@gmail.com

Anam Latif
Human Nutrition and Dietetics, National Institute of Food Science and Technology, Faculty of Food,
Nutrition and Home Sciences, University of Agriculture, Faisalabad-38040, Pakistan,
Tel: +92 (0) 312 9947820, E-mail: anamlatif101@gmail.com

Hilal A. Makroo
PhD Research Scholar, Department of Food Engineering and Technology, Tezpur University, Napaam,
Sonitpur, Assam – 784028, India, Mobile: +91-9435285672, E-mail: hilalmakroo@gmail.com

Ida Idayu Muhamad
Professor, Department of Bioprocess and Polymer Engineering, University Teknologi Malaysia, 81310
Johor Bahru, Johor, Mobile: +60-75535503, E-mail: idaidayu@utm.my

Reshma B. Nambiar
PhD Research Fellow, Department of Food Process Engineering, School of Bioengineering, SRM
University, Kattankulathur – 603203, Tamil Nadu, India, Mobile: +91 9600532156,
E-mail: reshma.reshbn@gmail.com

Radhakrishnan Preetha
Professor, Department of Food Processing, School of Bioengineering, SRM University, Kattankulathur,
Tamil Nadu – 603203, India, E-mail: preetha.r@ktr.srmuniv.ac.in

Kundeti Saranya Chandana Priya
Trainee in MTR Foods Pvt. Ltd., School of Food Technology, Jawaharlal Nehru Technological
University Kakinada (JNTUK), Kakinada, East Godavari District 533003, Andhra Pradesh, India,
Mobile: +91-7893668316, E-mail: saranyabiotech21@gmail.com

Keerthi Thalakattil Raghavan
Professor & Head, School of Biosciences, Mahatma Gandhi University, Priyadarsini Hills P.O.,
Kottayam – 686560, Kerala, India, Mobile: +91-9497655293, E-mail: keerthisureshbabu@gmail.com

Emmanuel Rotimi Sadiku
Professor, Department of Chemical, Metallurgical and Materials Engineering (Polymer Technology
Division), Tshwane University of Technology, Staatsartillerie Road, 0183, Pretoria West Campus, South
Africa. Mobile: +2-7128413255, +2-7722598629, E-mail: sadikur@tut.ac.za and ogidiolu@gmail.com

Sasikanth Sarangam
Assistant Professor, School of Food Technology, Jawaharlal Nehru Technological University Kakinada
(JNTUK), Kakinada, East Godavari (Dist.)-533003, Andhra Pradesh, India, Mobile: +91-8008175975,
E-mail: sarangamsasikanth@gmail.com

Juhi Saxena
PhD Research Scholar, Department of Food Engineering and Technology, Tezpur University, Napaam,
Sonitpur, Assam- 784028, India, Mobile: +91-8876482471, E-mail: juhi.saxena167@gmail.com

S. Periyar Selvam
Assistant Professor, Department of Food Process and Engineering, SRM Institute of Science and
Technology, Kattankulathur – 603203, Tamil Nadu, India, Mobile: +91-9444821490,
E-mail: periyar.india@gmail.com

Vijay Singh Sharanagat
Assistant Professor, Department of Food Engineering, National Institute of Food Technology Entrepreneurship and Management (NIFTEM), Kundli, Sonepat- 131028, Haryana, India, Tel: +91-1302281228, E-mail: vijaysinghs42@gmail.com

Aamir Shehzad
Assistant Professor, National Institute of Food Science and Technology, Faculty of Food, Nutrition and Home Sciences, University of Agriculture, Faisalabad-38040, Pakistan, Tel: +92 (0) 41 9201105, Fax: +92 (0) 41 9201439, E-mail: draamir@uaf.edu.pk

Ramesh Shruthy
Department of Food Processing, School of Bioengineering, SRM University, Kattankulathur, Tamil Nadu-603203, India

Karuna Singh
Assistant Professor, Amity Institute of Food Technology, Amity University, Noida, Sector-125, U.P. 201303, India, Mobile: +91-9810450875, E-mail: ksingh11@amity.edu

Lochan Singh
PhD Research Scholar, National Institute of Food Technology Entrepreneurship and Management (NIFTEM), Kundli, Sonepat – 131028, Haryana, India, E-mail: lochan2626@gmail.com

Vinti Singla
Student, Department of Food Technology Management, National Institute of Food Technology Entrepreneurship and Management (NIFTEM), Kundli, Sonepat – 131028, Haryana, India, Tel: +91-9467850831, E-mail: vintisingla2506@gmail.com

Hafiz Ansar Rasul Suleria
McKenzie Fellow, Department of Agriculture and Food Systems,
The University of Melbourne, Level 3, 780 Elizabeth Street, Parkville, Victoria 3010 Australia; E-mail: hafiz.suleria@unimelb.edu.au

Monika Thakur
Assistant Professor, Amity Institute of Food Technology, Amity University, Noida, Sector-125, U.P. 201303, India, Mobile: +91-9810495426, E-mail: mthakur1@amity.edu

Asna Zahid
Human Nutrition and Dietetics, National Institute of Food Science and Technology, Faculty of Food, Nutrition and Home Sciences, University of Agriculture, Faisalabad-38040, Pakistan, Tel: +92 (0) 323 7755001, E-mail: asna.zahid52@gmail.com

ABBREVIATIONS

AAS	acrylic acid sodium
ACE	angiotensin converting the enzyme
Ad	adjuvant
AGF	antibiotic growth promoter
AIDS	acquired immune deficiency syndrome
ALA	alpha-linolenic acid
ANFs	antinutritional factors
AOAC	Association of Official Analytical Chemists
APF	automatic plate freezers
ARA	arachidonic acid
ATP	adenosine triphosphate
ATPase	adenosine triphosphatase
BHA	butylated hydroxyanisole
BLIS	bacteriocin Like Inhibitory Substance
BOD	biological oxygen demand
BSE	bovine spongiform encephalopathy
Ce	cerium
CFR	code of federal regulations
CFU	colony forming units
CH-chitin	chemically treated chitin
CMC	carboxymethyl chitosan
CMCTS	chitosan carboxymethyl chitosan
CMCTS-g-AAS	chitosan carboxymethyl chitosan-graft-acrylic acid sodium
CMCTS-g-MAAS	chitosan carboxymethyl chitosan-graft-methacrylic acid sodium
CO	cinnamon oil
COD	chemical oxygen demand
$COOCH_3$	carboxymethyl group
COS	chitosan oligosaccharides
cP	centipoise
CS	chitosan
CS-GEO	chitosan ginger essential oil
CS-g-PVCL	chitosan-graft-poly (N-vinylcaprolactam)

CSIRO	Commonwealth Scientific and Industrial Research Organization
DA	degree of *N*-acetylation
DD	degree of deacetylation
DGDG	di glycosyl diacylglycerol
DHA	docosahexaenoic (DHA) acids
DNA	deoxyribonucleic acid
DPA	docosapentaenoic acid
DSW	deep sea water
EAE	enzyme assisted extraction
EPA	eicosapentaenoic acid
EPS	extracellular polymeric substances
EST	expression sequence tag
EU	European Union
Eu	europium
FAO	Food and Agriculture Organization of United States
FCS	fucosylated chondroitin sulphate
FDA	Food and Drug Administration
FFA	free fatty acids
FOG	fat-oil-grease
FOS	fructose oligosaccharides
FOSHU	food for specific health use
FPH	fish protein hydrolysis
GAG	glycosaminoglycan
GIT	gastrointestinal tract
GLA	gamma-linolenic acid
GlcNH2	2-amino-2-deoxy-D-glucose glucosamine
GOS	galactooligosaccharides
GRAS	generally regarded as safe
H_2O_2	hydrogen peroxide
HAV	hepatitis A virus
HCl	hydrochloric acid
HE	hydroxyethyl
HIV	human immunodeficiency virus
HPC	hydroxypropyl chitosan
HPF	horizontal plate freezers
HTCC	N-(2-hydroxyl) propyl-3-trimethyl ammonium chitosan chloride
HTEC	N-(2-hydroxyl) propyl-3-triethyl ammonium chitosan chloride

IARC	International Agency for Research on Cancer
IHNV	hematopoietic necrosis virus
ISAV	infectious salmon anemia disease
IU	International Units
IUPAC	International Union of Pure and Applied Chemistry
KWE	kelp waste extract
La	lanthanum
LAB	lactic acid bacteria
LCPUFA	long-chain polyunsaturated fatty acids
LDCs	low digestible oligosaccharides
LMW	low molecular weight
LNF	liquid nitrogen freezers
M	molecular weight (g/mol)
M.O.	microorganisms
MAAS	methacrylic acid sodium
MAE	microwave assisted extraction
MGDG	monoglycosyl diacylglycerol
MMT	metric million tons
MO-chitin	microbiologically treated chitin
MOP	marine oligosaccharide preparation
MOP	method of preparation
MOS	mannan oligosaccharides
mRNA	messenger ribonucleic acid
MUFA	monounsaturated fatty acids
MW	molecular weight
NAG	N-acetyl glucosamine
NaOH	sodium hydroxide
NDO	non-digestible oligosaccharides
NHPDCS	N-(2-hydroxyphenyl) – N, N-dimethyl chitosan
NO	nitric oxide
NoV G	norovirus genogroup
NP	nanoparticles
OMP38	outer membrane protein38
ompK	outer membrane protein K
P, R	shape factors in plank's model
P-CS	phosphorylated chitosan
pDNA	plasmid DNA
pEGFP-N2-OMPK	eukaryotic expression vector and outer membrane protein K

PLE	pressurized liquid extraction
PLs	phospholipids
Pr (III)	praseodymium
PS	potassium sorbate
PUFAs	polyunsaturated fatty acids
pVAOMP38	DNA vaccine-pcDNA 3.1 vector and outer membrane protein38
RAAS	renin-angiotensin–aldosterone system
RBL	rhamnose-binding lectin
RNA	ribonucleic acid
SC-CO$_2$	supercritical carbon dioxide extraction
SCFE	supercritical fluid extraction
SCO	short chain oligosaccharides
SDS-PAGE	sodium dodecyl sulfate-polyacrylamide gel electrophoresis
SFA	saturated fatty acids
SL	sodium lactate
SMF	submerged fermentation
SOS	soybean oligosaccharides
SSF	solid-state fermentation
SVCV	spring viremia of carp virus
SWBC	seaweed waste biomass
SWWP	surimi wash water protein
T	temperature (K or °C)
Tcf	trillion cubic feet
TGase	transglutaminase
TMA-N	trimethylamine nitrogen
TMC	N,N,N-trimethyl chitosan
TSS	total suspended solids
TVB-N	total volatile basic nitrogen
UAE	ultrasound assisted extraction
UNIDO	United Nation Industrial Development Organization
USD	United States Dollar
VS	volatile solids
WHC	water holding capacity
WHO	World Health Organization
WSSV	white spot syndrome virus
XOS	xylooligosaccharide
Yb	ytterbium

SYMBOLS

γ_w	activity coefficient of water
τ_j	molecular dissociation of compound j in food material
$°C$	Degree Celsius
ΔH_{av}	average latent of fusion
ΔT_F	freezing point depression
a_w	thermodynamic activity of water
B	ratio of the mass of unfreezable water to mass of dry solids in the food
C	specific heat (kJ/kg°C)
j	the jth component of the food material
j_0	the jth component of the food material at the initial freezing point
k	thermal conductivity (W/m°C)
m_x	weight fraction of x-component in food material
n	number of components in a food
R	universal gas constant (= 8.314 kJ/kg mol K)
w_0	water at the initial freezing point of a food material
X	mass fraction
λ_w	latent heat of fusion (J/g)
τ	molecular dissociation
ω-3	omega-3 fatty acid
ω-6	omega-6 fatty acid

PREFACE

To be healthy, it is our moral responsibility,
Towards Almighty God, ourselves, and our family;
Eating fruits and vegetables makes us healthy,
Believe and have faith;
Reduction of food waste can reduce world hunger and
can make our planet eco – friendly.

—Megh R. Goyal

Marine ecosystems are an excellent source of foods and nutraceuticals nowadays. The marine source is an attractive feedstock for the development of drugs, potential foods, and biomaterials. There are notable marine products, such as omega-3 polyunsaturated fatty acids and chitin/chitosan, which are used as drugs and in the development of biomaterials and tissue engineering scaffolds in biomedical engineering. There is a need for substantial scientific evidence regarding efficacy, dosage, and safety for traditional medicine to have a place in modern medicine.

Fourteen scientific reports on therapeutic values of different marine sources against diseases are presented in this book volume, titled *Technological Processes for Marine Foods, From Water to Fork: Bioactive Compounds, Industrial Applications, and Genomics*. The book covers the bioactive compounds and health-promoting potentials of marine sources against different ailments such as diabetes, hypertension, and microbial infections. Mechanisms of health benefits of bioactive compounds are also discussed. This book aims to further encourage the need for the development of marine-based drugs through innovative and groundbreaking research studies. The understanding of the therapeutic values of these bioactive molecules will also help us to improve their sustainability, as people and governments will be encouraged to preserve and conserve the marine sources for future generations. This book can be a valuable reference for postgraduate students in food biotechnology and food processing.

The contribution of all cooperating authors to this book volume has been most valuable in the compilation. Their names are mentioned in each chapter and in the list of contributors. We appreciate you all for having patience with our editorial skills. This book would not have been written without the valuable cooperation of these investigators, many of whom are renowned scientists who have worked in the field of food engineering and food science throughout their professional careers.

We will like to thank the editorial staff, Sandy Jones Sickels, Vice President, and Ashish Kumar, Publisher and President, at Apple Academic Press, Inc., for making every effort to publish the book when the diminishing water resources are a significant issue worldwide. Special thanks are due to the AAP Production Staff for their work as well.

We request the reader to offer their constructive suggestions that may help to improve the next edition.

We express our admiration to our families and colleagues for their understanding and collaboration during the preparation of this book volume. There is a piece of advice to one and all in the world: *"Permit that our almighty God, our Creator, provider of all and excellent Teacher, feed our life with Healthy Food Products and His Grace—; and Get married to your profession."*

—Megh R. Goyal, PhD, PE
Hafiz Ansar Rasul Suleria, PhD
Shanmugam Kirubanandan, MTech, MASc
Editors

PART I

Marine-Based Bioactive Compounds and Biomaterials

CHAPTER 1

PHARMACOLOGICAL APPLICATIONS OF MARINE-DERIVED COMPOUNDS: A PREVENTIVE APPROACH

SANA KHALID, MUNAWAR ABBAS, HUMA BADER-UL-AIN, and HAFIZ ANSAR RASUL SULERIA

ABSTRACT

Marine foods are rich sources of bioactive compounds. Polyunsaturated fatty acids, peptides, minerals, and vitamins are found in these species in sufficient amounts. The primary and most promising source of protein is fish and its byproducts play an essential role as functional and medicinal foods for the prevention and treatments of many chronic diseases. Several species of marine microalgae produce high quantities of lipids; some produce long-chain ω-3 and ω-6, like arachidonic acid (ARA), docosahexaenoic acid (DHA), eicosapentaenoic acid (EPA), gamma-linolenic acid (GLA). Marine polysaccharides consist of chitin, chitosan, alginate, agar, carrageenan, fucoidan, fucosylated chondroitin sulfate (FCS). These polysaccharides also confer potential health benefits. The functional and medicinal aspects of marine-based bioactive compounds are the limelight of this chapter.

1.1 INTRODUCTION

Globally, half of the marine species exist on oceans thus covering about 70% of the Earth's surface [50]. Due to high taxonomic and ecosystem diversity, the marine world provides an excellent resource of functional ingredients with health benefits, including polyunsaturated fatty acids (PUFAs), proteins/peptides, polysaccharides, polyphenols, saponins, sterols, and pigments. Investigation of new bioactive components from the ocean seems to be a challenge [88]. These naturally occurring compounds may be obtained from different categories, such as: microorganisms (microalgae, bacteria), plants (brown algae, red algae, and green algae), invertebrates (crustaceans, sponges, sea cucumbers, ascidians, etc.) as well as some vertebrates [36]. In recent times, marine micro-organisms have captured more significant interest in producing biologically active substances, owing to their property of adaptation of specific and extreme environmental conditions. These microbes from marine sources have also gained interest for being an innovative and promising source of an array of biological compounds, which are used as functional ingredients for the development of many medicinal foods and pharmaceuticals [142].

Globally, the prevalence of obesity, cancer, diabetes, and cardiovascular diseases is increasing day by day. In 2001, uproots of approximately 59% of 56.5 million total reported deaths and 46% of the reported diseases are chronic aberrations. Therefore, in the era of food and pharmaceuticals, their combined form (i.e., functional or medicinal foods) is capturing increasing attention. These functional foods and nutraceuticals have many important and potent functional ingredients such as: anti-oxidants, dietary fibers, omega-3 polyunsaturated fatty acids, vitamins, and minerals. These functional bioactive moieties along with physical exercise are recommended for the prevention and cure of such chronic diseases [71].

Seafood consumption is encouraged and accepted around the world as shown by a lower incidence of metabolic-related disorders in seafood consumption populations [81, 122]. Thousands of compounds with unique structures and bioactivities have been isolated from marine organisms so far [36].

This chapter highlights the importance of marine-based bioactive compounds for their health benefits.

1.2 BIOACTIVE COMPOUNDS

1.2.1 PROTEINS

Among the bioactive compounds of marine algae, enzymes, and their building blocks, i.e., proteins are of considerable importance. The major and most promising source of protein is fish and its byproducts. But owing to hygienic conditions, the fish is not of food grade and remain underutilized. However, through the filleting processes, the hygienic conditions can be improved. And fish and its by-products can be made safe for human consumption and present an array of functional perspectives. Among marine organisms, the byproducts of herring present a functional protein and therefore it plays an important role as functional and medicinal foods for the prevention and treatments of many chronic diseases. Uproot for the functional properties of a protein is the presence of amino acids, their composition, sequence, and size of peptide [50].

Amount and content of protein vary significantly within the same population's species [72]. Micro-algae have gained greater interest due to the presence of a considerable number of nitrogenous compounds such as: amino acids and proteins. However, the content of protein in brown seaweeds is less than 15% [28]. Whereas, the content of functional proteins in red seaweeds is high. Conclusively, there are major differences among red, brown, and green seaweeds. Research study on protein percentage in red algae (*Gracilaria changii*) demonstrated that about 34% of dry weight is a relatively high amount of protein that is less in green peas [78]. Whereas, red alga (*Corallina officinalis*) has approximately 7% of protein. Therefore, we can say that protein and amino acids content in micro-algae varies with the species [69]. Proteins from seaweed source are not high-quality proteins owing to the presence of a low quantity of some amino acids. However, these are considered functional owing to the presence of all essential amino acids in significant amounts. This property is the main pillar for their nutritional superiority among all terrestrial plants [91].

Most promising functional micro-algae are *Arthrospira* and *Chlorella* owing to the presence of a greater amount of protein and amino acids in them and this micro-algae are used for the treatment and prevention of many diseases [74]. Arya and Gupta referred the peptide microcolin-A, produced by *Lyngbyamajuscula*, as having immunosuppressive activity, inhibiting the innate immune system. Furthermore, metalloenzymes (such as superoxide dismutases and carbonic anhydrase) are functional enzymes owing to the presence of iron, manganese, copper or zinc. These functional ingredients play an important role against oxidative damages due to their antioxidant activity. Carbonic anhydrase is present in blood cells, and it catalyzes the reaction of conversion of CO into carbonic acid and bicarbonate ions in the tissues. Many other functional enzymes are produced by marine micro-algae such as: *Porphyridium*, *Anabaena*, and *I. galbana* [22].

1.2.2 FATTY ACIDS

The fatty acids (FAs) are promising functional ingredients in many medicinal and functional foods owing to their health endorsing perspectives (such as the treatment and prevention of many chronic diseases like cancer, diabetes, cardiovascular diseases, and osteoporosis) [73]. Several species of marine microalgae produce high quantities of lipids; some produce long-chain ω-3 and ω-6 FAs (like arachidonic acid (ARA), docosahexaenoic acid (DHA), eicosapentaenoic acid (EPA), gamma-linolenic acid (GLA)). Some examples are: *Arthrospira platensis, Isochrysisgalbana, Odontella, and Porphyridiumcruentum.* Microalgae of the genus *Phaeodactylum*, and diatoms in general, and *Pavlova lutheri* are particularly rich in EPA (20:5 ω-3). On the other hand, *P. lutheri* presents high quantities of DHA (22:6 ω-3). Guedes observed that *P. lutheri* presented the highest content in PUFAs among all the microalgae. *A. platensis* accumulates high amounts of GLA, a ω-6 fatty acid [22].

Moreover, there are many other rich sources of FAs, such as fish and fish oil are rich in PUFAs [138]. In seaweeds, the composition and content of FAs and lipids vary widely due to differences among species and groups. The average lipid range in seaweeds is 0.4 to 4.5% of the dry matter. Among lipids and FAs, palmitic acid is present in the highest amount in all seaweed groups and species. Furthermore, the considerable and main source of DHA is brown microalgae, i.e., *Schizochytrium* sp. [73]. Past research confirms the beneficial uses of dietary fish oils for humans. The oil of tuna and other highly migratory fish are considered useful marine sources of DHA [6, 98]. The presence of high levels of cholesterol causes an unpleasant odor in many oils; but in microalgae oil, there is no pleasant odor and no high levels of cholesterol. In addition, this microalgae oil presents many health benefits due to the presence of phytosterols and squalene [96].

1.2.3 POLYSACCHARIDES

Marine polysaccharides consist of chitin, chitosan, alginate, agar, carra-geenan, and fucosylated chondroitin sulfate (FCS), etc. Chitin, a long-chain polymer of an N-acetylglucosamine, is a naturally abundant polysaccharide; and is found in shells of sea crab and shrimp as a component of Crustaceans exoskeleton. Besides, chitin is distributed in the cell wall of green algae (*Chlorella sp.*), protozoa, and fungi (*Zygomycetes*) [94]. Chitosan, obtained by n-deacetylation of chitin, is the abundant secondary polysaccharide in nature [57]. Fucoidan refers to a heavily sulfated polysaccharide consisting of high proportions of L-fucose and is primarily found in brown algae. The molecular size, sulfation patterns, as well as monosaccharide composition vary in different sulfated polysaccharides, which accounts for varying biological actions and responses [36].

Various aquatic sources have been explored for bioactive polysaccharides, and a wide range of biological activities including anticoagulant, anti-inflammatory, antiviral, and antitumor activities have been reported [108, 137]. Marine microorganisms secrete structurally and functionally diverse extracellular polymeric substances (EPS) to defend against biotic as well as abiotic pressure in the ocean. EPS form a layer surrounding the marine microbial cells to help them to resist adverse and extreme conditions. Nichols and his research team reported that EPS extracted from six closely related marine *Pseudomonas* showed a great structural variation despite the close phylogenetic relationships of the strains [77, 89].

Marine-derived *Bacillus, Halomonas, Planococcus, Enterobacter, Alteromonas, Rhodococcus, Zoogloea*, and *Cyanobacteria* strains have been reported as common producers of extractable EPS [104]. Compared to other EPS-producing marine microbes, *Cyanobacteria* is known to produce a high amount of EPS in the marine environment. Acidic polysaccharides that contain a high amount of uronic acid, fucose, and sulfate are prominent in marine microbial EPS [139]. Acidic EPS are well-known immune boosters as they have the ability to induce secretion of inflammatory cytokines in living systems. Moreover, together with the increased level of nitric oxide production and inflammatory cytokines, acidic EPS potently defend against tumor progression and metastasis [48]. Marine filamentous fungal species (*Keissleriella sp., Penicillin* sp. and *Epicoccum* sp.) have also been reported as promising producers of bioactive EPS. Recent advanced research studies have proved that the anti-oxidative activity of marine fungal EPS is comparatively higher than the reported EPS [59]. It has been noticed that fungal EPS are homo or heteropolysaccharide of simple sugars such as: galactose, glucose, rhamnose, and mannose. Some marine *Lactobacillus* sp. is potent producers of EPS that can potentially be used as prebiotics with high bifidogenic effect [95, 115].

In addition to a microbial polysaccharide, marine microbes have been explored for bioactive monosaccharides. Due to several adaptations to the hard marine environment, marine-derived novel microbial species have been reported as high yield producers of both glucosamine and chitosan compared to the terrestrial species. These strains have potential as an alternative source for production of vegetarian glucosamine, which has growing demand in modern nutraceuticals [63]. Bioactive compounds from marine environment are summarized in Table 1.1.

TABLE 1.1 Bioactive Compounds from Marine Environment

Group of compounds	Bioactive compounds	Microorganisms	Reference
Polyunsaturated fatty acids	Eicosapentaenoic acid	*Phaeodactylum tricornutum; Porphyridium cruentum; Crypthecodinium,* and *Pavlova lutheri*	[24, 32, 90]
	γ-Linolenic acid	*Arthrospira; Porphyridium*	[18]
	Arachidonic acid	*Porphyridium cruentum*	[3, 18]
	Docosahexaenoic acid	*Crypthecodinium; Schizochytrium; Isochrysis galbana; Pavlova lutheri*	[90, 109]
Protein/enzymes	Proteins	*Dunaliella; Phaeodactylum tricornutum; Arthrospira platensis; P. cruentum*	[93]
	Superoxide dismutase	*P. tricornutum; Porphyridium; Anabaena; Synechococcus*	[33]
	Carbonic anhydrase	*I. galbana; Amphidinium carterae; Prorocentrum minimum*	[116]
Polysaccharides	Chitin and chitosan	*Chlorella sp., Absidia coerulea, Mucorrouxii, Gongronella butieri,* and *Absidiablakesleeana*	[5, 38]
	Fucoidan and Fucosylated chondroitin sulfate	*Acaudinamolpadioides*	[134]
	Alginates	*Pseudomonas aeruginosa* and *Azotobacter vinelandii*	[40]

1.3 TARGET MALADIES

1.3.1 OBESITY

Obesity is the main etiology for chronic aberrations and mortality [47]. Obesity has the inverse relationship with soluble fibers. Soluble and insoluble fibers are sufficiently found in seaweeds [25]. Fish oil from salmon, tuna, and mackerel is the most familiar and traditional source of marine lipid accepted by consumers and researchers. Dietary intake of fish oil has been reported to reduce the fat pad in many studies [75]. Fish oil supplementation was suggested to decrease body fat by affecting appetite. Dietary fish oil up-regulated ghrelin mRNA in the gastric fundus and duodenum, which might affect feeding behavior and energy intake [97]. Fucoidan isolated from sea cucumber showed strong anti-obesity effects both *in vitro* and *in vivo*. The novel fucoidan isolated from *Acaudinamolpadioides* consisting of 1→3-linked tetrafucose repeating units disturbed adipocytes differentiation. FCS isolated from *Acaudinamolpadioides* exhibited remarkable anti-adipogenic effects at early and later stages of differentiation [134].

Phlorotannins seemed to alleviate obesity and obesity-related disorders mediated by multiple mechanisms. Several studies demonstrated that phlorotannins may prevent fat absorption through the inhibition of pancreatic lipase. For example, phlorotannins extracted from *Eiseniabicyclis* and *Ecklonia cava* showed various inhibitory effects on pancreatic lipase [26]. Astaxanthin supplementation significantly decreased body weight and lipid levels in adipose tissue and skeletal muscle. Besides, astaxanthin alleviated insulin resistance through activation of post-receptor insulin signaling [7]. Anti-obesity compounds are derived from various marine species, including microorganisms, phytoplankton, invertebrates, and vertebrates. So far, algae exploration is the main focus; large numbers of algae compounds were isolated and tested on anti-obesity effects, especially brown algae. In addition, marine-derived lipids are also well documented for their anti-obese properties [36].

1.3.2   DIABETES MELLITUS

Globally, *Diabetes mellitus* is known as fatal malady [68]. Phycocolloids, fucoidans, phlorotannins, and pigments are biologically active compounds that are present in brown algae. Anti-diabetic effect of phlorotannins to promote consumer health has been reported in many studies. Dieckolextracted from *Ecklonia cava* and diphlorethohydroxycarmalol from *Ishigeokamurae* have demonstrated antidiabetic effects by inhibiting the digestive enzymes active for carbohydrate metabolism [62].

Various marine micro-algae species including *Chlorella pyrenoidosa*, *Chlorella sorokiniana* and *Chlorella vulgaris* of the genus *chlorella* have found to be effective against the hyperglycemia [17]. These all species have shown strong potential against the *diabetes mellitus* in animal models by improving the insulin resistance and reducing the insulin sensitivity, blood glucose, plasma cholesterol, and weight. These also showed potent hypoglycemic and renoprotective effects [4, 105, 128]. The hyperglycemia was controlled by many other phytochemicals in micro-algae. The most important compound was aquastatin that was derived from a marine-derived fungus, i.e., *Cosmospora sp.* and this compound showed a strong inhibitory effect against the protein tyrosine phosphate [106]. This, in turn, reduced the insulin sensitivity, increased the insulin resistance, improved the glycemic control and reduced the weight [53, 124].

Carotenoids in marine algae have strong potential against the abnormal blood glucose and insulin levels. Among carotenoids, fucoxanthin from *Phaeodactylum tricornutum* has considerable effects against the blood glucose and plasma insulin levels [51]. It significantly reduced the plasma glucose and improved the insulin levels and therefore increased the insulin resistance [86]. Scientist found that the consumption of *Isochrysisgalbana* in an animal model reduced the glucose level in blood and increases the insulin level and insulin resistance in blood. This consumption also resulted in weight loss of diabetic and obese rats and therefore reduced the cholesterol values and increased the lactic acid bacteria counts [19]. Table 1.2 shows the anti-diabetic screening of marine organisms, their extracts, and mechanisms of action.

TABLE 1.2   Anti-Diabetic Screening of Marine Organisms, Their Extracts, and Mechanisms of Action

Marine organisms	Extracts	Mechanism of action	Reference
Brown algae *Ecklonia cava*	Methanolic extracts	Reduce plasma glucose levels in rats	[45]
Corals *Sinularia firma* and *Sinularia erecta*	Methanolic extracts	Reduce plasma glucose levels in rats	[120, 123]
Red algae *Palmaria sp.*	Phenolic extracts	α-amylase inhibition	[61, 79]
Seagrass *Posidonia oceanica*	Raw extracts	Reduce plasma glucose levels in rats	[15, 31]
Seaweed *Laminaria angustata*	Raw extracts	Reduce plasma glucose levels in rats	[52, 56]
Sponge *Xetospongiamuta*	Aqueous extracts	Dipeptidyl peptidase IV inhibition	[84]

1.3.3   CARDIOVASCULAR DISEASES

The functional foods have captured greater importance owing to the presence of important functional ingredients such as: antioxidants, dietary fiber, carotenoids, vitamins, and minerals. Cardiovascular diseases and their related complications and risk factors can be prevented and treated through the antioxidant effect of functional and medicinal foods. Among functional foods, marine microalgae have played an important role in reducing the risk of cardiovascular diseases owing to the presence of various functional and phytochemical compounds and their respective antioxidant effect.

Among phytochemical compounds of marine micro-algae, chitosan has strong potential in reducing the total cholesterol and triacylglycerol levels in the plasma and liver through their excretion in feces. This is possible through their anti-oxidant, anti-lipidemic, and membrane stabilizing effects [27, 44]. Other functional ingredients are fucoidan from *Cladosiphon okamuranus* [121], and Laminaran from *Laminaria* spp. [20]. Both EPA and DHA decrease the risk of cardiovascular disease. Detailed reviews regarding enrichment of foods with EPA/DHA including epidemiological studies evaluating EPA and DHA cardio-protective effects have recently been published [30, 64]. EPA and DHA may exert their cardioprotective functions by influencing plasma triacylglycerol and cholesterol levels and modulation of the chronic inflammation in the vascular wall, which is one of the hallmarks of atherosclerosis [54].

Carotenoids are fat-soluble pigments produced by plants and microorganisms, and are present in fruits [37], vegetables, seaweeds, and some seafood. These are known to decrease the incidence and prevalence of cardiovascular events, perhaps by means of their antioxidant action on free radicals or by acting as anti-inflammatory molecules [29].

1.3.4 CANCERS

The mechanism behind the anti-tumor effect of marine algae phytochemicals is the apoptotic cell death [35, 111]. A broad spectrum of anticancer phytochemicals is present in marine algae. Marine algae mainly include cyanotoxins (i.e., anatoxin-A and microcystins/nodularin), Lagunamides A and B (filamentous marine cyanobacterium, *Lyngbya majuscule*) having a positive effect against bone cancer [67, 125]. Moreover, some other functional ingredients in marine algae include scytonemin, caracin-A, and borophycin. Formation of mitotic cell spindle and enzyme kinesis is involved in the control of cell cycle and is regulated by scytonemin. In addition to this property, scytonemin is involved in the inhibition of human fibroblasts and endothelial cells [114].

The main mechanism behind the anticancer effect of curacin-A is the inhibition of polymerization of the tubulin [14]. Another compound, borophycin (metabolite containing boron), has been isolated from marine cyanobacterial strains of *Nostoclinckia* and *N. spongiaeforme*. The mode of action of this compound against cancer is its potent cytotoxicity against human epidermoid carcinoma [9, 21]. In addition, derivatives of arachidonic acids (ARAs) also have strong potential against cancer due to the maintenance of homeostasis in mammalian systems. Okadaic acid, derived from Procentrum, has high toxicity against leukemia by inhibiting the protein phosphatase [119, 129]. Moreover, dinochromes A and B, isolated from the freshwater species *Peridiniumbipes*, are carotenoids and show effective anti-proliferative activity against the human tumor cell lines, such as neuroblastoma, osteosarcoma, and cervical cancer.

Green alga *Caulerpa sp.* has caulerpenyne, which is a promising phytochemical for the treatment and prevention of cancer, tumor, and for its effect against proliferation. Another species of green algae, i.e., *Cystophora sp.* is famous for antitumor activity owing to the presence of two functional ingredients including meroterpenes and usneoidone [10]. As far as the anticancer activities of brown seaweeds are concerned, there are many species such as: *Laminaria japonica, Porphyratenera, Gelidiumamansii, Saccharina japonica, Undariapinnatifida*, and *Euchemacottonii*. These species have potent phytochemicals such as sulfated polysaccharides, fucoidans, glucans, some secondary metabolites, and respective derivatives. Owing to these functional ingredients, brown seaweeds have considerable potential against gastric, colon, and breast cancers and their anti-proliferation activity [141, 112].

While referring to the anticancer activity of red algae, there are various functional species such as *Laurencia microcladia, Corallina pilulifera, Acanthospora spicifera*, and *Porphyra tenera*. These species have many important functional ingredients such as: elatol, ethanolic extract, carotene, lutein, chlorophyll-related compounds, secondary metabolites, and their halogenated derivatives. These show significant positive effects against cancer through various mechanisms such as: induction of cell cycle arrest in the G1 and the sub-G1 phases leading cells to apoptosis [13, 80, 127, 136]. Conclusively, marine algae are promising for pharmacological research towards the development of anticancer therapy.

1.4 SUMMARY

Marine products provide many health benefits mainly due to their characteristic bioactive compounds. The radical scavenging compounds (such as bioactive peptides, sulfated polysaccharides, phlorotannins, and pigments from marine foods and their by-products) exhibit anticancer, antidiabetic, antiobesity, and cardiovascular activities. This opens the door to produce new medicinal foods having strong potential to manage and treat chronic malfunctions. However, further research is required to identify the active ingredients and uncover mechanistic insights into the therapeutic effects of marine bioactive compounds.

KEYWORDS

- algae
- antioxidant
- bioactive
- cancer
- cardiovascular disease
- cholesterol
- fatty acids
- fish
- fucoidans
- functional foods
- linoleic acid
- nutraceuticals
- nutrients
- obesity
- phenolics
- pigments
- proteins
- seaweeds

REFERENCES

1. Ackman, R., (1989). Fatty acids. In: *Marine Biogenic Lipids, Fats and Oils* (Vol. 1, pp. 103–137). CRC-Taylor & Francis, Boca Raton, FL, USA.
2. Ackman, R., & Hooper, S., (1973). Non-methylene-interrupted fatty acids in lipids of shallow-water marine invertebrates: A comparison of two mollusks (*Littorina littorea* and *Lunatia triseriata*) with the sand shrimp (*Crangon septemspinosus*). *Comparative Biochemistry and Physiology Part B: Comparative Biochemistry, 46*(1), 153–165.
3. Ahern, T. J., Katoh, S., & Sada, E., (1983). Arachidonic acid production by the red alga Porphyridium cruentum. *Biotechnology and Bioengineering, 25*(4), 1057–1070.

4. Aizzat, O., Yap, S. W., Sopiah, H., Madiha, M., & Hazreen, M., (2010). Modulation of oxidative stress by *Chlorella vulgaris* in streptozotocin (STZ) induced diabetic Sprague-Dawley rats. *Advances in Medical Sciences, 55*(2), 281–288.

5. Alimentarius, C., (2007). Standard for infant formula and formulae for special medical purposes intended for infants. *Codex Stan Publication number 72–198* (pp. 113), FAO, Rome.

6. Arts, M. T., Ackman, R. G., & Holub, B. J., (2001). Essential fatty acids in aquatic ecosystems: A crucial link between diet and human health and evolution. *Canadian Journal of Fisheries and Aquatic Sciences, 58*(1), 122–137.

7. Arunkumar, E., Bhuvaneswari, S., & Anuradha, C., (2012). An intervention study in obese mice with astaxanthin, a marine carotenoid effects on insulin signaling and pro-inflammatory cytokines. *Food and Function, 3*(2), 120–126.

8. Azorkelsson, G., & Kristinsson, H. G., (2009). Bioactive peptides from marine sources. *State of art Report to the NORA Fund* (pp. 112–118). Gudjon Thorkelsson and Hordur G Kristinsson.

9. Banker, R., & Carmeli, S., (1999). Inhibitors of serine proteases from a water bloom of the cyanobacterium *Microcystis sp. Tetrahedron, 55*(35), 10835–10844.

10. Barbier, P., & Guise, S., (2001). Caulerpenyne from *Caulerpa taxifolia* has an antiproliferative activity on tumor cell line SK-N-SH and modifies the microtubule network. *Life Sciences, 70*(4), 415–429.

11. Bays, H. E., (2004). Current and investigational antiobesity agents and obesity therapeutic treatment targets. *Obesity, 12*(8), 1197–1211.

12. Burtin, P., (2003). Nutritional value of seaweeds. *Electronic Journal of Environmental, Agricultural and Food Chemistry, 2*(4), 498–503.

13. Campos, A., & Souza, C. B., (2012). Antitumor effects of elatol, a marine derivative compound obtained from red algae *Laurencia microcladia*. *Journal of Pharmacy and Pharmacology, 64*(8), 1146–1154.

14. Carte, B. K., (1996). Biomedical potential of marine natural products. *Bioscience, 46*(4), 271–286.

15. Celikler, S., & Tas, S., (2009). Anti-hyperglycemic and antigenotoxic potential of Ulva rigida ethanolic extract in the experimental *diabetes mellitus*. *Food and Chemical Toxicology, 47*(8), 1837–1840.

16. Cho, E. J., & Rhee, S. H., (1997). Antimutagenic and cancer cell growth inhibitory effects of seaweeds. *Preventive Nutrition and Food Science, 2*(4), 348–353.

17. Chou, Y. C., & Prakash, E., (2008). Bioassay-guided purification and identification of PPARα/γ agonists from Chlorella sorokiniana. *Phytotherapy Research, 22*(5), 605–613.

18. Cohen, Z., & Heimer, Y., (1992). Production of polyunsaturated fatty acids (EPA, ARA, and GLA) by the microalgae *Porphyridium* and *Spirulina*. In: *Industrial Applications of Single Cell Oils* (pp. 243–273). AOCS Press, Champaign - IL.

19. Costa, M., Rodrigues, J., & Fernandes, M., (2012). Marine cyanobacteria compounds with anticancer properties: A review on the implication of apoptosis. *Marine Drugs, 10*(10), 2181–2207.

20. Dumelod, R., (1999). Carbohydrate availability of Arroz Caldo with lambda-carrageenan. *International Journal of Food Science and Nutrition, 50*(4), 283–289.

21. Davidson, B. S., (1995). New dimensions in natural products research cultured marine microorganisms. *Current Opinion in Biotechnology, 6*(3), 284–291.

22. Raposo, M. F., & Morais, R. M., (2013). Health applications of bioactive compounds from marine microalgae. *Life Sciences, 93*(15), 479–486.

23. Debbab, A., & Aly, A. H., (2010). Bioactive compounds from marine bacteria and fungi. *Microbial Biotechnology*, *3*(5), 544–563.

24. Desbois, A. P., & Mearns, A., (2009). Fatty acid from the diatom *Phaeodactylum tricornutum* is antibacterial against diverse bacteria including multi-resistant *Staphylococcus aureus* (MRSA). *Marine Biotechnology*, *11*(1), 45–52.

25. Dettmar, P. W., & Strugala, V., (2011). The key role alginates play in health. *Food Hydrocolloids*, *25*(2), 263–266.

26. Eom, S. H., & Lee, M. S., (2013). Pancreatic lipase inhibitory activity of phlorotannins isolated from *Eisenia bicyclis*. *Phytotherapy Research*, *27*(1), 148–151.

27. Filipovic, J., & Skalko, N., (2001). Mucoadhesive chitosan-coated liposomes: Characteristics and stability. *Journal of Microencapsulation*, *18*(1), 3–12.

28. Fleurence, J., (1999). Seaweed proteins: Biochemical, nutritional aspects, and potential uses. *Trends in Food Science and Technology*, *10*(1), 25–28.

29. Giordano, P., & Scicchitano, P., (2012). Carotenoids and cardiovascular risk. *Current Pharmaceutical Design*, *18*(34), 5577–5589.

30. Givens, D. I., & Gibbs, R. A., (2008). Current intakes of EPA and DHA in European populations and the potential of animal-derived foods to increase them: Symposium on How can the n–3 content of the diet be improved? *Proceedings of the Nutrition Society*, *67*(3), 273–280.

31. Gokce, G., & Haznedaroglu, M. Z., (2008). Evaluation of antidiabetic, antioxidant and vasoprotective effects of Posidonia oceanica extract. *Journal of Ethnopharmacology*, *115*(1), 122–130.

32. Guedes, A., Amaro, H. M., & Malcata, F. X., (2011). Microalgae as sources of high added-value compounds: A brief review of recent work. *Biotechnology Progress*, *27*(3), 597–613.

33. Guzman, M. A., & Lopez, C. C., (2007). Effects of fertilizer-based culture media on the production of exocellular polysaccharides and cellular superoxide dismutase by *Phaeodactylum tricornutum* (Bohlin). *Journal of Applied Phycology*, *19*(1), 33–41.

34. Han, L., & Kimura, Y., (1999). Reduction in fat storage during chitin-chitosan treatment in mice fed a high-fat diet. *International Journal of Obesity & Related Metabolic Disorders*, *23*(2), 174–179.

35. Harada, H., & Noro, T., (1997). Selective antitumor activity *in vitro* from marine algae from Japan coasts. *Biological and Pharmaceutical Bulletin*, *20*(5), 541–546.

36. Hu, X., & Tao, N., (2016). Marine-derived bioactive compounds with anti-obesity effect: A review. *Journal of Functional Foods*, *21*, 372–387.

37. Ikeda, K., & Kitamura, A., (2013). Effect of *Undaria pinnatifida* (Wakame) on the development of cerebrovascular diseases in stroke-prone spontaneously hypertensive rats. *Clinical and Experimental Pharmacology and Physiology*, *30*(1–2), 44–48.

38. Je, J. Y., & Kim, S. K., (2012). Chitosan as potential marine nutraceutical. *Advances in Food and Nutrition Research*, *65*, 121.

39. Jiao, G., & Yu, G., (2011). Chemical structures and bioactivities of sulfated polysaccharides from marine algae. *Marine Drugs*, *9*(2), 196–223.

40. Johnson, F. A., & Craig, D. Q., (1997). Characterization of the block structure and molecular weight of sodium alginates. *Journal of Pharmacy and Pharmacology*, *49*(7), 639–643.

41. Joseph, J. D., (1989). Distribution and composition of lipids in marine invertebrates. *Marine Biogenic Lipids, Fats and Oils*, *2*, 49–143.

42. Kaats, G. R., & Michalek, J. E., (2006). Evaluating efficacy of a chitosan product using a double-blinded, placebo-controlled protocol. *Journal of the American College of Nutrition, 25*(5), 389–394.

43. Kadam, S., & Prabhasankar, P., (2010). Marine foods as functional ingredients in bakery and pasta products. *Food Research International, 43*(8), 1975–1980.

44. Kanauchi, O., & Deuchi, K., (1994). Increasing effect of a chitosan and ascorbic acid mixture on fecal dietary fat excretion. *Bioscience, Biotechnology, and Biochemistry, 58*(9), 1617–1620.

45. Kang, C., & Jin, Y. B., (2010). Brown alga *Ecklonia cava* attenuates type 1 diabetes by activating AMPK and Akt signaling pathways. *Food and Chemical Toxicology, 48*(2), 509–516.

46. Kaur, G., & Cameron, D., (2011). Docosapentaenoic acid (22: 5n–3): A review of its biological effects. *Progress in Lipid Research, 50*(1), 28–34.

47. Kelishadi, R., (2007). Childhood overweight, obesity, and the metabolic syndrome in developing countries. *Epidemiologic Reviews, 29*(1), 62–76.

48. Kim, G. Y., & Oh, Y. H., (2003). Acidic polysaccharide isolated from Phellinus linteus induces nitric oxide-mediated tumoricidal activity of macrophages through protein tyrosine kinase and protein kinase C. *Biochemical and Biophysical Research Communications, 309*(2), 399–407.

49. Kim, S. K., & Pangestuti, R., (2011). Potential role of marine algae on female health, beauty, and longevity. *Advances in Food and Nutrition Research, 64*, 41–55.

50. Kim, S. K., & Wijesekara, I., (2010). Development and biological activities of marine-derived bioactive peptides: A review. *Journal of Functional Foods, 2*(1), 1–9.

51. Kim, S. M., & Jung, Y. J., (2012). A potential commercial source of fucoxanthin extracted from the microalga *Phaeodactylum tricornutum*. *Applied Biochemistry and Biotechnology, 166*(7), 1843–1855.

52. Kimura, Y., & Watanabe, K., (1996). Effects of soluble sodium alginate on cholesterol excretion and glucose tolerance in rats. *Journal of Ethnopharmacology, 54*(1), 47–54.

53. Klaman, L. D., & Boss, O., (2000). Increased energy expenditure, decreased adiposity, and tissue-specific insulin sensitivity in protein-tyrosine phosphatase 1B-deficient mice. *Molecular and Cellular Biology, 20*(15), 5479–5489.

54. Komprda, T. A., (2012). Eicosapentaenoic and docosahexaenoic acids as inflammation-modulating and lipid homeostasis influencing nutraceuticals: A review. *Journal of Functional Foods, 4*(1), 25–38.

55. Kraffe, E., & Soudant, P., (2004). Fatty acids of serine, ethanolamine, and choline plasmalogens in some marine bivalves. *Lipids, 39*(1), 59–66.

56. Krish, S., & Das, A., (2014). *In-vitro* bioactivity of marine seaweed, *Cladophora rupestris*. *International Journal of Pharmacy and Biological Sciences, 5*(1), 898–908.

57. Kumar, M. N., (2000). Review of chitin and chitosan applications. *Reactive and Functional Polymers, 46*(1), 1–27.

58. Lahaye, M., (1991). Marine algae as source of fibers: Determination of soluble and insoluble dietary fiber contents in some sea vegetables. *Journal of the Science of Food and Agriculture, 54*(4), 587–594.

59. Laurienzo, P., (1991). Marine polysaccharides in pharmaceutical applications: An overview. *Marine Drugs, 8*(9), 2435–2465.

60. Lauritano, C., & Ianora, A., (2016). Marine organisms with anti-diabetes properties. *Marine Drugs, 14*(12), 220.

61. Lee, H. J., & Kim, Y. A., (2007). Screening of Korean marine plants extracts for inhibitory activity on protein tyrosine phosphatase 1B. *Journal of Applied Biological Chemistry*, *50*(2), 74–77.

62. Lee, S. H., & Jeon, Y. J., (2013). Anti-diabetic effects of brown algae-derived phlorotannins, marine polyphenols through diverse mechanisms. *Fitoterapia*, *86*, 129–136.

63. Logesh, A., & Thillaimaharani, K., (2012). Production of chitosan from endolichenic fungi isolated from mangrove environment and its antagonistic activity. *Asian Pacific Journal of Tropical Biomedicine*, *2*(2), 140–143.

64. Lopez, E., (2009). Health effects of oleic acid and long chain omega-3 fatty acids (EPA and DHA) enriched milks: A review of intervention studies. *Pharmacological Research*, *61*(3), 200–207.

65. Lordan, S., & Ross, R. P., (2011). Marine bioactives as functional food ingredients: Potential to reduce the incidence of chronic diseases. *Marine Drugs*, *9*(6), 1056–1100.

66. Maeda, H., & Hosokawa, M., (2007). Dietary combination of fucoxanthin and fish oil attenuates the weight gain of white adipose tissue and decreases blood glucose in obese/diabetic KK-Ay mice. *Journal of Agricultural and Food Chemistry*, *55*(19), 7701–7706.

67. Mahdi, E., & Fariba, K., (2012). Cancer treatment with using cyanobacteria and suitable drug delivery system. *Annals of Biological Research*, *3*(1), 622–627.

68. Malviya, N., & Jain, S., (2010). Antidiabetic potential of medicinal plants. *Acta Poloniae Pharmaceutica*, *67*(2), 113–118.

69. Marsham, S., & Scott, G. W., (2007). Comparison of nutritive chemistry of a range of temperate seaweeds. *Food Chemistry*, *100*(4), 1331–1336.

70. Martinez, M., & Ichaso, N., (2010). The desaturation pathway for DHA biosynthesis is operative in the human species: Differences between normal controls and children with the Zellweger syndrome. *Lipids in Health and Disease*, *9*(1), 98.

71. Mayakrishnan, V., & Kannappan, P., (2013). Cardioprotective activity of polysaccharides derived from marine algae: An overview. *Trends in Food Science and Technology*, *30*(2), 98–104.

72. McDermid, K. J., & Stuercke, B., (2003). Nutritional composition of edible Hawaiian seaweeds. *Journal of Applied Phycology*, *15*(6), 513–524.

73. Misurcova, L., & Ambrozova, J., (2011). Seaweed lipids as nutraceuticals. *Advances in Food and Nutrition Research*, *64*, 339–355.

74. Morais, R. M., (2001). *Functional Foods: An Introductory Course*. http://www.ift.org/knowledge-center/read-ift-publications/science-reports/scientific-status-summaries/functional-foods/introduction-to-functional-food.aspx. Accessed on December 31 of 2017.

75. Nakatani, T., & Kim, H. J., (2003). A low fish oil inhibits SREBP–1 proteolytic cascade, while a high-fish-oil feeding decreases SREBP–1 mRNA in mice liver relationship to anti-obesity. *Journal of Lipid Research*, *44*(2), 369–379.

76. Namvar, F., & Mohamed, S., (2012). Polyphenol-rich seaweed (*Eucheuma cottonii*) extract suppresses breast tumor via hormone modulation and apoptosis induction. *Food Chemistry*, *130*(2), 376–382.

77. Nichols, C. M., & Guezennec, J., (2005). Bacterial exopolysaccharides from extreme marine environments with special consideration of the southern ocean, sea ice, and deep-sea hydrothermal vents: A review. *Marine Biotechnology*, *7*(4), 253–271.

78. Norziah, M. H., & Ching, C. Y., (2000). Nutritional composition of edible seaweed *Gracilaria changgi*. *Food Chemistry*, *68*(1), 69–76.

79. Nwosu, F., & Morris, J., (2011). Anti-proliferative and potential anti-diabetic effects of phenolic-rich extracts from edible marine algae. *"Food Chemistry,"* *126*(3), 1006–1012.

80. Okai, Y., & Higashi-Okai, K., (1996). Identification of antimutagenic substances in an extract of edible red alga, *Porphyra tenera* (Asakusa-nori). *Cancer Letters*, *100*(1/2), 235–240.

81. Olsen, S. O., (2003). Understanding the relationship between age and seafood consumption: The mediating role of attitude, health involvement, and convenience. *Food Quality and Preference, 14*(3), 199–209.

82. Panlasigui, L. N., & Baello, O. Q., (2003). Blood cholesterol and lipid-lowering effects of carrageenan on human volunteers. *Asia Pacific Journal of Clinical Nutrition*, *12*(2), 209–214.

83. Park, H., & Choi, J., (2000). Docosahexaenoic acid-rich fish oil and pectin have a hypolipidemic effect, but pectin increases risk factor for colon cancer in rats. *Nutrition Research*, *20*(12), 1783–1794.

84. Pascual, I., & Lopez, A., (2007). Screening of inhibitors of porcine dipeptidyl peptidase IV activity in aqueous extracts from marine organisms. *Enzyme and Microbial Technology*, *40*(3), 414–419.

85. Pazos, A. J., & Roman, G., (1997). Lipid classes and fatty acid composition in the female gonad of Pecten maximus in relation to reproductive cycle and environmental variables. *Comparative Biochemistry and Physiology Part B: Biochemistry and Molecular Biology*, *117*(3), 393–402.

86. Peng, J., & Yuan, J. P., (2011). Fucoxanthin, a marine carotenoid present in brown seaweeds and diatoms: Metabolism and bioactivities relevant to human health. *Marine Drugs*, *9*(10), 1806–1828.

87. Pettigrew, R., & Hamilton, D., (1997). Obesity and female reproductive function. *British Medical Bulletin*, *53*(2), 341–358.

88. Plaza, M., & Cifuentes, A., (2008). In the search of new functional food ingredients from algae. *Trends in Food Science and Technology, 19*(1), 31–39.

89. Poli, A., & Anzelmo, G., (2010). Bacterial exopolysaccharides from extreme marine habitats: Production, characterization and biological activities. *Marine Drugs*, *8*(6), 1779–1802.

90. Pulz, O., & Gross, W., (2004). Valuable products from biotechnology of microalgae. *Applied Microbiology and Biotechnology*, *65*(6), 635–648.

91. Qasim, R., (1991). Amino acid composition of some common seaweeds. *Pakistan Journal of Pharmaceutical Sciences*, *4*(1), 49–54.

92. Qiu, X., & Hong, H., (2001). Identification of a fatty acid denaturize from *Thraustochytrium sp.* involved in the biosynthesis of docosahexaenoic acid by heterologous expression in *Saccharomyces cerevisiae* and *Brassica juncea. Journal of Biological Chemistry*, *276*(34), 31561–31566.

93. Rasmussen, R. S., & Morrissey, M. T., (2007). Marine biotechnology for production of food ingredients. *Advances in Food and Nutrition Research*, *52*, 237–292.

94. Rinaudo, M., (2006). Chitin and chitosan: Properties and applications. *Progress in Polymer Science*, *31*(7), 603–632.

95. Ruas, P., & Hugenholtz, J., (2002). Overview of the functionality of exopolysaccharides produced by lactic acid bacteria. *International Dairy Journal, 12*(2), 163–171.

96. Rodriguez, N., & Beltran, S., (2010). Production of omega-3 polyunsaturated fatty acid concentrates: A review. *Innovative Food Science and Emerging Technologies*, *11*(1), 1–12.

97. Saidpour, A., & Kimiagar, M., (2012). The modifying effects of fish oil on fasting ghrelin mRNA expression in weaned rats. *Gene, 507*(1), 44–49.

98. Saito, H., (2004). Lipid and FA composition of the pearl oyster *Pinctada fucata martensii*: Influence of season and maturation. *Lipids, 39*(10), 997–1005.

99. Saito, H., (2007). Identification of novel n–4 series polyunsaturated fatty acids in a deep-sea clam, *Calyptogena phaseoliformis*. *Journal of Chromatography A., 1163*(1), 247–259.

100. Saito, H., (2008). Unusual novel n–4 polyunsaturated fatty acids in cold-seep mussels (*Bathymodiolus japonicus* and *Bathymodiolus platifrons*), originating from symbiotic methanotrophic bacteria. *Journal of Chromatography A., 12*(2), 242–254.

101. Saito, H., & Marty, Y., (2010). High levels of icosapentaenoic acid in the lipids of oyster *Crassostrea gigas* ranging over both Japan and France. *Journal of Oleo Science, 59*(6), 281–292.

102. Saito, H., & Seike, Y., (2005). High docosahexaenoic acid levels in both neutral and polar lipids of a highly migratory fish, *Thunnus tonggol* (Bleeker). *Lipids, 40*(9), 941–953.

103. Santhosh, S., & Sini, T., (2006). Effect of chitosan supplementation on antitubercular drugs-induced hepatotoxicity in rats. *Toxicology, 219*(1), 53–59.

104. Satpute, S. K., & Banat, I. M., (2010). Biosurfactants, bioemulsifiers, and exopolysaccharides from marine microorganisms. *Biotechnology Advances, 28*(4), 436–450.

105. Senthilkumar, T., & Ashokkumar, N., (2012). Impact of *Chlorella pyrenoidosa* on the attenuation of hyperglycemia-mediated oxidative stress and protection of kidney tissue in streptozotocin-cadmium induced diabetic nephropathic rats. *Biomedicine and Preventive Nutrition, 2*(2), 125–131.

106. Seo, C., & Sohn, J. H., (2009). Isolation of the protein tyrosine phosphatase 1B inhibitory metabolite from the marine-derived fungus Cosmospora sp. SF–5060. *Bioorganic and Medicinal Chemistry Letters, 19*(21), 6095–6097.

107. Shahidi, F., (2007). Nutraceuticals and healthful products from aquatic resources. *The Journal of Ocean Technology, 2*(2), 36–48.

108. Sheng, G. P., & Yu, H. Q., (2010). Extracellular polymeric substances (EPS) of microbial aggregates in biological wastewater treatment systems: A review. *Biotechnology Advances, 28*(6), 882–894.

109. Sijtsma, L., & Swaaf, M., (2004). Biotechnological production and applications of the ω-3 polyunsaturated fatty acid docosahexaenoic acid. *Applied Microbiology and Biotechnology, 64*(2), 146–153.

110. Simpson, B., & Nayeri, G., (1998). Enzymatic hydrolysis of shrimp meat. *Food Chemistry, 61*(1), 131–138.

111. Sithranga, N., & Kathiresan, K., (2010). Anticancer drugs from marine flora: An overview. *Journal of Oncology*. http://dx.doi.org/10.1155/2010/214186.

112. Stepanenko, O. V., (2011). Modern fluorescent proteins: From chromophore formation to novel intracellular applications. *Biotechniques, 51*(5), 313.

113. Stevenson, C., & Capper, E., (2002). Scytonemin-a marine natural product inhibitor of kinases key in hyperproliferative inflammatory diseases. *Inflammation Research, 51*(2), 112–114.

114. Stevenson, C. S., & Capper, E. A., (2002). The identification and characterization of the marine natural product scytonemin as a novel antiproliferative pharmacophore. *Journal of Pharmacology and Experimental Therapeutics, 303*(2), 858–866.

115. Sun, C., & Wang, J. W., (2004). Free radical scavenging and antioxidant activities of EPS2, an exopolysaccharide produced by a marine filamentous fungus *Keissleriella sp.* YS 4108. *Life Sciences, 75*(9), 1063–1073.

116. Sun, Y., & Wang, C., (2009). The optimal growth conditions for the biomass production of *Isochrysis galbana* and the effects that phosphorus, Zn, CO_2, and light intensity have on the biochemical composition of *Isochrysis galbana* and the activity of extracellular CA. *Biotechnology and Bioprocess Engineering, 14*(2), 225–231.

117. Sun, Z., & Liu, J., (2011). Astaxanthin is responsible for antiglycoxidative properties of microalga *Chlorella zofingiensis*. *Food Chemistry, 126*(4), 1629–1635.

118. Sun, Z., & Peng, X., (2010). Inhibitory effects of microalgal extracts on the formation of advanced glycation endproducts (AGEs). *Food Chemistry, 120*(1), 261–267.

119. Takai, A., & Bialojan, C., (1987). Smooth muscle myosin phosphatase inhibition and force enhancement by black sponge toxin. *FEBS Letters, 217*(1), 81–84.

120. Tamrakar, A. K., & Tiwari, P., (2008). Antihyperglycaemic activity of *Sinularia firma* and *Sinularia erecta* in streptozotocin-induced diabetic rats. *Medicinal Chemistry Research, 17*(2–7), 62–73.

121. Thomes, P., & Rajendran, M., (2010). Cardioprotective activity of *Cladosiphon okamuranus* fucoidan against isoproterenol-induced myocardial infarction in rats. *Phytomedicine, 18*(1), 52–57.

122. Thorsdottir, I., & Tomasson, H., (2007). Randomized trial of weight-loss-diets for young adults varying in fish and fish oil content. *International Journal of Obesity, 31*(10), 1560–1566.

123. Tiwari, P., & Rahuja, N., (2008). Search for antihyperglycemic activity in few marine flora and fauna. *Indian Journal of Science and Technology, 1*(5), 1–5.

124. Tonks, N. K., (2006). Protein tyrosine phosphatases: From genes, to function, to disease. *Nature Reviews Molecular Cell Biology, 7*(11), 833–846.

125. Tripathi, A., & Fang, W., (2012). Biochemical studies of the lagunamides, potent cytotoxic cyclic depsipeptides from the marine cyanobacterium *Lyngbya majuscula*. *Marine Drugs, 10*(5), 1126–1137.

126. Bussel, B. C., & Henry, R. M., (2011). Fish consumption in healthy adults is associated with decreased circulating biomarkers of endothelial dysfunction and inflammation during a 6-year follow-up. *The Journal of Nutrition, 141*(9), 1719–1725.

127. Vasanthi, H., & Rajamanickam, G., (2004). Tumoricidal effect of the red algae *Acanthophora spicifera* on Ehrlich as cites carcinoma in mice. *Seaweed Research and Utilization Journal, 26*, 217–224.

128. Vecina, J. F., & Oliveira, A. G., (2014). Chlorella modulates insulin signaling pathway and prevents high-fat-diet-induced insulin resistance in mice. *Life Sciences, 95*(1), 45–52.

129. Walker, T. R., & Watson, S. P., (1992). Okadaic acid inhibits activation of phospholipase C in human platelets by mimicking the actions of protein kinases A and C. *British Journal of Pharmacology, 105*(3), 627–631.

130. Ward, O. P., & Singh, A., (2005). Omega-3/6 fatty acids: Alternative sources of production. *Process Biochemistry, 40*(12), 3627–3652.

131. Washida, K., & Koyama, T., (2006). Karatungiols A and B, two novel antimicrobial polyol compounds, from the symbiotic marine dinoflagellate *Amphidinium sp.* *Tetrahedron Letters, 47*(15), 2521–2525.

132. Wijesekara, I., & Pangestuti, R., (2011). Biological activities and potential health benefits of sulfated polysaccharides derived from marine algae. *Carbohydrate Polymers, 84*(1), 14–21.

133. Xie, W., & Xu, P., (2001). Antioxidant activity of water-soluble chitosan derivatives. *Bioorganic and Medicinal Chemistry Letters, 11*(13), 1699–1701.

134. Xu, H., & Wang, J., (2015). Inhibitory effect of fucosylated chondroitin sulfate from the sea cucumber Acaudina molpadioides on adipogenesis is dependent on Wnt/β-catenin pathway. *Journal of Bioscience and Bioengineering*, *119*(1), 85–91.

135. Yaguchi, T., & Tanaka, S., (1997). Production of high yields of docosahexaenoic acid by *Schizochytrium sp.* strain SR21. *Journal of the American Oil Chemists' Society*, *74*(11), 1431–1434.

136. Yamamoto, I., & Maruyama, H., (1986). The effect of dietary or intraperitoneally injected seaweed preparations on the growth of sarcoma–180 cells subcutaneously implanted into mice. *Cancer Letters*, *30*(2), 125–131.

137. Yang, L., & Zhang, L. M., (2009). Chemical structural and chain conformational characterization of some bioactive polysaccharides isolated from natural sources. *Carbohydrate Polymers*, *76*(3), 349–361.

138. Yap, C., & Chen, F., (2001). Polyunsaturated fatty acids: Biological significance, biosynthesis, and production by microalgae and microalgae-like organisms. *Algae and Their Biotechnological Potential* (pp. 1–32). Springer.

139. Ye, S., & Liu, F., (2012). Antioxidant activities of an exopolysaccharide isolated and purified from marine Pseudomonas PF–6. *Carbohydrate Polymers*, *87*(1), 764–770.

140. Yim, J. H., & Son, E., (2005). Novel sulfated polysaccharide derived from red-tide microalga Gyrodinium impudicum strain KG03 with immunostimulating activity in vivo. *Marine Biotechnology*, *7*(4), 331–338.

141. Yubin, J., & Guangmei, Z., (1998). *Pharmacological Action and Application of Available Antitumor Composition of Traditional Chinese Medicine* (p. 214). Heilongjiang Science and Technology Press, Heilongjiang, China.

142. Zhao, C., & Wu, Y., (2015). Hypotensive, hypoglycaemic and hypolipidaemic effects of bioactive compounds from microalgae and marine microorganisms. *International Journal of Food Science & Technology*, *50*(8), 1705–1717.

CHAPTER 2

BIOACTIVE COMPOUNDS FROM MARINE SOURCES

VIJAY SINGH SHARANAGAT, VINTI SINGLA, and LOCHAN SINGH

ABSTRACT

Peptides, polysaccharides, lipids, chitooligosaccharides, lectins, phlorotannins, pigments, and diterpenes are major bioactive compounds found in marine foods, which may be effectively extracted through techniques like microwave and enzyme assisted extraction, supercritical fluid extraction, etc. This chapter thus provides an insight on specific bioactive compounds present in marine foods and the technological advancements in their extraction. Authors have also discussed health benefits and advantages of these extraction techniques.

2.1 INTRODUCTION

Bioactive word is made of two words: bio and active. The word 'Bio' is derived from the Greek "bios," which refers to 'life' while 'active' is obtained from the Latin meaning 'full of energy.' Bioactive compounds refer to extra nutritional constituents present in small quantities in foods. Research has shown that these compounds can act as antibiotics, anti-fouling agents, fungicides, biosensors, immune-modulators, and UV protective compounds and have therapeutic effects on our health. There is a growing interest in bioactive compounds owing to their wide range of applications in plant science, geo-medicine, agrochemicals, modern pharmacology, cosmetics, nano-bioscience, food industry, etc. It is thus a very promising area for research and development aiming at diversification of the resources of bioactive compounds and improving their salvage or synthesis pathways.

Identifying the bioactive compounds, their source and health effects are active focus areas for scientific inquiry. The vast seawaters and the immense microflora present in it are the sources of bioactive compounds.

Most bioactive compounds are secondary metabolites produced by marine microbes like micro-algae, diatoms, bacteria, fungi, actinomycetes, and others. Numerous novel bioactive compounds from marine sources have been screened for anti-diabetic, anti-hypertensive, anti-microbial, anti-tumor, anti-arthritic, anti-cellulite, contraceptive, hemolytic, and anti-inflammatory effects [3]. A certain group of microorganisms that grow in an environment of high pressure and salt concentrations can be found only in sea waters. It is believed that sea animals develop self-defense mechanisms, which result in the release of these secondary metabolites [19]. Particularly microalgae have gained focus because of their ability to synthesize such complex bioactive compounds, which under lab conditions could not be manufactured by chemical means. Another advantage of a microalga is that they could culture easily in comparison to higher order sea animals.

The functional foods must be safe, authentic, and traceable. In this regard, marine products provide a new alternative for the isolation of biocompounds to meet the increasing demands of consumers. Main functional ingredients from marine foods include bioactive peptides (fish, invertebrates, and seaweeds), chitooligosaccharides (crustaceans), sulfated polysaccharides (seaweeds and invertebrates), phlorotannins (seaweeds), lectins (fish), fucoxanthins (seaweeds) and others, which have been discussed in later sections of this chapter.

This chapter focuses on the bioactive compounds from a marine source.

2.2 MARINE BIOACTIVE COMPOUNDS

Bioactive compounds obtained from the marine resource can be classified under topics that are listed in Figure 2.1.

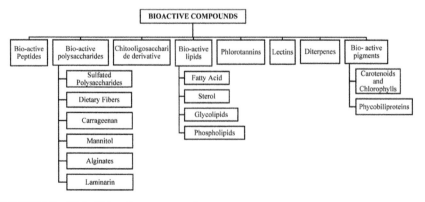

FIGURE 2.1   List of bioactive compounds.

2.2.1 BIOACTIVE PEPTIDES

Bioactive peptides are specific protein fragments, which are synthesized when large prepropeptides are broken down and modified to give a positive impact on the physiological functions of the body. Marine organisms constitute the best sources of structurally diverse bioactive peptides [166]. These peptides possess numerous health beneficial effects. Some important functions of these peptides include: ACE inhibitory and anti-hypertensive, antioxidative, anticoagulant, and antimicrobial effects. Thus they become the integral components of nutraceuticals, functional foods and pharmaceuticals owing to their disease prevention and treatment potential [24, 65]. Table 2.1 summarizes different bioactive peptides purified from sea animals and their therapeutic effects.

When all amino acids of the bioactive peptide act in cooperation with each other, then this exerts a certain degree of antioxidant activity. The use of peptides as a food additive is prevented due to the bitter taste of hydrolysates [98], but on treating with activated carbon, this bitterness is removed [163]. It is challenging for the food industry to properly utilize these bioactive compounds, as the processing leads to undesirable alterations at the molecular level or interactions with rest of the constituents of food resulting in low bioactivities [129]. Anti-microbial peptides work selectively against pathogenic microbes without any toxicity to the host mammals leading to its enhanced pharmacological applications for treating infections [45].

2.2.2 BIOACTIVE POLYSACCHARIDES

Simple sugars containing glycosidic linkages form polysaccharides; and are used in a large number of commercial products such as: stabilizers, emulsifiers, thickeners, food & beverages, animal feed, etc. [123, 179]. With a huge reserve of storage polysaccharides, cell wall polysaccharides and mucopolysaccharides algae are recognized as an important source of these sugar polymers [100, 131]. Mainly cellulose, hemicellulose, and neutral polysaccharides present in the cell wall of the algal act as a support system for the thallus in water. These celluloses and hemicelluloses are weak in comparison to those found inland plants and trees. In seaweed species (like *Fucus sp., Laminaria sp., Chondrus crispus, Ascophyllum nodosum, Ulva sp., Porphyra sp., Sargassum sp., Palmaria palmate and Gracilaria sp.*),4 to 76% dry weight content is total polysaccharide, the cellulose content from 2–10% of dry weight and the hemicellulose content is about 9%. Maximum

TABLE 2.1 Therapeutic Effects and Mode of Action of Bioactive Peptides Purified from Sea Animals

Source of Bioactive peptide	Purified from sea animal	Mode of action	Therapeutic effects	Reference
ACE inhibitory peptide	Marine zooplankton (*Brachionus rotundiformis*)	Curtail the activity of the renin-angiotensin-aldosterone system (RAAS) by inhibiting the angiotensin-converting enzyme.	Treatment of hypertension and congestive heart failure.	[27]
Antimicrobial peptides	Marine invertebrates	Interact with bacterial membranes (can be cytotoxic) and pass through it to reach the target inside the cell.	Treat pathogenic infections.	[45]
Gelatin peptides	Fish skin	Accelerate calcium absorption by increasing calcium bioavailability.	Antioxidants, anti-hypertensive	[24, 84, 125]
Marine anti-coagulant proteins	Blood ark shell, starfish, yellowfin sole.	Inhibit the intrinsic coagulation pathway.	Prevent blood clotting.	[73, 97, 153]
Marine oligosaccharide preparation (MOP)	Chum salmon	By killing malignant cells and increasing the lymphokines secretion.	Immuno-modulatory effects.	[198]
Marine protein hydrolysate	Hoki, tuna, scad, conger eel, cod.	Accelerate calcium absorption. Metabolize calcium by decreasing the number of osteoclasts.	Antioxidants, cardiovascular disease, osteoporosis, obesity, and others.	[65–67, 90, 126, 154, 169, 176]
Peptides derived from collagen	Starfish (*Asterias amurensis*).	Stimulates collagen synthesis when absorbed by joint cartilage.	Anti-wrinkle effects, cartilage regeneration, improve joint health.	[102, 140, 141]
Protein hydrolysate	Bigeye tuna	Strong suppression of systolic blood pressure.	Anti-hypertensive	[108]

polysaccharide concentrations are seen in species *Ascophyllum*, *Porphyra*, *Palmaria*, and *Ulva*. However, only *Ulva sp.* exhibits the presence of lignin and at low concentration (3% on a dry weight basis).

2.2.2.1   SULFATED POLYSACCHARIDES

Sulfated polysaccharides are sugar polymers with a certain degree of sulfation. Many species of marine algae are an excellent source of sulfated polysaccharides like fucoidans and sulfate galactans [145]. Dextran sulfate, pentosan polysulfate, fucoidan, and carrageenans are some of the sulfated polysaccharides, which have shown to be potent inhibitors of some enveloped viruses, e.g., herpes simplex virus, human cytomegalovirus, Sindbis virus, vesicular stomatitis virus, and human immunodeficiency virus [12]. Table 2.2 shows the advantages of sulfated polysaccharides and their therapeutic effects.

TABLE 2.2   Therapeutic Effects and Advantages of Sulfated Polysaccharides

Therapeutic effect of sulfated polysaccharide	Source	Advantage over other drugs	Reference
Antipyretic, analgesic, anti-inflammatory	Brown algae (*Sargassum fulvellum*)	Potential scavenger of NO than commercially available BHA and α-tocophorol (antioxidants).	[90]
Immuno-modulators	Green algae (*Ulva rigida*)	Expression of chemokines and interleukins doubled. Stimulation of macrophage secretion of prostaglandin.	[109]
Sulfated fucoidans (anti-coagulant)	Marine brown algae	Higher anti-coagulant activity than heparin. Anti-viral activity against diseases like HIV.	[33, 35, 36, 42, 81, 121, 139, 191]
Sulfated galactans (anti-coagulant)	Marine red algae (*carrageenan*)	Novel anti-coagulant	[28, 94, 162]
Vaginal anti-viral formulations	Seaweed	Low cost, a broad spectrum of anti-viral property, more acceptance and safe	[18]

In general, polysaccharides from algal source with a low degree of sulfation do not work actively against viruses [37]. The anti-viral activity rather depends on many other factors such as: molecular weight, dynamic stereochemistry and constituent sugars [2, 37, 115]. Marine algae are also an important source of antioxidant compounds [122, 159] and their antioxidant activity depends on the structure of sulfated polysaccharides like the type of major sugar, the degree of sulfating and glycosidic branching [150, 199].

Fucans is a type of sulfated polysaccharide, which is present in all brown algae but is absent in green algae, red algae, freshwater algae, and terrestrial plants [165]. The extracellular matrix of brown algae has only two types of polysaccharides: sulfated fucans and alginic acid. Fucans is divided into three main categories: fucoidans, xylofucoglycuronans, and glycorunogalactofucans [52]. Fucoidans have a solubility in acid and water solutions [160]. The bioactive functions of fucoidan include: immune-stimulating effects, viral inhibition, and anti-inflammatory effects. This bioactive compound can modify cell surface and cell wall-reinforcing properties [182]. In a research study, it was found that intake of fucoidan lowers the severity of symptoms in the initial phase of arthritis triggered by *S. aureus* in mice [184].

2.2.2.2   DIETARY FIBERS

Dietary fibers constitute plant cell wall polysaccharides and lignin. They have laxative functions resulting in human large bowel functioning, causing a surge in stool output, easy dilution of colonic contents, a quicker rate of passage through the gut and thus, resultantly improvement in the colonic metabolism of minerals, nitrogen, and bile acids. These alterations are brought about by fiber passing through the gut remaining undigested and retaining water in the gut lumen [173].

Dietary fibers are beneficial due to their ability to stimulate intestinal microflora. They have a diverse chemical composition and biological properties [104]. Marine plants have higher proportions of soluble dietary fibers than terrestrial plants [103]. Marine alga has soluble dietary fibers (like alginic acid, agars, furonan, laminaran, and porphyrin) and insoluble dietary fibers (like xylan, cellulose, and mannans). Soluble dietary fibers are similar to prebiotics and offer resistance to breakdown in the upper digestive tract of the body because of lack of suitable microflora, but the gut microbiomes break them down and in the process, derive their energy from them. They also suppress the growth of pathogenic microbes like *Escherichia coli, Campylobacter jejuni*, and others. The dietary fiber content in different marine algae is as follows: *Undaria* (58% dry weight), *Fucus* (50% dry weight), *Porphyra* (30% dry weight) and *Saccharina* (29% dry weight) [131].

The highest content of soluble dietary fiber is in *Undaria pinnatifida* (wakame), *Chondrus*, and *Porphyra* (15–22%) and insoluble dietary fiber are in *Fucus* (40%) and *Laminaria* (27%) [47]. Also, studies indicate that undigested polysaccharides of seaweed are dietary fibers, but they modify dietary protein and mineral digestibility in the body. For example, the apparent digestibility and retention coefficients of minerals such as potassium (K), calcium (Ca), sodium (Na), iron (Fe) and magnesium (Mg) were found to be less in vivo studies on rats fed with seaweeds [181].

Marine polysaccharides are important in functional foods and nutraceuticals because of their therapeutic effects like anti-tumor, anti-herpetic, anticoagulant, hypocholesterolemic (as indicated by in vivo studies on rats), anti-diabetic, anti-viral, and anti-cellulite (prevent obesity) [5, 11, 49, 107, 131, 196]. Use of soluble fiber delays availability of glucose and proximal small intestine's absorption due to which glucose levels at the postprandial stage are lowered down. Therefore, they are recommended for diabetic patients [68]. Insoluble dietary fibers (celluloses) decrease digestive tract transit time [116].

2.2.2.3   CARRAGEENAN

Carrageenans belong to a biomolecular group made of linear chains of a polysaccharide having sulfate half-esters and linked to the sugar moiety, obtained through extraction from edible red seaweeds. They are water-soluble, increase the viscosity of solutions and pH stable. Carrageenan has three forms (kappa, iota, and lambda) with different gel-forming properties [155]. In the food industry, carrageenan is used as a thickener and stabilizer (E–407) in canned food products, salad dressings, dessert mousses, bakery fillings, instant desserts, ice cream, and canned pet foods. It is commonly found in non-dairy milk such as almond milk and coconut milk. Similarly, the industrial usage of pure extract of carrageenan is also diverse. It is a clarifier in beer, wines, and honey, and has therapeutic effects on human health. Due to its hypoglycemic effect, it can be used for the prevention and control of diabetes [44]. It exhibits both anti-tumor and antiviral properties [168, 186, 195, 200]. In its native form, it is indigestible polysaccharide, but its refined versions pose considerable health risks. International Agency for Research on Cancer (IARC) also categorized degraded carrageenan as one of the probable human carcinogens, whereas carrageenan in native form still enjoys GRAS status (21 CFR 182.7255) and is utilized as a food additive [29, 177]. In a study with 0.5–2% concentration of degraded carrageenan when fed to rhesus monkeys, it resulted in diarrhea, hemorrhage, and ulcerations [16]. In a human study un-degraded carrageenan, in concentrations lower than what is present in a typical diet, resulted in higher cell death, reduced cell multiplication and cell cycle arrest [20]. Carrageenan has higher anti-thrombotic activity than κ-carrageenan because of more sulfate content. However, the toxicity is independent of sulfate content and depends on the molecular weight of the carrageenan.

2.2.2.4 MANNITOL

Mannitol is commercially manufactured from fructose. After starch hydrolysis, hydrogenation occurs where hydrogen molecules are added to fructose to restructure for forming sugar mannose. Among terrestrial sources, mannitol is present in fruits, leaves, stem, and various other parts of plants, with strawberries, celery, onions, pumpkins, and mushrooms being rich sources. With 60% sweetness of sucrose, the caloric value of mannitol is 1.6 calories per gram.

Under the marine environment, it is present in brown algae sp. (*Laminaria* and *Ecklonia*), and the content varies with seasonal fluctuations and atmospheric conditions. This sugar alcohol is used in candies, chocolate-flavored compound coatings, chewing gums, the pharmaceutical industry, paints, and varnishes, leather, and paper making, plastics, and explosives. As a sucrose replacer, mannitol is used in making sugar free confectionaries for diabetic people. It is also used as a flavor enhancer, non-hygroscopic, and chemically inert additive in food products. It has a sweet, pleasant taste and gives the desired mouth feel. It is non-carcinogenic and hence is safe for consumption by children and aged persons. Unlike other sweeteners, mannitol does not lead to dental caries [133]. As a fat replacer, mannitol has nutraceutical applications. It is heat stable, laxative, and provides less than 4 kcal per gram.

2.2.2.5 ALGINATES

Alginates are present in two forms: acid and salt. The acid form called alginic acid comprises apolymeri of uronic acids, where uronic acid occurs as two different units (L-glucoronic acid and D-mannuronic acid) and is a polysaccharide found in the extracellular matrix of brown algae while the salt form forms 40–47% dry weight of all brown seaweeds as a component in cell wall [8, 155]. The sources of alginic acid include: *Saccharina latissimi*, *Laminaria digitata*, *Laminaria hyperborea*, *Laminaria sp.*, *Fucus vesiculosus*, *Ascophyllum nodosum*, *Undaria pinnatifida*, *Sargassum vulgare*, Brown seaweed species, and *Ulva* species.

Commercial applications of alginic acid are in food industries, pharmaceuticals, feed, and cosmetics. Bioactive effects of alginic acid include: antihypertensive effects, hypocholesterolemic, hypolipidemic, and hypoglycemic responses, protection against carcinogens and as a prebiotic dietary fiber [92, 105, 144]. The hypocholesterolemic response is generated due to the enhanced elimination of cholesterol from the digestive system and increased fecal cholesterol content [44, 64, 138]. Other effects include: prevention of uptake of toxic chemical substances and protection of the surface membranes of the intestine and stomach [82, 131, 137]. Marine sources are prominent, because alginic acid cannot be derived from any land plants.

2.2.2.6   LAMINARIN

Laminarin is a marine polysaccharide, which is a rich and readily available carbon source for marine prokaryotes. It can stimulate growth, provides tolerance to thermal stress and induces disease resistance. Derived mainly from the fronds of Laminaria/Saccharina (maximum 32% of dry weight), Laminarin has commercial prebiotic applications [41]. Due to its non-gelling nature, it can also be used in the medical and pharmaceutical industries. It is also under study as a cancer curing and tumor-inhibiting agent [127].

On sulphation, reduction or oxidation, it shows blood anticoagulant properties, also the anticoagulation is more with a higher degree of sulphation [165]. The Laminarin provides immunity by enhancing the number B-cells and helper T-cells in the body as well as by protecting from bacterial pathogens and severe irradiation. Its ability to lower serum cholesterol levels and systolic blood pressure make it effective against coronary heart diseases by lowering the plaque formation in the arteries [61]. Laminarin has therapeutic effects on the patients of hypercholesterolemia and hyperlipidemia as it lowers the cholesterol absorption in the intestine [92, 105, 144]. In France, commercial utilization of Laminarin in the form of dietary fiber has gained popularity [41].

2.2.3   CHITOOLIGO-SACCHARIDE DERIVATIVES

Industrial utilization of Chitin is because of a broad spectrum of properties, e.g., gelling ability poly-electrolyte properties, reactive functional groups, and high adsorption capacity.

It is biodegradable, bacteriostatic, and fungistatic. Industrially derived from fungal mycelia and crustacean shells, chitin possesses anti-tumor effects. When chitin is deacetylated in concentrated NaOH solution, then chitosan is produced [174]. In the food industry, chitinous biopolymers are used as preservative agents [31], edible films, fat replacers and flavor ingredients. Kitozyme (Brussels), Primex (Norway), and United Chittechnologies, Inc. (United States) are key global players involved in the chitin and chitosan' production, and new entrants are expected to come in the near future [180].

2.2.4   BIOACTIVE LIPIDS

Vitamins (A, D, E, and K) and carotenoids with fat-soluble nature are carried in the cell through the lipid compounds that consist of waxes, fats, phospholipids (PLs), sterols, mono-, di-, and triglycerides, fat-soluble vitamins, etc. The designation of 'lipid' is determined by their unique solubility characteristics. Simple lipids are fatty acid esters with alcohol and consist of fats and waxes (fatty acid esters with long chain alcohols other than glycerol). Compound lipids contain groups in addition to an alcohol attached fatty acid ester. The third category of lipids consists of derived lipids (which are obtained from neutral and/or compound lipids) like fat-soluble vitamins, sterols, and hydrocarbons. About 70% of the dietary lipids come from vegetable oils, 28% of the fat of land animals and the remaining 2% from marine oils. Approximately one g day^{-1} is the consumption of marine oils (per capita) throughout the world. Differences between vegetable oils and marine oils lie in the length of the carbon chain. Vegetable oils (unsaturated and saturated fatty acids (SFA)) have 16- and 18-carbon chain (C16 and C18), while marine oils have 14–22 carbon chain (C14-C22) [135]. The maximum lipid content has been reported in the fronds of *Laminaria, Saccharina, and A. esculenta* species [55]. Also, the aquatic species found in cold waters generally have higher amounts of polyunsaturated fatty acids (PUFAs) [135].

2.2.4.1   FATTY ACIDS

Marine-based long-chain polyunsaturated fatty acids (LCPUFA) possess twenty and above carbon atoms with double bonds (two or more) from the methyl (omega) end of the fatty acid. Omega-3 (also known as α-linolenic acid) and omega-6 (also called as linoleic acid) are important bioactive fatty acids. Both omega-3 and omega-6 fatty acids exhibit affinity for same enzymes in the body including Δ–6-desaturase, elongase, Δ–5-desaturase and Δ–4-desaturase thus maintaining the proper ratio between omega-6 and omega-3 becomes indispensable. Particularly, omega-3 fatty acids hold importance is owing to their role in functional foods [155].

2.2.4.1.1 Omega- 3 Fatty Acids

LCPUFA called as Omega-3 fatty acids are considered as good fat. These fatty acids have the first double bond at the third carbon atom from the methyl end. Their therapeutic functions include: brain development, nerve development and alleviation of cardiovascular diseases, autoimmune disorders and atherosclerosis. There are three types of omega-3 fatty acids: Docosahexaenoic acid (DHA), Eicosapentaenoic acid (EPA), and Alpha-linolenic acid (ALA).

EPA and DHA are mainly found in cold water fish. They not only prevent heart diseases but also slow down the onset of inflammatory conditions like arthritis. American Heart Association recommended individuals with coronary heart disease to use 1 g/day combined consumption of EPA and DHA. In the western diet system, the consumption of omega-6 fatty acids results in a higher intake ratio of omega-6/omega-3 fatty acids favoring the onset of many diseases. Marine algae and fish are good sources of omega-3 fatty acids. Only some bacteria and fungi have been related to the production of these PUFAs in considerable amounts [197]. Table 2.3 highlights processing of various marine sources for omega-3 fatty acids.

Production of a long chain of omega-3 fatty acids in fishes in noteworthy quantities does not take place. They assimilate them by consuming zooplankton (that feed on algae) through their diet [1]. Also, there are taste and odor problems connected with fish. Therefore, algal-based technologies have gained popularity because they can eliminate these problems. Initial attempts that were made in the early 1980s involved photosynthetic microalgae strains for outdoor pond production. However here also, a large amount of water was required for harvesting and early self -shading resulted in low culture densities. Commercial fermentation technologies were developed as a substitute to pond production for heterotrophic production of microalgae. Heterotrophic processes, where fungi or bacteria were used for the production of a long chain of omega-3 fatty acids, were less competitive compared to the algal-based technologies, because of lower lipid production by the bacterial systems and long fermentation times in fungal systems to obtain considerable omega-3 fatty acid production [14].

TABLE 2.3 Processing of Various Marine Sources for Omega-3 Fatty Acids

Source	Production technology	Advantages	Limitations	Reference
Bluegreen *Spirulina* and green *Dunaliella*	Outdoor pond production (photosynthetic approach).	Tolerance of *Dunaliella* for high salinity and of *Spirulina* for high pH.	A large amount of water required for harvesting and low culture densities before self -shading starts.	[13, 50]
Fish	Pressing fish flesh.	Oil contains 20–30% of omega 3 fatty acids.	Stability, odor, and taste problems. Fish do not produce omega-3 fatty acids on their own and rather acquire them by eating zooplankton.	[1, 14]
Isochrysis galbana	Pond cultivation.	Highest Omega-3 production from photosynthetic strain at lab scale.	Average production limits to 4 to 8 mg/L/day only in outdoor ponds.	[113]
Microalgae, fungi, or bacteria (production organisms)	Heterotrophic production in fermenters.	High degree of control and rapid growth. Light cell densities, lower the costs for biomass harvesting.	Setting up large-scale fermentation facility is costly.	[14, 197, 151]

2.2.4.2   STEROLS

Sterols are found naturally in fungi, plants, and animals with cholesterol being the common animal sterol while phytosterols and phytostanols (reduced sterols) trace back their origin to plants. Cholesterol is necessary for cellular functionality as it alters the membrane fluidity of the animal cell, facilitates signaling by acting as a secondary messenger and is a predecessor to steroid hormones and fat-soluble vitamins. Important phytosterols include: sitosterol, campesterol, stigmasterol, and brassicasterol. The amount and type of sterol will depend on the seaweed species. Different sterols are their marine sources (Table 2.4).

TABLE 2.4   Sterols in Various Seaweed Species

Marine source	Sterols present as	Reference
Green seaweed species	Cholesterol, 28isofucocholesterol, β-sitosterol, and 24-methylene-cholesterol.	[161, 189]
Brown seaweed species	Cholesterol, fucosterol, and brassicasterol.	
Red seaweed species	Cholesterol, desmosterol, fucosterol, sitosterol, and chalinasterol	

In brown seaweed species (*Laminaria* and *Undaria*), 662–2,320 µg/ g of dry weight, i.e., 83–97% of total sterol content is because of fucosterol. The prominent sterol in red seaweed (*Palmaria* and *Porphyra*) is desmosterol with a sterol content of 75.69–313.41 µg/g of dry weight. *C. crispus,* the red seaweed, possesses cholesterol in the highest amount [100, 161].

2.2.4.3   GLYCOLIPIDS

Glycolipids are lipids attached with energy-giving carbohydrates. The glycolipid content, particularly glycosphingolipids, which are found in mammalian cells, differ with cell location and type, and organelle (Table 2.5). Glycolipids cause numerous neurological disorders and metabolic disorders like lack of insulin resistance. They may also control the reaction of key membrane proteins like insulin receptors, growth factor receptors, and some ion channels. They play a major role in the forming lipid rafts, resultantly influencing the dynamics and structure of membrane arrangements within cells [167].

TABLE 2.5   Individual Glycolipids in Red, Green, and Brown Algae

Marine algae	Total glycolipid content	Individual glycolipids present as	Reference
Red algae	16 to 32 µmol g−1 dry weight	MGDG, DGDG, and sulphaquinovosyldiacyl-glycerol are present in the same quantity.	[39, 40]
Green algae	11 to 32µmol g−1 dry weight	MGDG and DGDG are main glycolipids.	
Brown algae	—	MGDG (26% to 47%), DGDG (20% to 44%), sulphaquinovosyl-glycerol (18% to 52%).	

Major glycolipids present in marine algae are: monoglycosyldiacyl glycerol (MGDG), diglycosyldiacyl glycerol (DGDG) and sulphaquinovo-syldiacyl-glycerol (Table 2.5). The main acyl lipids in *F. vesiculosus* and *A. nodosum* are sulphoquinovosyldiacyl glycerol, monogalactosyldiacyl glycerol, digalactosyldiacyl glycerol, and trimethyl-betaalaninediacyl glycerol [72]. The maximum concentration of SFA and mono-unsaturated fatty acids (MUFA) was found in sulfoquinovosyldiacyl-glycerides for algae *Ulva, Undaria, and Palmaria* [80].

2.2.4.4   PHOSPHOLIPIDS

PLs are made of 2 fatty acids and a phosphate compound bound to a glycerol head with hydrophobic and hydrophilic regions. They are not considered as true fat because one fatty acid of triglyceride is replaced by a phosphate group. Cell membranes utilize the dual fat-soluble and water-soluble property of PLs to retain structure and transport materials.

Glycerol containing PLs include: phosphatidylethanolamine, phosphatidylcholine, phosphatidylinositol, phosphatidic acid, and phosphatidylserine. Marine phospholipids are important bioactive compounds. As they resist oxidation (do not undergo rancidity easily), they have higher contents of omega-3 fatty acids (EPA and DHA) with greater bioavailability, therapeutic effects on humans and animals [171]. Also, they are natural emulsifiers, allow absorption of lipophilic nutrients (such as fatty acids, cholesterol, and similar compounds), facilitate lipoprotein synthesis in the liver, conduct nerve signals, protect the nervous system and control the muscles. Choline, which is a bioactive molecule, acts as a precursor to the neurotransmitter acetylcholine is present in phosphatidylcholine and sphingomyelin [128]. In red seaweed species, phosphatidylcholine (62–78%), phosphatidylglycerol

(10–23%), unidentified polar PLs (2.7–10%) and minor PLs (phosphatidyl-ethanolamine and diphosphatidylglycerol) are present.

Generally, the phospholipid content in red seaweeds varies with species and makes 10–21% of the total lipid, i.e., 0.5–2.6 mg g^{-1}of dry weight [38]. *F. vesiculosus* and *A. nodosum* have phosphatidylethanolamine as the major PLs, but form less than 10% of the total acyl lipids [72]. Because of their tight intermolecular packing arrangement at *sn*–2 positions and their synergy with α-tocopherol, they present stable liposomes; thus they have food and feed applications [56].

2.2.5 PHLOROTANNINS

Phlorotannins are polyphenolic secondary metabolites of brown algae that act as antibacterial agents and protect the algae from predators. They are formed from the polymerization of phloroglucinol and synthesis occurs primarily by the acetate-malonate pathway. Ranging from various molecular sizes, they show hydrophilic behavior [152]. Therapeutic effects of these phlorotannins include: antioxidant activity, anti-HIV [10] and various others like anti-proliferative, anti-hypertensive, and anti-inflammatory [74, 76, 95]. Their main role is in cell wall synthesis and defense mechanisms [9]. Fucoid species of brown algae have higher concentrations of phlorotannins than other species obtained from tropical Pacific regions [175].

2.2.6 LECTINS

Lectins consist of a varied group of carbohydrate binding proteins present in marine organisms and have non-immune origin [62]. Bioactive lectins are present in macro-algal species such as: *Eucheuma spp.*, *Ulva sp.*, and *Gracilaria sp.* They are commercially isolated from green marine algae (*Codium fragile subsp. tomentosoides*) [170]. The major role of lectins is their antibiotic activity against fish pathogen bacteria. Algal lectins have also been screened for other bioactive properties including cytotoxic, mito-genic, anti-inflammatory, anti-nociceptive, anti-adhesive, and anti-HIV activities [22, 130, 170]. Soluble and membrane-bound glycoconjugates linked to glycan structures interact specifically with Lectins. Red algae (*Pterocladia capillacea* and *E. serra*) in aqueous and saline ethanol media at low concentrations stopped *V. vulnificus* from growing [111]. Novel calcium-dependent intelectin is present in the skin mucus of catfish *Silurus asotus*. Intelectin has a structure similar to Lactoferrin, which is a receptor present in the intestine. RBL family lectins, whose physiological functions are not dependent on the carbohydrate-recognition ability, are useful for the diagnosis of various pathological conditions such as Burkitt's lymphoma, which is very virulent [178].

2.2.7 DITERPENES

Diterpenes are non-volatile halogenated compounds, which are found in excess in brown algae (genus *Dictyota*). Its different carbon structure includes: dolabellane, xenicane, and prenylated guaiane skeletons [23]. Typical algal terpenes found and isolated from different species of *Dictyota* are dictyol C, dictyodial, and dictyol H. [120].

2.2.8 BIOACTIVE PIGMENTS

2.2.8.1 CAROTENOIDS AND CHLOROPHYLLS

Carotenoids represent red, orange or yellow wavelengths of photosynthetic pigments in all algae, higher plants, and many photosynthetic bacteria. Because of a high degree of unsaturation, they are prone to oxidation and thus prevent the rest of the molecules. Therefore, carotenoid pigments are anti-oxidative in nature [136]. Green algae Chlorella, when propagated under mass culture conditions by using a synthetic nutrient medium like potassium nitrate, resulted in low yields of carotene and xanthophyll in both light and darkness [77].

Green seaweeds get their color from pigments Chlorophyll-A and Chlorophyll-B. Thylakoids (light-harvesting structures) use Chlorophyll-A for carrying out photosynthesis [112, 155]. During processing, Chlorophyll gets converted to pheophytin, pyropheophytin, and pheophorbide. These compounds have antimutagenic and anticancer properties. In studies on seaweeds containing chlorophyll, it was found that seaweed cultivated and obtained from harbor districts relatively has three times higher content than the seaweed harvested from an open sea locality [106]. The brown color of seaweeds is because of xanthophyll pigment (fucoxanthin). In red seaweeds, the green pigment (Chlorophyll *a*) and yellow pigment (beta-carotene) are masked by red pigments (phycoerythrin and phycocyanin). Fucoxanthin is a functional pigment derived from brown seaweeds. This pigment has anti-inflammatory, anti-oxidant, and anti-tumor effects [52, 119].

2.2.8.2 PHYCOBILIPROTEINS

Unlike chlorophyll and carotenoids, phycobiliproteins are not oil-soluble and consist of linear tetrapyrroles type of pigmented phycobilins. They are not embedded in the membranes but rather form particles (phycobilisomes) on the thylakoidal surface. Combination of the two different and principal phycobilins - phycoerythrobilin (red) and phycocyanobilin (blue), absorb at wavelengths of the visible light to give varied absorption spectra [112]. Phycobiliproteins are divided into phycoerythrin, phycocyanin, allophycocyanin, and phycoerythrocyanin. Inside the phycobilisomes, water-soluble phycobiliproteins are important in the photosynthetic process of Cyanophyta, Rhodophyta, and Cyptophyta algae [7, 34]. Other photosynthetic pigments that are present in algae are responsible for harvesting light in deep waters [187].

2.3 SIGNIFICANT BIOACTIVE COMPOUNDS FROM ALGAL SOURCES

With a high content of polysaccharides, minerals, PUFA, vitamins, and bioactive compounds, marine algae constitute an important source of nutrition. In algae, the process of nourishment occurs via osmosis in the absence of roots and vascular system. Their chemical composition categorizes them into brown alga (*phaeophyta*), a green alga (*chlorophyta*), and a red alga (*rhodophyta*) [52]. Their broad spectrum of functions includes: defense mechanisms, antioxidative, anti-microbial, and anti-tumor properties (Table 2.6).

TABLE 2.6 Functions of Bioactive Compounds from Algal Sources

Algal sources	Bioactive compound	Function	Reference
Brown algae	Fucoxanthin (most abundant carotenoid in nature).	Anti-oxidant and anti-tumor. Recently studied for controlling weight gain by enhancing metabolism	[119]
Brown algae (*Fucus vesiculosus* and *Leathesia nana)*	Bromophenols	Anti-microbial and grazing deterrents.	[193, 194]
Marine microalgae	Oxylipins	Defense mechanisms.	[101]
Sargassum vulgare	Alginic acid	Anti-tumor	[172]

2.4   BIOACTIVE COMPOUNDS FROM MARINE BYPRODUCTS

The aquatic by-products are not considered as waste anymore, and present research aims at exploring their potential applications [123]. Various technologies can be used for the extraction and purification of marine bioactive compounds like biopeptides, fatty acids, oligosaccharides, biopolymers, and other water-soluble minerals [86]. Commercial applications of bioactive compounds isolated from aquatic by-products are summarized in Table 2.7.

After deproteinization, demineralization, and decoloration, shellfish and crustacean shell waste can be processed to produce chitin [60]. This chitin after partial hydrolysis or deacetylation will lead to chitin oligomers or chitosan oligomers, which have nutraceutical applications. Shrimp processing waste on mild enzymatic treatment isolates approximately 90% of the protein and carotenoids. The carotenoprotein produced after treatment has applications in feed supplementation [174].

2.5   PROCESSING TECHNOLOGY FOR BIOACTIVE COMPOUNDS

Novel bioprocessing technologies are being developed to isolate important bioactives from marine sources for their further applications in therapeutic foods (Table 2.8). Some bio-transformation processes via enzyme-mediated hydrolysis in batch reactors are involved in the extraction of these bioactive ingredients. Based on the target compound, different techniques are currently used like to isolate with maximum yield and purity. Membrane bioreactor technology installed with ultrafiltration membranes has come up for the bioprocessing and is considered as an efficient technique for complete utilization of marine source products [86, 89, 134]. Ultrafiltration membrane system results in indesirable size and bioactive property of the isolates.

For obtaining biopeptides, a three-enzyme system is used. Here enzymatic digestion is done sequentially. Ultrafiltration membrane system in combination with a multistep recycling membrane reactor aids in serial enzymatic digestions [71]. Recently novel technologies have also been reviewed for extracting bioactive compounds from marine algae since high temperatures and solvents employed during conventional extraction methods affect the functionality of these bioactive compounds. Other limitations of conventional methods include their energy-intensive nature, mass transfer resistance due to the involvement of multi-phases, longer extraction times, reproducibility challenges and lastly difficulty in disposing of the solvents because of harmful effects on the environment.

TABLE 2.7 Bioactive Compounds from Marine By-products: Their Advantages and Applications

Byproduct	Compound isolated	Advantages	Commercial applications	Reference
Fish bone	Hydroxyapatite [$Ca_{10}(PO_4)_6(OH)_2$]	Thermodynamically stable at physiological pH. Participates in bone bonding.	Hydroxyapatite is used for faster bone repair after surgeries.	[69]
Fish eggs	Lectins (natural glycoproteins).	Lectin-pathogen complex doesn't allow pathogens to cause disease (thus acting as an antibiotic).	Nutraceutical effects of lectin.	[15, 75]
Fish frames and cutoffs	Muscle proteins	Balanced amino acid composition and easily digestible.	Enzymatically hydrolyzed to recover protein biomass. Protein hydrolysates show functional properties.	[17, 183]
Fish internal organs	Enzymes (pepsin, trypsin, chymotrypsin and collagenases)	Catalyst even at low concentrations, low temperatures and stable over a wide range of pH.	Enzymes can be used for commercial production of bioactive at large scale.	[25, 53, 54, 85, 146].
Fish skin	Collagen and gelatin.	Pose a low risk of unknown pathogens like BSE (Bovine Spongiform Encephalopathy)	Commercial alternative of bovine-derived collagen and gelatin.	[51, 87]
Shellfish and crustacean shells.	Chitin (biologically active polysaccharide).	Bioactive compounds derived from chitin are non-toxic, biocompatible, and biodegradable.	Chitin oligomers and chitosan oligomers have used in nutraceuticals. Chitosan has also got food applications.	[88, 70]

TABLE 2.8 Principles and Advantages of Novel Techniques for Extraction of Bioactive Marine Compounds

Novel extraction technique	Principle of extraction	Advantages	Reference
Enzyme-assisted extraction	Breaking down algal cell walls and cuticle (complex biomolecules) using enzymes such as proteases and carbo-hydrases to marine algae at desired pH and temperature conditions to release the desired bioactive compounds.	• Eco-friendly • Nontoxic • High bioactive yielding technology • No issue of water solubility and insolubility • Low-cost technique with food grade enzymes • Preserves the original efficacy of compounds.	[4, 57, 71, 147, 190]
Microwave-assisted extraction	Microwave energy is transferred to the solution and heating occurs by twin mechanisms of dipole rotation and ionic conduction. The rotation of molecules is proportionate to the frequency of radiation. The absorbed energy is redistributed in the media, and homogeneous heating takes place. Due to disruption of H-bonds, solvent penetrates more in the matrix and extracts the bioactive compounds.	• Better extraction rate • Less solvents required • Higher extraction yield • Less costly than SFE • Closed-vessel microwave-assisted extraction is particularly being used for those compounds that need conditions like high temperature and pressure.	[78, 99, 158, 188]
Pressurized liquid extraction	Under moderate conditions (50– 200°C and 3.5–20 MPa), the solvents get heated to above their boiling point due to which the diffusion and penetration increases. At higher temperature, the solubility is more, viscosity, and surface tension are much lower. The reduced viscosity gives the solvent a better distribution over the algal matrix, resulting in a better yield.	• Less quantity of solvent required for extraction. • In pressurized hot water extraction, no solvent is required. • Faster method than other solvent extraction techniques. • When compared with SFE, flexibility in the use of solvents.	[63, 132, 146]

TABLE 2.8 (Continued)

Novel extraction technique	Principle of extraction	Advantages	Reference
Supercritical fluid extraction	Above the critical limits, the density of the fluid used is equivalent to that of a liquid and the viscosity to that of a gas. However, the diffusivity lies between a liquid and a gas. Higher diffusivity allows better penetration and faster extraction of the bioactive compounds. This technique is feasible for the extraction of functional compounds (nutraceuticals).	• No loss in the volatile behavior of bioactive compounds. • Faster extraction rate • Greater yield • Eco-friendly • Little or no requirement of organic solvents. • Fluid density can be changed by varying temperature and pressure.	[58, 59, 78]
Ultrasound-assisted extraction	Ultrasonic waves are mechanical waves that traverse through solid, liquid, and gas media through rarefactions and compression resulting in negative pressure in the liquid as a result of which bubble formation takes place at high-pressure values. These bubbles result in cavitation and subsequently macro turbulence which causes particle breakdown and the extraction of bioactive compounds from the algal matrix.	• Simple and inexpensive • High extraction yield and fast kinetics • Wide range of solvents • Heat sensitive bioactive compounds can be extracted with very less deterioration	[30, 114, 185]

Novel technologies include: microwave-assisted extraction (MAE), enzyme-assisted extraction (EAE), supercritical fluid extraction (SCFE), ultrasound-assisted extraction (UAE) and pressurized liquid extraction (PLE). In a study using pressurized liquid and solid phase extraction, it was found that marine algal products contained bioactive compounds like phenolic acids (vanillic, caffeic, salicylic acid, p-hydroxybenzoic, chlorogenic, protocatechuic, 2,3-dihydroxybenzoic) in microgram levels per unit gram of lyophilized sample [143]. Also, metabolites can be effectively isolated by high-speed counter-current chromatography. General pretreatment given in most of the extraction processes involves washing of the harvested algae to remove salts and impurities followed by drying and milling. Some pretreatments specifically prevent co-extraction of undesired functional compounds showing nearly the same solubility.

2.5.1 APPLICATIONS OF NOVEL EXTRACTION TECHNIQUES

- *Microwave-assisted extraction*: Sulfated polysaccharides (fucoidan) were extracted from brown seaweed (*Fucus vesiculosus*) [156]. Optimum fucoidan recovery occurred at 800 kPa, with a treatment time of 1 min, using 1 g of alga / 25 mL of water. Carotenoids were extracted from *Dunaliella tertiolecta* at 56°C and atmospheric pressure [148]. Iodine from *Palmaria* (dulse), *Porphyra* (nori), *Laminaria ochroleuca* (kombu), *Ulva rigida* (sea lettuce), *Himanthalia elongata* (sea spaghetti) and *Undaria pinnatifida* (wakame) at 200°C, 1000 W and 0–5 min holding time [157].
- *Ultrasound-assisted extraction*: Isoflavones were extracted from algal species (*Porphyra sp., Sargassum muticum, Halopytis incurvus, Sargassum vulgare, Undaria pinnatifida, Hypnea spinella,* and *Chondrus crispus*) by using sonication for 30 min [93]. At 17 kHz, 65°C minerals were isolated from *Laminaria ochroleuca* (brown marine algae), *Palmaria* (red marine algae), *Porphyra, Undaria pinnatifida,* and *Himanthalia elongate* using ultrasound extraction [43].
- *Enzyme-assisted extraction*: Using alginase lyase enzymes at 37°C and pH 6.2, EAE was used to isolate fucoxanthin from *Undaria pinnatifida* [21]. Using carbohydrases and proteases to degrade the algal cell walls, antioxidant-rich extracts were recovered by this technique from *Sargassum horneri* and brown seaweed species [57, 147].
- *Pressurized liquid extraction*: At 110°C and 90% ethanol, fucoxanthin was isolated using PLE from *Eisenia bicyclis* [164]. Optimum

extraction of zeaxanthin from *Chlorella ellipsoidea* occurred at 115.4°C and 23.3 min [96]. *Himanthalia elongata* was used as the source of bioactive phenols (extraction at 50, 100, 150 and 200°C for 20 min) [149].

- **Supercritical fluid extraction**: Anti-oxidants (such as polyphenols, chlorophyll, carotenoids, and others) have been isolated from marine sources using supercritical fluid. Using ethanol and acids as solvents and *Haematococcus pluvialis* as the source, astaxanthin was extracted [48]. From *Scenedesmus almeriensis*, carotenoids were isolated at 40 MPa and 60°C [117]. Chlorophyll was extracted from *Dunaliella salina* using methanol [118]. Antioxidants like polyphenols and β-carotene from *Sargassum muticum and Botryococcus braunii, Chlorella vulgaris, Dunaliella salina* respectively. *Sargassum muticum* was subject to modified CO_2, 12% ethanol, 15.2 MPa, 60°C for 90 min extraction while *Botryococcus braunii* and other sources of β-carotene were exposed to of 30 MPa and 40°C [6, 124]. A 40–50°C temperature range and 24.1–37.9 MPa pressure range were employed for PUFAs extraction from *Hypneacharoides sp.* [32].

2.6 FUTURE TRENDS

Bioactive compounds hold myriad of applications in the food industry. The market for bioactive compounds is continuously growing with significant demand across various food verticals. Also, the marine environment acts as a reservoir for the generation of new bioactive compounds. Greater efforts are required to fully exploit the potential uses of these compounds and deliver a safe end product to consumers. Nutrition could be a target market for marine biopeptides, which were until now used primarily in dairy products only. Though the market for sports nutrition is growing annually at an average rate of 5–7%, but ensuring good quality of raw materials and constant batch compositions is a challenge ahead [180].

2.7 SUMMARY

Consumption of high-calorie diets and sedentary lifestyles leads to the onset of diseases such as cancers, obesity, cardiovascular diseases, and diabetes. Use of functional foods (the concept of which developed in Japan) has gained popularity to prevent and cure physiological and metabolic disorders.

A functional food excluding pills and capsules must retain the shape of the food and impart a physiological effect apart from any observed nutritional effects. The most common method of producing novel functional foods is by incorporating functional ingredients (bioactive compounds) to traditional foods. Apart from terrestrial plants and animals, vast seas and oceans are underutilized sources of functional ingredients. Here it is essential to consider how bioactive compounds can be obtained from new matrices and the appropriate combination of low cost-intensive, selective, and environment-friendly extraction techniques with fulfilling legal requirements at the same time (like using food-grade solvents with minimum health implications). Synthesis of new value-added ingredients from bioactive compounds has been identified as a key research area. But constraints of insufficient supply, high costs of productions, end- product quality and harmful effects if any need to be considered while creating new formulations.

Screening of bioactives using novel technologies like MAE, UAE, enzyme-assisted extraction, use of supercritical fluid or pressurized liquid provides better extraction rate and yield than previously used methods of hydro-distillation, maceration, and soxhlet extraction. Also, the ultrafiltration membrane system results in desirable size and bioactive property of the functional isolates. Apart from the processing technology various aspects of bioactive peptides, polysaccharides, chitooligosaccharides, lipids, pigments, phlorotannins, lectins, and diterpenes have been discussed in this chapter. Also, a brief layout of marine by-product screening for functional compounds has been given. Now the major challenge that awaits food scientists is to develop selective formulations for use as functional foods and nutraceuticals that do not hold back the consumer from buying because of sensory related issues but rather deliver these marine functional ingredients in their bioavailable forms with least possible side effects on health.

KEYWORDS

- algae
- anti-hypertensive
- carotenoid
- enzyme
- marine food

- **membrane bioreactor**
- **peptides**
- **pigments**
- **polysaccharides**
- **seaweed**
- **supercritical fluid extraction**

REFERENCES

1. Ackman, R. G., Jangaard, P. M., Hoyle, R. J., & Brockerhoff, H., (1964). Origin of marine fatty acids. I. Analysis of the fatty acids produced by the diatom *Skeletonema costatum*. *J. Fish. Res. Bd. Canada, 21*, 747–756.

2. Adhikari, U., Mateu, C. G., Chattopadhyay, K., Pujol, C. A., Damonte, E. B., & Ray, B., (2006). Structure and antiviral activity of sulfated fucans from *Stoechospermum marginatum*. *Phytochemistry, 67*, 2474–2482.

3. Ahmed, A. B. A., Taha, R. M., Mohajer, S., Elaagib, M. E., & Kim, S. K., (2012). Preparation, properties, and biological applications of water-soluble chitin oligosaccharides from marine organisms. *Russian Journal of Marine Biology, 38*, 349–356.

4. Ahn, C. B., Jeon, Y. J., Kang, D. S., Shin, T. S., & Jung, B. M., (2004). Free radical scavenging activity of enzymatic extracts from a brown seaweed Scytosiphon lomentaria by electron spin resonance spectrometry. *Food Res. Int., 37*, 253–258.

5. Amano, H., Kakinuma, M., Coury, D. A., Ohno, H., & Hara, T., (2005). Effect of a seaweed mixture on serum lipid level and platelet aggregation in rats. *Fisheries Science, 71*, 1160–1166.

6. Anaëlle, T., Serrano Leon, E., Laurent, V., Elena, I., Mendiola, J. A., Stéphane, C., et al., (2013). Green improved processes to extract bioactive phenolic compounds from brown macroalgae using *Sargassum muticum* as model. *Talanta, 104*, 44–52.

7. Aneiros, A., & Garateix, A., (2004). Bioactive peptides from marine sources: Pharmacological properties and isolation procedures. *J Chromatogr B Analyt Technol Biomed Life Sci., 803*, 41–53.

8. Arasaki, S., & Arasaki, T., (1983). *Low Calorie, High Nutrition Vegetables from the Sea to Help You Look and Feel Better* (p. 196). Japan Publications, Japan - Tokyo.

9. Arnold, T. M., & Targett, N. M., (2003). To grow and defend: Lack of trade-offs for brown algal phlorotannins. *Oikos*, 406–408.

10. Artan, M., Li, Y., Karadeniz, F., Lee, S. H., Kim, M. M., & Kim, S. K., (2008). Anti-HIV-1 activity of phloroglucinol derivative, 6, 6'- Bieckol, from *Ecklonia cava*. *Bioorganic and Medical Chemistry, 16*, 7921–7926.

11. Athukorala, Y., Lee, K. W., Kim, S. K., & Jeon, Y. J., (2007). Anticoagulant activity of marine green and brown algae collected from Jeju Island in Korea. *Bioresource Technology, 98*(9), 1711–1716.

12. Baba, M., Snoeck, R., Pauwels, R., & Clercq, E., (1988). Sulfated polysaccharides are potent and selective inhibitors of various enveloped viruses, including herpes simplex virus, cytomegalovirus, vesicular stomatitis virus, and human immunodeficiency virus. *ASM Journals, 32*(11), 1742–1745.

13. Barclay, W. R., Terry, K. L., Nagle, N. J., Weissman, J. C., & Goebel, R. P., (1987). Potential of new strains of marine and inland saline-adapted microalgae for aquaculture. *J. World Aquacult. Soc., 18*, 218–228.

14. Barclay, W. R., Meager, K. M., & Abril, J. R., (1994). Heterotrophic production of long chain omega-3 fatty acids utilizing algae and algae-like microorganisms. *Journal of Applied Phycology, 6*, 123–129.

15. Bazil, V., & Entlicher, G., (1999). Complexity of lectins from the hard roe of perch (*Perca fluviatilis* L.). *International Journal of Biochemistry and Cell Biology, 31*, 431–442.

16. Benitz, K. F., Golberg, L., & Couston, F., (1973). Intestinal effects of carrageenans in the rhesus monkey. *Food Cosmet. Toxicol., 11*, 565–575.

17. Benkajul, S., & Morrissey, M. T., (1997). Protein hydrolysates from Pacific whiting solid wastes. *Journal of Agricultural and Food Chemistry, 45*, 3423–3430.

18. Beress, A., Wassermann, O., Bruhn, T., & Beress, L., (1993). A new procedure for the isolation of anti-HIV compounds (polysaccharides and polyphenols) from the marine alga *Fucus vesiculosus. Journal of Natural Products, 56*, 478–488.

19. Bhatnagar, I., & Kim, S. K., (2012). Pharmacologically prospective antibiotic agents and their sources: A marine microbial perspective. *Environmental Toxicology and Pharmacology, 34*(3), 631–643.

20. Bhattacharyya, S., Borthakur, A., Dudeja, P. K., & Tobacman, J. K., (2008). Carrageenan induces cell cycle arrest in human intestinal epithelial cells *in vitro. J. Nutr., 138*(3), 469–475.

21. Billakanti, J. M., Catchpole, O., Fenton, T., & Mitchell, K., (2012). Extraction of fucoxanthin from *Undaria pinnatifida* using enzymatic pre-treatment followed by DME and EtOH co-solvent extraction. In: King, J., (ed.), *10thInternational Symposium on Supercritical Fluids* (pp. 13–16). CASSS: Emeryville – CA.

22. Bird, K. T., Chiles, T. C., Longley, R. E., Kendrick, A. F., & Kinkema, M. D., (1993). Agglutinins from marine macroalgae of the Southeastern United States. *J. Appl. Phycol., 5*, 213–218.

23. Blunt, J. W., Copp, B. R., Hu, W. P., Munro, M. H. G., Northcote, P. T., & Prinsep, M. R., (2009). Marine natural products. *Natural Products Reports, 26*, 170–244.

24. Byun, H. G., & Kim, S. K., (2001). Purification and characterization of angiotensin I converting enzyme (ACE) inhibitory peptides from Alaska Pollack (*Theragra chalcogramma*) skin. *Proc. Biochem., 36*, 1155–1162.

25. Byun, H. G., Park, P. J., Sung, N. J., & Kim, S. K., (2003). Purification and characterization of a serine proteinase from the tuna pyloric caeca. *Journal of Food Biochemistry, 26*, 479–494.

26. Byun, H. G., Kim, Y. T., Park, P. J., Lin, X., & Kim, S. K., (2005). Chitooligosaccharides as a novel β-secretase inhibitor. *Carbohydr. Polym., 61*, 198–202.

27. Byun, H. G., Lee, J. K., Park, H. G., Jeon, J. K., & Kim, S. K., (2009). Antioxidant peptides isolated from the marine rotifer, *Brachionus rotundiformis. Process Biochemistry, 44*, 842–846.

28. Carlucci, M. J., Pujol, C. A., Ciancia, M., Noseda, M. D., Matulewicz, M. C., & Damonte, E. B., (1997). Antiherpetic and anticoagulant properties of carrageenans from the red

seaweed *Gigartina skottsbergii* and their cyclized derivatives: Correlation between structure and biological activity. *International Journal of Biological Macromolecules*, *20*, 97–105.

29. Carthew, P., (2002). Safety of carrageenan in foods. *Environ Health Perspect*, *110*, 166–176.

30. Chemat, F., Zill-e-Huma, & Khan, M. K., (2011). Applications of ultrasound in food technology: Processing, preservation, and extraction. *Ultrason. Sonochem.*, *18*, 813–835.

31. Chen, C., Liau, W., & Tsai, G., (1998). Antibacterial effects of N-sulfonated and N-sulfobenzoyl chitosan and application to oyster preservation. *Journal of Food Protection*, *61*, 1124–1128.

32. Cheung, P. C. K., (1999). Temperature and pressure effects on supercritical carbon dioxide extraction of n–3 fatty acids from red seaweed. *Food Chem.*, *65*, 399–403.

33. Chevolot, L., Foucault, A., Chaubet, F., Kervarec, N., Sinquin, C., & Fisher, A. M., (1999). Further data on the structure of brown seaweed fucans: Relationships with anticoagulant activity. *Carbohydrate Research*, *319*, 154–165.

34. Chronakis, I. S., Galatanu, A. N., Nylander, T., & Lindman, B., (2000). The behavior of protein preparations from blue-green algae (*Spirulina platensis* strain Pacifica) at the air/water interface. *Colloid Surface Physicochem. Eng. Aspect*, *173*, 181–192.

35. Church, F. C., Meade, J. B., Treanor, E. R., & Whinna, H. C., (1989). Antithrombin activity of fucoidan: The interaction of fucoidan with heparin cofactor II, antithrombin III, and thrombin. *Journal of Biological Chemistry*, *264*, 3618–3623 .

36. Colliec, S., Fischer, A. M., Tapon-Bretaudiere, J., Boisson, C., Durand, P., & Jozefonvicz, J., (1991). Anticoagulant properties of a fucoidan fraction. *Thrombosis Research*, *64*(2), 143–154.

37. Damonte, E. B., Matulewicz, M. C., & Cerezo, A. S., (2004). Sulfated seaweed polysaccharides as antiviral agents. *Current Medicinal Chemistry*, *11*, 2399–2419.

38. Dembitsky, V. M., & Rozentsvet, O. A., (1990). Phospholipid composition of some marine red algae. *Phytochemistry*, *29*, 3149–3152.

39. Dembitsky, V. M., Rozentsvet, O. A., & Pechenkina, E. E., (1990). Glycolipids, phospholipids, and fatty-acids of brown-algae species. *Phytochemistry*, *29*, 3417–3421.

40. Dembitsky, V. M., Pechenkinashubina, E. E., & Rozentsvet, O. A., (1991). Glycolipids and fatty-acids of some seaweeds and marine grasses from the Black Sea. *Phytochemistry*, *30*, 2279–2283.

41. Deville, C., Damas, J., Forget, P., Dandrifosse, G., & Peulen, O., (2004). Laminarin in the dietary fiber concept. *J. Sci. Food Agric.*, *84*, 1030–1038.

42. Dobashi, K., Nishino, T., Fujihara, M., & Nagumo, T., (1989). Isolation and preliminary characterization of fucose-containing sulfated polysaccharides with blood-anticoagulant activity from the brown seaweed *Hizikia fusiforme*. *Carbohydrate Research*, *194*, 315–320.

43. Domínguez-González, R., Moreda-Piñeiro, A., Bermejo-Barrera, A., & Bermejo-Barrera, P., (2005). Application of ultrasound-assisted acid leaching procedures for major and trace elements determination in edible seaweed by inductively coupled plasma-optical emission spectrometry. *Talanta*, *66*, 937–942.

44. Dumelod, B. D., Ramirez, R. P. B., Tiangson, C. L. P., Barrios, E. B., & Panlasigui, L. N., (1999). Carbohydrate availability of arroz caldo with *lambdacarrageenan*. *Int. J. Food Sci. Nutr.*, *50*, 283–289.

45. Epand, R. M., & Vogel, H. J., (1999). Diversity of antimicrobial peptides and their mechanisms of action. *BBA Biomembranes*, *1462*, 11–28.

46. Eskilsson, C. S., Hartonen, K., Mathiasson, L., & Riekkola, M. L., (2004). Pressurized hot water extraction of insecticides from process dust – comparison with supercritical fluid extraction. *J. Sep. Sci., 27,* 59–64.

47. Fleury, N., & Lahaye, M., (1991). Chemical and physicochemical characterization of fibers from *Laminaria digitata* (Kombu Breton) - A physiological approach. *Journal of the Science of Food and Agriculture, 55*(3), 389–400.

48. Fujii, K., (2012). Process integration of supercritical carbon dioxide extraction and acid treatment for astaxanthin extraction from a vegetative microalga. *Food Bio. Prod. Process, 90,* 762–766.

49. Ghosh, T., Chattopadhyay, K., Marschall, M., Karmakar, P., Mandal, P., & Ray, B., (2009). Focus on antivirally active sulfated polysaccharides: From structure-activity analysis to clinical evaluation. *Glycobiology, 19*(1), 2–15.

50. Gladue, R. M., (1991). Heterotrophic microalgae production: Potential for application to aquaculture feeds. In: Fulks, W., & Main, K. L., (eds.), *Rotifer and Microalgae Culture Systems* (pp. 275–286). Oceanic Institute, Honolulu.

51. Gomez-Gillen, M. C., Turnay, J., Fernandez-Diaz, M. D., Ulmo, N., Lizarbe, M. A., & Montero, P., (2002). Structural and physical properties of gelatin extracted from different marine species: A comparative study. *Food Hydrocolloids, 16,* 25–34.

52. Gupta, S., & Abu-Ghannam, N., (2011). Bioactive potential and possible health effects of edible brown seaweeds. *Trends in Food Science and Technology, 22,* 315–326.

53. Haard, N. F., (1998). Specialty enzymes from marine organisms. *Food Technology, 52,* 64–67.

54. Haard, N. F., & Simpson, B. K., (1994). Protease from aquatic organisms and their uses in the seafood industry. In: Martin, A. M., (ed.), *Fisheries Processing Biotechnological Applications* (pp. 132–154). Chapman & Hall, London.

55. Haug, A., & Jensen, A., (1954). Seasonal variation in the chemical composition of *Alaria esculenta, Laminaria saccharina, Laminaria hyperborea* and *Laminaria digitata* from Northern Norway, Norwegian Institute of Seaweed Research. *Akademisk Trykkningssentral* (Vol. 4, pp. 1–14). Blindern, Norway.

56. Henna Lu, F. S., Nielsen, N. H., Heinrich, M. T., & Jacobson, C., (2011). Oxidative stability of marine phospholites in the liposomal form and their applications. *AOCS Journal, 46*(1), 3–23.

57. Heo, S. J., Park, E. J., Lee, K. W., & Jeon, Y. J., (2005). Antioxidant activities of enzymatic extracts from brown seaweeds. *Bioresour. Technol., 96,* 1613–1623.

58. Herrero, M., Cifuentes, A., & Ibañez, E., (2006). Sub- and supercritical fluid extraction of functional ingredients from different natural sources: Plants, food-by-products, algae, and microalgae: A review. *Food Chem., 98,* 136–148.

59. Herrero, M., Mendiola, J. A., Cifuentes, A., & Ibáñez, E., (2010). Supercritical fluid extraction: Recent advances and applications. *J. Chromatogr., A., 1217,* 2495–2511.

60. Hines, B. M., Suleria, H. A. R., & Osborne, S. A., (2016). Automated screening potential thrombin inhibitors using the epMotion® 5075. *Eppendorf., 377,* 1–6.

61. Hoffmane, R., Paper, D. H., Donaldson, J., Alban, S., & Franz, G., (1995). Characterization of a laminarin sulfate which inhibits basic fibrgrowth factor-factor bindiendothelial cellal-cell proliferation. *J. Cell Sci., 108,* 3591–3598.

62. Hori, K., Matsubara, K., & Miyazawa, K., (2000). Primary structures of two hemagglutinins from the marine red alga, *Hypnea japonica. Biochim. Biophys. Acta Gen. Subj., 147,* 226–236.

63. Ibañez, E., Herrero, M., Mendiola, J., & Castro-Puyana, M., (2012). Extraction and characterization of bioactive compounds with health benefits from marine resources: Mamicroalgaero algae, cyanobacteria, and invertebrates. In: Hayes, M., (ed.), *Marine Bioactive Compounds* (pp. 55–98). Springer, New York.

64. Ito, K., & Tsuchida, Y., (1972). The effect of algal polysaccharides on depressing of plasma cholesterol level in rats. *Proceedings of the 7th International Seaweed Symposium* (pp. 451–455). Tokyo – Japan.

65. Je, J. Y., Park, J. Y., Jung, W. K., Park, P. J., & Kim, S. K., (2005). Isolation of angiotensin I converting enzyme (ACE) inhibitor from fermented oyster sauce, *Crassostrea gigas*. *Food Chemistry, 90*, 809–814.

66. Je, J. Y., Qian, Z. J., Byun, H. G., & Kim, S. K., (2007). Purification and characterization of an antioxidant peptide obtained from tuna backbone protein by enzymatic hydrolysis. *Process Biochemistry, 42*, 840–846.

67. Je, J. Y., Qian, Z. J., Lee, S. H., Byun, H. G., & Kim, S. K., (2008). Purification and antioxidant properties of bigeye tuna (*Thunnus obesus*) dark muscle peptide on free radical-mediated oxidation systems. *Journal of Medicinal Food, 11*(4), 629–637.

68. Jenkins, D. J. A., Wolever, T. M. S., Leeds, A. R., Gassull, M. A., Haisman, P., & Dilawari, J., (1978). Dietary fibers, fiber analogs, and glucose-tolerance - importance of viscosity. *British Medical Journal, 1*, 1392–1394.

69. Jensen, S. S., Aaboe, M., Pinhold, E. M., Hjrting-Hansen, Z., Melsen, F., & Ruyter, I. E., (1996). Tissue reaction and material characteristics of four bone substitutes. *International Journal of Oral & Maxillofacial Implants, 11*, 55–66.

70. Jeon, Y. J., & Kim, S. K., (2000). Continuous production of chitooligosaccharides using a dual reactor system. *Process Biochemistry, 35*, 623–632.

71. Jeon, Y. J., Wijesinghe, W. A. J. P., & Kim, S. K., (2011). Enzyme-assisted extraction and recovery of bioactive components from seaweeds. In: *Handbook of Marine Macroalgae* (pp. 221–228). Wiley, New York.

72. Jones, A. L., & Harwood, J. L., (1992). Lipid composition of the brown algae *Fucus vesiculosus* and *Ascophyllum nodosum*. *Phytochemistry, 31*, 3397–3403.

73. Jung, W. K., Je, J. Y., & Kim, S. K., (2001). A novel anticoagulant protein from *Scapharca broughtonii*. *J. Biochem. Mol. Biol., 35*, 199–205.

74. Jung, H. A., Hyun, S. K., Kim, H. R., & Choi, J. S., Angiotensin-convertingnverting enzyme I inhibitory activity of phlorotannins from *Ecklonia stolonifera*. *Fisheries Science, 72*, 1292–1299.

75. Jung, W. K., Park, P. J., & Kim, S. K., (2003). Purification and characterization of a new lectin from the hard roe of skipjack tuna, *Katsuwonus pelamis*. *International Journal of Biochemistry and Cell Biology, 35*, 255–265.

76. Jung, W. K., Ahn, Y. W., Lee, S. H., Choi, Y. H., Kim, S. K., Yea, S. S., et al., (2009). *Ecklonia cavaethanolic* extracts inhibit lipopolysaccharide-induced cyclooxygenase–2 and inducible nitric oxide synthase expression in BV2 Microglia via the MAP kinase and NF-kB pathways. *Food Chemical Toxicology, 47*, 410–417.

77. Kathrein, H. R., (1960). Production of carotenoids by the cultivation of algae. *U.S. Patent 2, 949*, 700.

78. Kaufmann, B., & Christen, P., (2002). Recent extraction techniques for natural products: Microwave-assisted extraction and pressurized solvent extraction. *Phytochem. Anal., 13*, 105–113.

79. Khosravi-Darani, K., (2010). Research activities on supercritical fluid science in food biotechnology. *Crit. Rev. Food Sci. Nutr., 50*, 479–488.

80. Khotimchenko, S. V., (2003). The fatty acid composition of glycolipids of marine macrophytes. *Russ. J. Mar. Biol.*, *29*, 126–128.

81. Killing, H., (*1913*). Zur biochemie der Meersalgen. *Zeitschrift für Physiologische Chemie, 83*, 171–197.

82. Kim, I. H., & Lee, J. H., (2008). Antimicrobial activities methicillin-resistantesistant *Staphylococcus aureus* from macroalgae. *J. Ind. Eng. Chem.*, *14*, 568–572.

83. Kim, S. H., Choi, D. S., Athukorala, Y., Jeon, V., Senevirathne, M., & Rha, C. K., (2007). Antioxidant activity of sulfated polysaccharides isolated from *Sargassum fulvellum*. *Journal of Food Science and Nutrition*, *12*, 65–73.

84. Kim, S. K., Kim, Y. T., & Byun, H. G., (2001). Isolation and characterization of antioxidative peptides from gelatin hydrolysate of Alaska pollack skin. *J. Agri. Food Chem.*, *49*, 1984–1989.

85. Kim, S. K., Park, P. J., Kim, J. B., & Shahidi, F., (2002). Purification and characterization of a collagenolytic protease from the filefish, *Novoden Modestrus. Journal of Biochemistry and Molecular Biology*, *35*, 165–171.

86. Kim, S. K., & Mendis, E., (2006). Bioactive compounds from marine processing byproducts: A review. *Food Res. Int.*, 39, 383–393.

87. Kim, S. K., Byun, H. G., & Lee, E. H., (1994). Optimum extraction conditions of gelatin from fish skins and its physical properties. *Journal of Korean Industrial and Engineering Chemistry*, *5*, 547–559.

88. Kim, S. K., Park, P. J., Yang, H. P., & Hanm, S. S., (2001). Subacute toxicity of chitosan oligosaccharide in Sprague–Dawly rats: Antibiotics - Arzneim-Forsch. *Drug Research*, *51*, 769–774.

89. Kim, S. K., & Rajapakse, N., (2005). Enzymatic production and biological activities of chitosan oligosaccharides (COS): A review. *Carbohydrate Polymers*, *62*, 357–368.

90. Kim, S. Y., Je, J. Y., & Kim, S. K., (2007). Purification and characterization of antioxidant peptide from hoki (*Johnius belengerii*) frame protein by gastrointestinal digestion. *J. Nut. Biochem.*, *18*, 31–38.

91. Kim, S. K., & Wijesekara, I., (2010). Development and biological activities of marine-derived bioactive peptides: A review. *Journal of Functional Foods*, *2*(1), 1–9.

92. Kiriyama, S., Okazaki, Y., & Yoshida, A., (1969). Hypocholesterolemic effect of polysaccharides and polysaccharide-rich foodstuffs in cholesterol-fed rats. *J. Nutr.*, *97*, 382–388.

93. Klejdus, B., Lojková, L., Plaza, M., Šnóblová, M., & Štěrbová, D., (2010). Hyphenated technique for the extraction and determination of isoflavones in algae: Ultrasound-assisted supercritical fluid extraction followed by fast chromatography with tandem mass spectrometry. *J. Chromatogr., A, 1217*, 7956–7965.

94. Kolender, A. A., Pujol, C. A., Damonte, E. B., Matulewicz, M. C., & Cerezo, A. S., (1997). The system of sulfated α-(1 → 3)-linked D-mannans from the red seaweed *Nothogenia fastigiata*: Structures, antiherpetic, and anticoagulant properties. *Carbohydrate Research*, *304*, 53–60.

95. Kong, C. S., Kim, J. A., Yoon, N. Y., & Kim, S. K., (2009). Induction of apoptosis by phloroglucinol derivative from *Ecklonia cava* in MCF–7 human breast cancer cells. *Food Chemical Toxicology*, *47*, 1653–1658.

96. Koo, S. Y., Cha, K. H., Song, D. G., Chung, D., & Pan, C. H., (2011). Optimization of pressurized liquid extraction of zeaxanthin from *Chlorella ellipsoidea. J. Appl. Phycol.*, *24*, 725–730.

97. Koyama, T., Noguchi, K., Aniya, Y., & Sakanashi, M., (1998). Analysis for sites of anticoagulant action of plancinin, a new anticoagulant peptide isolated from the starfish *Acanthaster planci*, in the blood coagulation cascade. *General Pharmacology, 31*, 277–282.

98. Kristinsson, H. G., & Rasco, B. A., (2000). Fish protein hydrolysates: Production, bio-chemical, and functional properties. *Critical Reviews in Food Science and Nutrition, 32*, 1–39.

99. Kubrakova, I. V., & Toropchenova, E. S., (2008). Microwave heating for enhancing efficiency of analytical operations (review). *Inorg. Mater., 44*, 1509–1519.

100. Kumar, C. S., Ganesan, P., Suresh, P. V., & Bhaskar, N., (2008). Seaweeds as a source of nutritionally beneficial compounds: A review. *J. Food Sci. Technol., 45*, 1–13.

101. Kupper, F. C., Gaquerel, E., Boneberg, E. M., Morath, S., Salaun, J. P., & Potin, P., (2006). Early events in the perception of lipopolysaccharides in the brown alga *Laminaria digitata* include an oxidative burst and activation of fatty acid oxidation cascades. *Journal of Experimental Botany, 57*, 1991–1999.

102. Kwon, M. C., Kim, C. H., Kim, H. S., Syed, A. Q., Hwang, B. Y., & Lee, H. Y., (2007). Anti-wrinkle activity of low molecular weight peptides derived from the collagen isolated from *Asterias amurensis. Korean Journal of Food Science and Technology, 39*, 625–629.

103. Lahaye, M., (1991). Marine-algae as sources of fibers - determination of soluble and insoluble dietary fiber contents in some sea vegetables. *Journal of the Science of Food and Agriculture, 54*(4), 587–594.

104. Lahaye, M., & Kaeffer, B., (1997). Seaweed dietary fibers: Structure, physicochemical, and biological properties relevant to intestinal physiology. *Sciences des Aliments, 17*(6), 563–584.

105. Lamela, M., Anca, J., Villar, R., Otero, J., & Calleja, J. M., (1989). Hypoglycemic activity of several seaweed extracts. *J. Ethnopharmacol., 27*, 35–43.

106. Larsen, B., & Haug, A., (1958). *The Influence of Habitat on the Chemical Composition of Ascophyllum Nodosum (L.)* (pp. 29–38). Le Jol.-Reprint of report 20. Norwegian Institute of Seaweed Research, Trondheim, Norway.

107. Lee, J. B., Hayashi, K., Hashimoto, M., Nakano, T., & Hayashi, T., (2004). Novel antiviral fucoidan from sporophyll of *Undaria pinnatifida* (Mekabu). *Chemical & Pharmaceutical Bulletin, 52*(9), 1091–1094.

108. Lee, S. H., Qian, Z. J., & Kim, S. K., (2010). A novel angiotensin I converting enzyme inhibitory peptide from tuna frame protein hydrolysate and its antihypertensive effect in spontaneously hypertensive rats. *Food Chemistry, 118*, 96–102.

109. Leiro, J. M., Castro, R., Arranz, J. A., & Lamas, J., (2007). Immunomodulating activities of acidic sulphated polysaccharides obtained from the seaweed *Ulva rigida* C AGARDH. *International Immunopharmacology, 7*, 879–888.

110. Li, Y., Qian, Z. J., Ryu, B. M., Lee, S. H., Kim, M. M., & Kim, S. K., (2009). Chemical components and its antioxidant properties in vitro: An edible marine brown alga, *Ecklonia cava. Bioorganic and Medicinal Chemistry, 1*, 1963–1973.

111. Liao, W. R., Lin, J. Y., Shieh, W. Y., Jeng, W. L., & Huang, R., (2003). Antibiotic activity of lectins from marine algae against marine vibrios. *J. Ind. Microbiol. Biotechnol., 30*, 433–439.

112. Lobban, C. S., & Harrison, P. J., (1994). *Seaweed Ecology and Physiology* (p. 384). Cambridge University Press, Cambridge.

113. López, A. D., Molina, G. E., Sánchez, P. J. A., Garcia, S. J. L., & Garcia, C. F., (1992). Isolation of clones of *Isochrysis galbana* rich in eciosapentaenoic acid. *Aquaculture*, *102*, 363–371.

114. Luque-García, J. L., & Luque de Castro, M. D., (2003). Ultrasound: A powerful tool for leaching. *Trends Anal. Chem.*, *22*, 41–47.

115. Luscher-Mattil, M., (2000). Polyanions: A lost chance in the fight against HIV and other virus diseases. *Antiviral Chemistry and Chemotherapy*, *11*, 249–259.

116. Mabeau, S., & Fleurence, J., (1993). Seaweed in food products: Biochemical and nutritional aspects. *Trends in Food Science & Technology*, *4*(4), 103–107.

117. Macías-Sánchez, M. D., Fernandez-Sevilla, J. M., Fernández, F. G. A., García, M. C. C., & Grima, E. M., (2010). Supercritical fluid extraction of carotenoids from *Scenedesmus almeriensis*. *Food Chem.*, *123*, 928–935.

118. Macías-Sánchez, M. D., Mantell, C., Rodríguez, M., Martínez de la Ossa, E., Lubián, L. M., & Montero, O., (2009). Comparison of supercritical fluid and ultrasound-assisted extraction of carotenoids and chlorophyll a from *Dunaliella salina*. *Talanta*, *77*, 948–952.

119. Maeda, H., Hosokawa, M., Sashima, T., & Miyashita, K., (2007). Dietary combination of fucoxanthin and fish oil attenuates the weight gain of white adipose tissue and decreases blood glucose in obese/diabetic KK-A^y mice. *Journal of Agricultural and Food Chemistry*, *55*, 7701–7706.

120. Manzo, E., Ciavatta, M. L., Bakkas, S., Villani, G., Varcamonti, M., Zanfardino, A., & Gavagnin, M., (2009). Diterpene content of the alga *Dictyota ciliolata* from a Moroccan lagoon. *Phytochemistry Letters*, *2*, 211–215.

121. Matsubara, K., (2004). Recent advances in marine algal anticoagulants. *Current Medicinal Chemistry*, *2*, 13–19.

122. Mayer, A. M. S., & Hamann, M. T., (2002). Marine pharmacology in 1999: Compounds with antibacterial, anticoagulant, antifungal, anthelmintic, anti-inflammatory, antiplatelet, antiprotozoal, and antiviral activities affecting the cardiovascular, endocrine, immune, and nervous systems, and other miscellaneous mechanisms of action. *Comparative Biochemistry and Physiology Part C.*, *132*, 315–339.

123. McHugh, D. J., (1987). *Production and Utilization of Products from Commercial Seaweeds* (pp. 1–189). FAO Fisheries Technical Report, Rome.

124. Mendes, R. L., Nobre, B. P., Cardoso, M. T., Pereira, A. P., & Palavra, A. F., (2003). Supercritical carbon dioxide extraction of compounds with pharmaceutical importance from microalgae. *Inorg. Chim. Acta*, *356*, 328–334.

125. Mendis, E., Rajapakse, N., Byun, H. G., & Kim, S. K., (2005). Investigation of jumbo squid (*Dosidicus gigas*) skin gelatin peptides for their in vitro antioxidant effects. *Life Sciences*, *77*, 2166–2178.

126. Mendis, E., Rajapakse, N., & Kim, S. K., (2005). Antioxidant properties of a radical-scavenging peptide purified from enzymatically prepared fish skin gelatin hydrolysate. *Journal of Agricultural Food Chemistry*, *53*, 581–587.

127. Miao, H. Q., Elkin, M., Aingorn, E., Ishai-Michaeli, R., Stein, C. A., & Vlodavsky, I., (1999). Inhibition of heparanase activity and tumor metastasis by laminarin sulfate and synthetic phosphorothioate oligodeoxynucleotides. *Int. J. Cancer*, *83*, 424–431.

128. Miraglio, A. M., (2009). *Phospholipids: Structure Plus Functionality*. http://www.naturalproductsinsider.com/articles/2006/03/phospholipids-structure-plus functionality. aspx Accessed 5 June of 2017.

129. Moller, N. P., Scholz-Ahrens, K. E., Roos, N., & Schrezenmeir, J., (2008). Bioactive peptides and proteins from foods: Indication for health effects. *European Journal of Nutrition, 47*, 171–182.

130. Mori, T. O., Keefe, B. R., Sowder, R. C., Bringans, S., Gardella, R., Berg, S., et al., (2005). Isolation and characterization of Griffiths in, a novel HIV inactivating protein, from the red alga *Griffithsia* sp. *J. Biol. Chem., 280*, 9345–9353.

131. Murata, M., & Nakazoe, J., (2001). Production and use of marine algae in Japan. *Jpn. Agr. Res. Q., 35*, 281–290.

132. Mustafa, A., & Turner, C., (2011). Pressurized liquid extraction as a green approach in food and herbal plants extraction: A review. *Anal. Chim. Acta, 703*, 8–18.

133. Nabors, L. O. B., (2004). Alternative sweeteners. *Agro Food Industry HiTech., 15*, 39–41.

134. Nagai, T., & Suzuki, N., (2000). Isolation of collagen from fish waste material — Skin, bone, and fins. *Food Chemistry, 68*, 277–281.

135. Narayan, B., Miyashita, K., & Hosakawa, M., (2002). Physiological effects of eicosapentaenoic acid (EPA) and docosahexaenoic acid (DHA)-A review. *Food Rev. Int., 22*, 291–307.

136. Ngo, D. H., Wijesekara, I., Vo, T. S., Ta, Q. V., & Kim, S. K., (2011). Marine food-derived functional ingredients as potential antioxidants in the food industry: An overview. *Food Research International, 44*, 523–529.

137. Nishide, E., & Uchida, H., (2003). Effects of Ulva powder on the ingestion and excretion of cholesterol in rats. In: Chapman, A. R. O., Anderson, R. J., Vreeland, V. J., & Davison, I. R., (eds.), *Proceedings of the 17th International Seaweed Symposium* (pp. 165–168). Oxford University Press, Oxford.

138. Nishide, E., Anzai, H., & Uchida, N., (1993). Effects of alginates on the ingestion and excretion of cholesterol in the rat. *J. Appl. Phycol., 5*, 207–211.

139. Nishino, T., Yamauchi, T., Horie, M., Nagumo, T., & Suzuki, H., (2000). Effects of fucoidan on the activation of plasminogen by u-PA and t-PA. *Thrombosis Research, 99*, 623–634.

140. Oesser, S., & Seifert, J., (2003). Stimulation of type II collagen biosynthesis and secretion in bovine chondrocytes cultured with degraded collagen. *Cell Tissue Res., 311*, 393–399.

141. Oesser, S., Adams, M., Babel, W., & Seifert, J., (1999). Oral administration of 14C labeled gelatin hydrolysate leads to an accumulation of radioactivity in cartilage of mice (C57/ BL). *J Nutr., 129*, 1891–1895.

142. Ogawa, T., Watanabe, M., Naganuma, T., & Muramoto, K., (2011). Diversified carbohydrate-binding lectins from marine resources. *Journal of Amino Acids* volume 2011, open access ID: 838914, p. 20.

143. Onofrejova, L., Vasickova, J., Klejdus, B., Stratil, P., Misurcova, L., Kracmar, S., Kopecky, J., & Vacek, J., (2010). Bioactive phenols in algae: The application of pressurized-liquid and solid-phase extraction techniques. *J. Pharm. Biomed. Anal., 51*, 464–470.

144. Panlasigui, L.N., Baello, O.Q., Dimatangal, J. M., & Dumelod, B. D., (2003). Blood cholesterol and lipid-lowering effects of carrageenan on human volunteers. *Asia Pac. J. Clin. Nutr., 12*, 209–214.

145. Painter, T. J., & Aspinall, G. O., (1983). *The Polysaccharides* (pp. 195–285). Academic Press, New York.

146. Park, P. J., Lee, S. H., Byun, H. G., Kim, S. H., & Kim, S. K., (2002). Purification and characterization of a collagenase from the mackerel, *Scomber japonicas. Journal of Biochemistry and Molecular Biology, 35,* 576–582.

147. Park, P. J., Shahidi, F., & Jeon, Y. J., (2004). Antioxidant activities of enzymatic extracts from an edible seaweed *Sargassum horneri* using ESR spectrometry. *J. Food Lipids, 11,* 15–27.

148. Pasquet, V., Chérouvrier, J. R., Farhat, F., Thiéry, V., Piot, J. M., Bérard, J. B., et al., (2011). Study on the microalgal pigments extraction process: Performance of microwave assisted extraction. *Process Biochem., 46,* 59–67.

149. Plaza, M., Santoyo, S., Jaime, L., García-Blairsy, R. G., Herrero, M., Señoráns, F. J., & Ibáñez, E., (2010). Screening for bioactive compounds from algae. *J. Pharm. Biomed. Anal., 51,* 450–455.

150. Qi, H., Zhang, Q., Zhao, T., Chen, R., Zhang, H., & Niu, X., (2005). Antioxidant activity of different sulfate content derivatives of polysaccharide extracted from *Ulva pertusa* (Chlorophyta) *in vitro. International Journal of Biological Macromolecules, 37,* 195–199.

151. Radwan, S., (1991). Sources of C20-polyunsaturated fatty acids for biotechnological use. *Appl. Microbio. Biotech., 35,* 421–430.

152. Ragan, M. A., & Glombitza, K. W., (1986). *Handbook of Physiological Methods* (pp. 129–241). Cambridge University Press, Cambridge, UK.

153. Rajapakse, N., Jung, W. K., & Mendis, E., (2005). A novel anticoagulant purified from fish protein hydrolysate inhibits factor XIIa and platelet aggregation. *Life Sci., 76,* 2607–2619.

154. Ranathunga, S., Rajapakse, N., & Kim, S. K., (2006). Purificatioantioxidativezation of antioxidantative peptide derived from muscle of conger eel (*Conger myriaster*). *European Food Research and Technology, 222,* 310–315.

155. Rasmussen, R. S., & Morrissey, M. T., (2007). Marine biotechnology for production of food ingredients. *Advances in Food and Nutrition Research, 52,* 237–292.

156. Rodriguez-Jasso, R. M., Mussatto, S. I., Pastrana, L., Aguilar, C. N., & Teixeira, J. A., (2011). Microwave-assisted extraction of sulfated polysaccharides (fucoidan) from brown seaweed. *Carbohydr. Polym., 86,* 1137–1144.

157. Romarís-Hortas, V., Moreda-Piñeiro, A., & BeMicrowave-assisted (2009). Microwave assisted extraction of iodine and bromine from edible seaweed for inductively coupled plasma-mass spectrometry determination. *Talanta, 79,* 947–952.

158. Routray, W., & Orsat, V., (2012). Microwave-assisted extraction of flavonoids: A review. *Food Bioprocess Technol., 5,* 409–424.

159. Ruperez, P., (2001). Antioxidant activity of sulphated polysaccharides from the Spanish marine seaweed Nori. In: *Proceedings of the COST 916 European Conference on Bioactive Compounds in Plant Foods: Health Effects and Perspectives for the Food Industry* (pp. 114–118). Tenerife, Canary Islands, *Spain.*

160. Rupérez, P., Ahrazem, O., & Leal, J. A., (2002). Potential antioxidant capacity of sulphated polysaccharides from edible brown seaweed *Fucus vesiculosus. Journal of Agricultural and Food Chemistry, 50,* 840–845.

161. Sanchez-Machado, D. I., Lopez-Hernandez, J., Paseiro-Losada, P., & Lopez, C. J., (2004). An HPLC method for the quantification of sterols in edible seaweeds. *Biomed. Chromatogr., 18,* 183–190.

162. Sen Sr., A. K., Das, A. K., Banerji, N., Siddhanta, A. K., Mody, K. H., & Ramavat, B. K., (1994). A new sulfated polysaanticoagulantpotent blood anti-coagulant activity from the red seaweed *Grateloupia indica. International Journal of Biological Macromolecules, 16*, 279–280.

163. Shahidi, F., Han, X. Q., & Synowiecki, J., (1995). Production and characteristics of protein hydrolysates from capelin (*Mallotus villosus*). *Food Chemistry, 53*, 285–293.

164. Shang, Y. F., Kim, S. M., Lee, W. J., & Um, B. H., (2011). Pressurized liquid method for fucoxanthin extraction from *Eisenia bicyclis* (Kjellman) Setchell. *J. Biosci. Bioeng., 111*, 237–241.

165. Shanmugam, M., & Mody, K. H., (2000). Heparinoid-active sulphated polysaccharides from marine algae as potential blood anticoagulant agents. *Current Science, 79*, 1672–1683.

166. Sharma, S., Singh, R., & Rana, S., (2011). Bioactive peptides: A review. *Int. J. Bioautomation, 15*(4), 223–250.

167. Shorthouse, D., Hedger, G., Koldso, H., & Sanson, M. S. P., (2016). Molecular simulations of glycolipids: Towards mammalian cell membrane models. *Biochimie, 120*, 105–109.

168. Skoler-Karpoff, S., Ramjee, G., Ahmed, K., Altini, L., Plagianos, M. G., Friedland, B., et al., (2008). Efficacy of carraguard for prevention of HIV infection in women in SPlacebo-controlleddomized, Placebo controlled trial. *The Lancet, 372*, 1977–1987.

169. Slizyte, R., Mozuraityte, R., Martinez-Alvarez, O., Falch, E., Fouchereau-Peron, M., & Rustad, T., (2009). Functional, bioactive, and antioxidative properties of hydrolysates obtained from cod (*Gadus morhua*) backbones. *Process Biochemistry, 44*, 668–677.

170. Smit, A.J., (2004). Medicinal and pharmaceutical uses of seaweed natural products: A review. *J. Appl. Phycol., 16*, 245–262.

171. Sørensen, H. O., (2009). Marine phospholipids. *Int. Aqua. Feed, 12*, 14–15.

172. De Sousa, A. P. A., Torres, M. R., Pessoa, C., De Moraes, M. O., Filho, F. D. R., Alves, A. P. N. N., & Costa-Lotufo, L. V., (2007). *In vivo* growth-inhibition of Sarcoma 180 tumor by alginates from brown seaweed *Sargassum vulgare. Carbohydrate Polymers, 69*, 7–13.

173. Suleria, H. A. R., Addepalli, R., Masci, P., Gobe, G., & Osborne, S. A., (2017). *In vitro* anti-blacklipory activities of black lip abalone (*Haliotis rubra*) in RAW 264.7 macrophages. *Food and Agricultural Immunology, 28*(4), 711–724.

174. Synowiecki, J., & Al-Khateeb, N. A., (2003). Production, properties, and some new application of chitin and its derivatives. *Food Science and Nutrition, 43*, 145–171.

175. Targett, N. M., & Arnold, T. M., (1998). Predicting the effects of brown algal phlorotannins on marine herbivores in tropical and temperate oceans. *Journal of Phycology, 36*, 195–205.

176. Thiansilakul, Y., Benjakul, S., & Shahidi, F., (2007). Antioxidative activity of protein hydrolysate from round scad muscle using alkalase and flavor enzyme. *Journal of Food Biochemistry, 31*, 266–287.

177. Tobacman, J. K., (2001). Review of harmful gastrointestinal effects of carrageenan in animal experiments. *Environ Health Perspect, 109*, 983–994.

178. Tomohisa, O., Watanabe, M., Naganuma, T., & Muramoto, K., (2011). Diversified carbohydrate-binding lectins from marine resources. *Journal of Amino Acids, 2011*, 20.

179. Tseng, C. K., (2001). Algal biotechnology industries and research activities in China. *J. Appl. Phycol., 13*, 375–380.

180. Rustad, T., & Hayes, M., (2012). Bioactive peptides and protein hydrolysates: Generation, isolation procedures, and biological and chemical characterizations. In: *Marine Bioactive Compounds* (pp. 99–113). Springer Science +Business Media, New York, USA.

181. Urbano, M. G., & Goñi, I., (2002). Bioavailability of nutrients in rats fed on edible seaweeds, Nori (*Porphyra tenera*) and Wakame (*Undaria pinnatifida*), as a source of dietary fiber. *Food Chemistry, 76*(3), 281–286.

182. Usov, A. I., Smirnova, G. P., & Klochkova, N. G., (2001). Polysaccharides of algae: 55. Polysaccharide composition of several brown algae from Kamchatka. *Russ. J. Bioorgan. Chem., 27*, 395–399.

183. Venugopal, V., Chawla, S. P., & Nair, P. M., (1996). Spray dried protein powder from threadfin beam: Preparation, properties, and comparison with FPC type B. *Journal of Muscle Foods, 7*, 55–58.

184. Verdrengh, M., Erlandsson-Harris, H., & Tarkowski, A., (2000). Role of Staphylococcus experimental staphylococcus aureus-induced arthritis. *Eur. J. Immunol., 30*, 1606–1613.

185. Vilkhu, K., Manasseh, R., Mawson, R., & Ashokkumar, M., (2011). Ultrasonic recovery and modification of food ingredients. In: Feng, H., Barbosa-Canovas, G., & Weiss, J., (eds.), *Ultrasound Technologies for Food and Bioprocessing* (pp. 345–368). Springer, New York.

186. Vlieghe, P., Clerc, T., Pannecouque, C., Witvrouw, M., De Clercq, E., Salles, J. P., & Kraus, J. L., (2002). Synthes kappa-carrageenanntly bound kappacarrageenan-AZT conjugates with improved anti-HIV activities. *J. Med. Chem., 45*, 1275–1283.

187. Voet, D., Voet, J., & Pratt, C. W., (1999). *Fundamentals of Biochemistry* (p. 1095). Wiley, New York.

188. Wang, L., & Weller, C. L., (2006). Recent advances in extraction of nutraceuticals from plants. *Trends Food Sci. Technol., 17*, 300–312.

189. Whittaker, M. H., Frankos, V. H., Wolterbeek, A. M. P., & Waalkens-Berendsen, D. H., (2000). Effects of dietary phytosterols on cholesterol metabolism and atherosclerosis: Clinical and experimental evidence. *Am. J. Med., 109*, 600–601.

190. Wijesinghe, W. A., & Jeon, Y. J., (2012). Enzyme-assistant extraction (EAE) of bioactive components: A useful approach for recovery of industrially important metabolites from seaweeds: A review. *Fitoterapia, 83*, 6–12.

191. Witvrouw, M., & De Clercq, E., (1997). Sulfated polysaccharides extracted from sea algae as potential antiviral drugs. *General Pharmacology, 29*, 497–511.

192. Wu, Y. R., Lin, Y. C., & Chuang, H. W., (2016). Laminarin modulates the chloroplast antioxidant system to enhance abiotic stress tolerance partially through the regulation of the defensin-like gene expression, *Elsevier Journal, 247*, 83–92.

193. Xu, X., Song, F., Wang, S., Li, S., Xiao, F., & Zhao, J., (2004). Dibenzyl bromophenols with diverse dimerization patterns from the brown alga *Leathesia nana*. *Journal of Natural Products, 67*, 1661–1666.

194. Xu, X. L., Fan, X., Song, F. H., Zhao, J. L., Han, L. J., & Yang, Y. C., (2004). Bromophenols from the brown alga *Leathesia nana*. *Journal of Asian Natural Products Research, 6*, 217–221.

195. Yan, X. J., Zheng, L., Chen, H. M., Lin, W., & Zhang, W. W., (2004). Enriched accumulation and biotransformation of selenium in the edible seaweed *Laminaria japonica*. *J. Agric. Food Chem., 52*, 6460–6464.

196. Ye, H., Wang, K., Zhou, C., Liu, J., & Zeng, X., (2008). Purification, antitumor, and antioxidant activities *in vitro* of polysaccharides from the brown seaweed *Sargassum pallidum. Food Chemistry, 111*(2), 428–432.
197. Ward, O. P., & Singh, A., (2005). Omega-3/6 fatty acids: Alternative sources of production. *Process Biochemistry, 40*(12), 3627–3652.
198. Yuan, H., Song, J., Li, X., Li, N., & Dai, J., (2006). Immunomodulation and antitumor activity of [kappa]-carrageenan oligosaccharides. *Cancer Letters, 243*(2), 228–234.
199. Zhang, Q., Li, N., Zhou, G., Lu, X., Xu, Z., & Li, Z., (2003). *In vivo* antioxidant activity of polysaccharide fraction from *Porphyra haitanensis* (Rhodephyta) in aging mice. *Pharmacological Research, 48*, 151–155.
200. Zhou, Y., Yang, H. S., Hu, H. Y., Liu, Y., Mao, Y. Z., Zhou, H., et al., (2006). Bioremediation potential of the macroalga *Gracilaria lemaneiformis* (Rhodophyta) integrated into fed fish culture in coastal waters of North China. *Aquaculture, 252*, 264–276.

CHAPTER 3

MARINE BIOACTIVE COMPONENTS: SOURCES, HEALTH BENEFITS, AND FUTURE PROSPECTS

MONIKA THAKUR

ABSTRACT

Marine organisms (like sponges, tunicates, bryozoans, mollusks, bacteria, micro-algae, macro-algae, cyanobacteria, fishes, crustaceans, seaweeds) have a valuable source of bioactive components in comparison to other terrestrial sources. Marine organisms have bioactive components such as: protein and peptides (collagen, gelatin, albumins), polysaccharides (carrageenan, agar-agar, fucans, fucanoids, chitin, chitosan, and derivatives), ω-3 fatty acids, phenolic compounds, pigments (phlorotannins, β-carotene, chlorophylls, and lutein), enzymes (gastric proteases, pepsins, gastricsins, chymosins, serine, cysteine, lipases, transglutaminase), fat, and water-soluble vitamins. These components have various nutraceutical properties such as: anti-viral, anti-bacterial, anti-coagulant, antioxidant, anti-hypertensive, radioprotective, anti-parasitic, anti-inflammatory, and anti-cancerous, etc. Marine novel foods also explore the unexplored sources of oceans, their characteristics, capabilities, challenges, and opportunities. This chapter summarizes marine bioactive components, their sources, health perspective, and future prospects for a healthy lifestyle.

3.1 INTRODUCTION

Our lifestyle has been continuously searching for natural foods that can improve biological functions and make people healthier, fitter, and to live longer. As a result of consumer awareness of the consociation between food, disease, and health, the consumption of functional foods has increased in recent decades. The role of food as a whole in the improvement of health has been well recognized, and thus they led to the development of functional foods [15, 21]. The concept of functional foods includes: They should be in the naturally occurring form; should be the essential part of our daily diet; and help in regulating/controlling diseases [43].

Marine organisms survive in distinct environmental conditions like - high concentration of ions, low levels of light, a variation of cold and warm temperatures variations, pressure, oxygen levels, water movements, density, and pH prevailing in their habitats. The marine- ecosystem because of its prodigious biodiversity is a rich natural resource of many biologically active compounds such as: polyunsaturated fatty acids, sterols, proteins, polysac-charides, antioxidants, and pigments. Since these marine organisms live in complex habitats, therefore they are exposed to extreme conditions and adapt to new environmental surroundings. These marine based bioactive components also have excellent nutraceutical potential with additional health benefits like anti-cancerous, antiviral, immune-potentiating, hypocholester-olemic, and hepato-protective and anti-inflammatory properties, etc.

Since the marine ecosystem has great taxonomic diversifications, the endeavors related to the search for new, magnificent bioactive compounds from the marine environment has been a source of unlimited potential [35, 36]. Marine-based ecosystems consist of large heterogeneity of marine organisms in comparison to terrestrial ecosystems [14]. The organisms are basically growing in diverse habitats, from the intertidal zone to the deep sea environment having organisms such as: Poriferans (sponges), Coelenterate, Cnidarians (corals, sea anemones, hydrozoans, jellyfish), Annelidans (Poly-chaetes, marine worms), Bryozoans (moss animals or sea mats), Arthropods (lobsters, crabs, shrimps, prawns, crayfish), Molluscans (oysters, abalone, clams, mussels, squid, cuttlefish, octopus etc.), and Echinodermata (sea stars, sea cucumbers, sea urchins) [44].

This divergent group also includes macro and microalgae, marine bacteria and fungi, cyanobacteria's, marine fishes and crustaceans, etc. Marine sources have recently received extravagant attention, and these sources have been searched out for new bioactive compounds. Hippocrates had also described the nutritional and therapeutic effects of various marine based bioactive components and the impact on the health of human beings [48]. Borresen [6] also explained that 40% of global fisheries and seafoods are globally produced from the marine ecosystems only.

Marine-based bioactive components have been isolated from a wide array of sources, which have unique and exclusive properties [36]. These components are also called as MNP's (Marine natural Products). The marine environment containing a vast array of marine organisms with unique biological properties has been one of the most underutilized biological resources. There are many marine species, still lying unexplored and under-exploited. Therefore, there has been focused attention by the researchers on exploiting underutilized natural resources for multifarious use.

The present chapter reviews the marine organisms, their bioactive components, various health perspectives, marine toxic compounds, and future prospects.

3.2   MARINE SOURCES OF BIOACTIVE MOLECULES

Marine-based species comprise of approximately one-half of the total universal biodiversity, the oceans and aquatic environment, and plenty of resource for novel bioactive components. Marine species comprises bioactive compounds, and much attention has been paid as they play a pivotal role in disease prevention and maintenance of human health. These bioactive compounds have been derived from a vast array of resources such as: microorganisms, marine sponges, macro, and microalgae, marine fungi, marine bacteria, etc. The host organism synthesizes the bioactive components as non-primary/secondary metabolites to conserve the homeostasis in their habitat and surroundings.

The marine-based biodiversity proves as an incomprehensive resource for the development of nutraceuticals [2]. Some of the marine sources are listed in Figure 3.1. Some other organisms have also been explored for the extraction of novel functional food ingredients from marine resources. Recently, seaweeds have been considered as a potential source of antioxidants due to their poly-phenolic compounds and carotenoid contents [27].

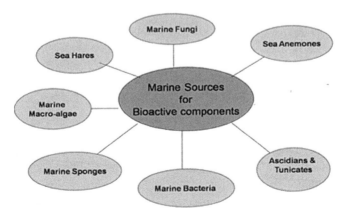

FIGURE 3.1   **(See color insert.)** Marine sources of bioactive components.

3.3 MARINE BIOACTIVE COMPONENTS

The marine-based resources, because of their unparallel biodiversity have been excellent sources of unlimited marine based bioactive compounds. A diverse array of bioactive components can be screened from the extracts of marine organisms. Many of them help in the development of functional foods and nutraceuticals [18, 25]. Table 3.1 presents the marine food ingredients with their bioactive components, sources, and related health benefits.

The peptide compounds isolated from protein hydrolysates isolated from fish, and fucans and galactans isolated from algae possess anticancer anticoagulant, and hypocholesterolemic properties, etc. In addition to these properties, the oil derived from marine fishes and bacteria are exceptional sources of ω-3 fatty acids, whereas seaweeds and the crustaceans are excellent sources of various antioxidants like carotenoids, phenolic constituents flavanoids, etc. Therefore, the present review focuses on the marine-based bioactive compounds as main functional food ingredients for the guardianship of health and prevention from various ailments [24].

Deep-sea water (DSW) has also recently been the focus as it has been enriched with various important minerals and nutrients. Kim et al. [19] discussed the inhibitory effects of DSW on breast cancer invasion. It has been reported that desalinated DSW have a much super-ion effect on preventing the development of atherosclerosis in rabbits fed with a high cholesterol diet in comparison to desalinated surface seawater with a similar profile of important minerals (Mg, K, Ca, Na, Cl, and sulfate ions) [17].

The natural marine sources play a very important role in human health as they have a wide range of nutraceutical properties such as: anti-microbial, antioxidant, anti-hypertensive, anti-coagulant, anti-cancerous, anti-inflammatory, immune-modulatory, and many other medicinal effects. Thus, these bioactive components are of utmost importance to us and give a vast array of health benefits.

3.4 MARINE NUTRACEUTICALS IN THE MARKET

The marine products have become significant for both pharmaceutical and health-food based industries. Many marine based nutraceuticals are available in the markets such as: fish oil (ω-3fatty acids); seal blubber, algal, and shark liver oil; squalene, shark cartilage; macroalgae; chitin, chitosan as well as their monomers and oligomers, enzymes, peptides, and related compounds, fat-soluble vitamins A, D, and E; protein hydrolysates and glucosamine.

TABLE 3.1   Marine-Based Functional Bioactive Components and Health Benefits

Functional food ingredients	Bioactive components	Source	Health benefits	Reference
Carotenoids	a,b,e-Carotene, lycopene, and xanthophylls (cryptoxanthin, lutein, zeaxanthin, rhodovibrin, capsanthin, rhodoxanthin, violaxanthin, flavoxanthin, luteochrome, bixin, and crocetin)	Salmon, Mussel, Shrimp, crab, trout, lobster, redfish, tuna, squid, octopus, sea cucumber, crayfish, Atlantic herring, pink salmon, algae, fungi, sponges, starfish, sea urchin, sea anemones, corals	Anti-inflammatory, antioxidant, immune-modulatory, anti-cancerous, prevent cardiovascular and neurodegenerative diseases	[26, 51]
DSW (Deep sea water)	Nutrient and minerals (Mg, K, Ca, Na, Cl, and Sulphate ions)	Red mold (*Monascus* sp.)	Effect of breast cancer, prevent tumor metastasis, hypolipidemic properties, anti-art hero-sclerosis effect	[17, 19, 49]
Marine enzymes	Catechol oxidase, gastric proteases, pepsins, gastricsins, chymosins, serine, cysteine, lipases, transglutaminase; tyrosinases; cresolase	Mackerel, Salmon, squid, seal, tuna, sardine, cod, and rainbow trout, etc.	Anti-oxidant potential and anti microbial activity	[7, 8, 31, 38]
Phenolic compounds	Phlorotannins	Polyphenolic compounds in the brown algae	Anti-oxidant potential and anti microbial activity	[22, 31, 40]
Polysaccharides	Phlorotannins, glutathione, alginates, carrageenan, fucoidan, agar, furcelleran, ascophyllan, laminarin, polyuronides carrageenan, agar-agar, fucans, fucanoids, LMW chitin, chitosan, and derivatives	Brown algae; Crabs, Krill; Macroalgae (Gelidium, Gracilaria, Hypnea, and Gigartina)	Anti-tumor, anti-coagulant, anti – HIV agents, cardioprotective, anti-inflammatory, anti-allergic, anti-oxidant, anti-diabetic, antibacterial, and protease inhibitor activities	[4, 10, 23, 39, 45–47]
Prebiotics	LMW COS, algal polysaccharides	Cyanobacterial biomass of *Spirulina sp.*. stimulate *Lactobacillus sp.* and *Bifidobacterium sp.*	Prebiotic effect	[16, 30, 34]

TABLE 3.1 (Continued)

Functional food ingredients	Bioactive components	Source	Health benefits	Reference
Protein & peptides	Protein hydrolysates, bioactive peptides collagen, gelatin, albumins	Fish, crustacean, and molluscs (Pacific hake, squid, Alaska pollock, oyster, blue mussel, capelin, scad, herring, conger eel, yellow stripe trevally, dried bonito, sardine, yellowfin sole, salmon, nori, wakame, mackerel, mussel, prawn, oyster), krill, algae; fish (cod, tuna, herring, pallock, haddock)	Antioxidant, anti-inflammatory, anticoagulant, antitumor, immune-modulatory, antibacterial, antithrombotic, and antihypertensive activity, anti-skin aging activity, prevent, and treat chronic atrophic gastritis; ACE inhibitor activity	[1, 9, 12, 20, 28, 29, 41, 51]
Vitamins & Minerals	Fat and water-soluble vitamins, Fe, I, Mg, Zn	All marine-based organisms and seaweeds	Minerals and vitamins are required for physiological functions of the body and also serve as co-factors from various biochemical reactions.	[33, 34]
ω-3 fatty acids	EP, DPA, and DHA	Fish (capelin, sardine, anchovy, mackerel, herring, freshwater fish, tuna, salmon, menhaden, halibut, cod), marine mammals (Seal blubber, whale blubber), algae, fungi, krill	Prevention and treatment of coronary artery disease, hypertension, diabetes, anti-arthritis, and other inflammatory, anti-cancerous, essential for growth and development of brain and retina, immune-modulatory.	[13, 32]

Various other nutraceuticals available in the markets are: ω-3 oils, Chitosan, chitosan oligosaccharides and glucosamine, carotenoids, and xanthophylls, Protein hydrolysates, Algal phenolics, Enzymes, and carbohydrates.

3.5   MARINE TOXINS

Marine-based products are important from all aspects both nutritionally and health-wise, but there have been some health hazards like food poisoning, etc. Through the food chain, many of the toxic compounds are added in the body. This toxic compound may cause a different kind of neurological and gastrointestinal disorders in the body. The Seafood poisoning because of ingestion of toxic compounds from marine sources is an underexploited causal agent. But this risk has been increased because of certain other factors like damages to coral reefs; changes in climate; and the spread of toxic - algal blooms, etc. Generally, toxins are passed from one organism to another by the food chain. Thus indirectly or sometimes directly by the food chain, the toxin is ingested by humans. Some of the very common marine toxins are:

- **Adverse interaction of marine algae** with various other nutrients (carbohydrates, proteins, fat, minerals, and vitamins): There has been the possibility of the formation of toxic, allergenic or carcinogenic substances.
- **Ciguatoxins** are poisoning because of toxic compound Ciguatera, which is because of the ingestion of coral-reef fishes. This toxic compound is accumulated in the body by the marine-based food web. The toxin of the poisoning is a heat-stable and lipid-soluble. Ciguatoxins or Maitotoxins are concentrated in the fish liver, intestines, roe, and heads.
- **Halitoxin** is a toxic complex of various marine sponges of the genus *Haliclona sp.* like *H. viridis; H. rubens* and *H. erina,* etc. Halitoxin is a complex mixture of salts of pyridiniums, which are toxic in nature and are toxic for mice and fishes.
- **Lophotoxin** is a neuro-muscular toxin isolated from several Pacific gorgonians like *Lophogorgia*. This toxin stops the nerve-stimulated contraction of the muscle and is highly dangerous.
- **Palytoxin** is a toxic compound generated from marine coelenterates (*Palythoa sp.*). The toxin is poisonous, water-soluble compound and has an influence on the Ca and K ion transport system in the nerves and the cardiovascular system of the body.

Therefore, with the generation of such toxins after consuming marine-based food products, the system should be established to understand the risks involved and their evaluation for the same [49]. Since the population is more aware of the importance and use of marine foods, the chances of the occurrence of these toxins also seem to be increasing in frequency and in the particular geographic distribution [32]. Thus, while focusing majorly on the use of marine sources as functional foods, the other side of the source as Marine toxins should also be addressed.

3.6   FUTURE PROSPECTS OF MARINE NUTRICEUTICALS

Marine nutraceuticals have a bright future and lucrative for both the pharmaceutical industries and food sector. In recent decades, there has been a growing interest in research and commercialization of the bioactive ingredients around the globe on these marine foods [37]. Within the short span of time, the Marine food recourses because of the plethora of bioactive components and their related health benefits are expected to command a huge market.

Till today, a limited number of bioactive components have been identified from marine sources and further research is needed to develop methods to apply them. Up till now, most of the Nutraceutical properties of marine-based bioactive components have been observed *in-vitro* or in animal model systems. But, since these bioactive components are having the potential for nutritional and health benefits, their activities are further to be analyzed in human beings also. Meanwhile, it has been reported that unexploited resources from the oceans offer many opportunities for future developments. These marine bioactive components resources are still underexploited and unexplored. There is a need for multi-disciplinary industrial and academic collaborations to develop new technologies and methodologies to exploit them for them for our healthy generations wisely.

3.7   SUMMARY

This chapter began with the introduction of consumer awareness on the conjunction between diet, health, and disease. Then introduction on marine foods was discussed with the examples of various marine organisms. The bioactive components derived from marine sources have also been discussed. These components have vast nutraceutical properties as antimicrobial, anticoagulant, antioxidant, anti-hypertensive, radioprotective, anti-parasitic, anti-inflammatory, anti-cancerous, and many more. The marine toxins were also discussed with the marine nutraceuticals. The chapter concludes with the rise in the demand for Marine novel foods also to explore the unexplored sources their characteristic, challenges, and opportunities.

KEYWORDS

- anti-cancerous
- antihypertensive
- antimicrobial
- antioxidant
- caretonoids
- chitosan
- ciguatoxin
- functional foods
- halitoxin
- health perspectives
- maitotoxin
- marine organisms
- marine toxins
- novel foods
- nutraceuticals
- palytoxin

REFERENCES

1. Agyei, D., & Danquah, K., (2011). Industrial-scale manufacturing of pharmaceutical-grade bioactive peptides. *Biotechnol. Adv., 29*, 272–277.
2. Aneiros, A., & Garateix, A., (2004). Bioactive peptides from marine sources: Pharmacological properties and isolation procedures. *J. Chromatogr. B. Anal. Technol. Biomed. Life Sci., 803*, 41–53.
3. Arct, J., & Pytkowska, K., (2008). Flavonoids as components of biologically active cosmeceuticals. *Clin. Dermatol., 26*, 347–357.

4. Berteau, O., & Mulloy, B., (2003). Sulfated fucans, fresh perspectives: Structures, functions, and biological properties of sulfated fucans and an overview of enzymes active toward this class of polysaccharide. *Glycobiology, 13*, 29–40.

5. Bhattacharya, S., & Shivaprakash, M. K., (2005). Evaluation of three Spirulina species grown under similar conditions for their growth and biochemicals. *J. Sci. Food Agric., 85*, 333–336.

6. Borresen, T., (2009). Seafood for improved health and wellbeing. *Food Technol., 63*, 88.

7. Chen, T., Embree, H. D., Brown, E. M., Taylor, M. M., & Payne, G. F., (2003). Enzyme-catalyzed gel formation of gelatin and chitosan: Potential for in situ applications. *Biomaterials, 24*, 2831–2841.

8. Diaz-Lopez, M., & Garcia-Carreno, F. L., (2000). Applications of fish and shellfish enzymes in food and feed products. In: Haard, N. F., & Simpson, B. K., (eds.), *Seafood Enzymes* (pp. 571–618). Marcel Dekker, Inc., New York, NY, USA.

9. Fouchereau-Peron, M., Duvail, L., Michel, C., Gildberg, A., Batista, I., & Gal, Y. I., (1999). Isolation of an acid fraction from a fish protein hydrolysate with calcitonin-gene-related-peptide-like biological activity. *Biotechnol. Appl. Biochem., 29*, 87–92.

10. Freile-Pelegrín, Y., & Murano, E., (2005). Agars from three species of *Gracilaria* (Rhodophyta) from the Yucatán Peninsula. *Bioresour. Technol., 96*, 295–302.

11. Fujiwara, S., (2002). Extremophiles: Developments of their special functions and potential resources. *J. Biosci. Bioeng., 94*, 518–525.

12. Gomez-Guillen, M. C., Turnay, J., Fernandez-Dıaz, M. D., Olmo, N., Lizarbe, M. A., & Montero, P., (2002). Structural and physical properties of gelatin extracted from different marine species: A comparative study. *Food Hydrocoll., 16*, 25–34.

13. Harris, K. A., Hill, A. M., & Kris-Etherton, P. M., (2010). Health benefits of marine-derived ω-3 fatty acids. *ACSMS Health Fit. J., 14*, 22–28.

14. Hill, R. T., & Fenical, W., (2010). Pharmaceuticals from marine natural products: Surge or ebb? *Curr. Opin. Biotechnol., 21*, 777–779.

15. Honkanen, P., (2009). Consumer acceptance of (marine) functional food. In: Luten, J., (ed.), *Marine Functional Food* (Vol. 1, pp. 141–154). Wageningen Academic Publishers, Wageningen, The Netherlands.

16. Honypattarakere, T., Cherntong, N., Wickienchot, S., Kolida, S., & Rastall, R. A., (2012). *In vitro* prebiotic evaluation of exopolysaccharides produced by marine isolated lactic acid bacteria. *Carbohydr. Polym., 87*, 846–852.

17. Hou, C. W., Tsai, Y. S., Jean, W. H., Chen, C. Y., Ivy, J. L., Huang, C. Y., & Kuo, C. H., (2013). Deep ocean mineral water accelerates recovery from physical fatigue. *J. Int. Soc. Sports Nutr., 10*, 1–7.

18. Kim, S., & Pallela, R., (2012). Medicinal foods from marine animals: Current status and prospects. *Adv. Food Nutr. Res., 65*, 1–9.

19. Kim, S., Chun, S. Y., Lee, D. H., Lee, K. S., & Nam, K. S., (2013). Mineral-enriched deep-sea water inhibits the metastatic potential of human breast cancer cell lines. *International Journal of Oncology, 43*, 1691–1700.

20. Lai, G., Yang, L., & Guoying, L., (2008). Effect of concentration and temperature on the rheological behavior of collagen solution. *Int. J. Biol. Macromol., 42*, 285–291.

21. Lakhanpal, T. N., & Rana, M., (2005). Medicinal and nutraceutical genetic resources of mushrooms. *Plant Genetic Resources: Characterization and Utilization, 3*, 288–303.

22. Lee, J. H., Seo, Y. B., Jeong, S. Y., Nam, S. W., & Kim, Y. T., (2007). Functional analysis of combinations in astaxanthin biosynthesis genes from *Paracoccushaeundaensis*. *Biotechnol. Bioprocess Eng., 12*, 312–317.

23. Li, L. Y., Sattler, I., Deng, Z. W., Groth, I., Walther, G., Menzel, K. D., et al., (2008). A seco-oleane-type triterpenes from *Phomopsis* sp. (strain HK10458) isolated from the mangrove plant *Hibiscus tiliaceus. Phytochemistry, 69*, 511–517.

24. Lordan, S., Ross, R. P., & Stanton, C., (2011). Marine bioactives as functional food ingredients: Potential to reduce the incidence of chronic diseases. *Marine Drugs, 9*, 1056–1100.

25. Luten, J. B., (2009). *Marine Functional Food* (pp. 166–167). Wageningen Academic, The Netherlands. http://www.wageningenacademic.com/doi/book/10.3920/978–90–8686–658–8Accessed on June 10.

26. Maeda, H., Sakuragi, Y., Bryant, D. A., & Dellapenna, D., (2005). Tocopherols protect *Synechocystis* sp. strain PCC 6803 from lipid peroxidation. *Plant Physiol., 138*, 1422–1435.

27. Miyashita, K., (2014). Marine antioxidants: Polyphenols and carotenoids from algae. In: Kristinsson, H. G., (ed.), *Antioxidants and Functional Components in Aquatic Foods* (pp. 219–229). John Wiley & Sons Ltd., West Sussex, UK.

28. Nicholson, J. P., Wolmarans, M. R., & Park, G. R., (2000). The role of albumin in critical illness. *Br. J. Anaesth., 85*, 599–610.

29. Noitup, P., Garnjanagoonchorn, W., & Morrissey, M. T., (2005). Fish skin type I collagen. *J. Aquat. Food Prod. Technol., 14*, 17–28.

30. O'Sullivan, L., Murphy, B., McLoughlin, P., Duggan, P., Lawlor, P. G., Hughes, H., & Gardiner, G. E., (2010). Prebiotics from marine macroalgae for human and animal health applications. *Mar. Drugs, 8*, 2038–2064.

31. Okada, T., & Morrissey, M. T., (2007). Marine enzymes from seafood by-products. In: Shahidi, F., (ed.), *Maximizing the Value of Marine Byproducts* (pp. 374–396). CRC Press: Boca Raton, FL, USA, and Woodhead Publishing Limited: Cambridge, UK.

32. Park, D. L., Guzman-Perez, S. E., & Lopez-Garcia, R., (1995). Aquatic biotoxins: Design and implementation of seafood safety monitoring programs. *Rev. Environ. Contam. Toxicol, 161*, 157–200.

33. Parr, R. M., Aras, N. K., & Iyengar, G. V., (2006). Dietary intakes of essential trace elements: Results from total diet studies supported by the IAEA. *J. Radioanal. Nucl. Chem., 270*, 155–161.

34. Pena-Rodriguez, A., Mawhinney, T. P., Ricque-Marie, D., & Cruz-Sua'rez, L. E., (2011). Chemical composition of cultivated seaweed *Ulvaclathrata* (Roth) C. Agardh. *Food Chem., 129*, 491–498.

35. Plaza, M., Cifuentes, A., & Ibáñez, E., (2010). In search of new functional food ingredients from algae. *Trends Food Sci. Technol., 19*, 31–39.

36. Rasmussen, R. S., (2007). *Morrissey.* Marine biotechnology for production of food ingredients. *Adv. Food Nutr. Res., 52*, 237–92.

37. Shahidi, F., & Ambigaipalan, P., (2015). Novel functional food ingredients from marine sources. *Current Opinion in Food Science, 2*, 123–129.

38. Shahidi, F., & Kamil, Y. V. A., (2001). Enzymes from fish and aquatic invertebrates and their application in the food industry. *Trends Food Sc. Technol., 12*, 435–464.

39. Shahidi, F., & Abuzaytoun, R., (2005). Chitin, chitosan, and co-products: Chemistry, production, applications, and health effects. *Adv. Food Nutr. Res., 49*, 93–135.

40. Shibata, T., Ishimaru, K., Kawaguchi, S., Yoshikawa, H., & Hama, Y., (2008). Antioxidant activities of phlorotannins isolated from Japanese Laminariaceae. *J. Appl. Phycol., 20*, 705–711.

41. Sibilla, S., Martin, G., Sarah, B., Anil, B. R., & Licia, G., (2015). An overview of the beneficial effects of hydrolyzed collagen as a nutraceutical on skin properties: Scientific background and clinical studies. *Open Nutraceuticals J., 8*, 29–42.

42. Sijtsma, L., & De Swaaf, M. E., (2004). Biotechnological production and applications of the ω-3 polyunsaturated fatty acid docosahexaenoic acid. *Appl. Microbiol. Biotechnol., 64*, 146–153.

43. Suleria, H. A. R., (2016). Marine processing waste - In search of bioactive molecules. *Nat. Prod. Chem. Res., 4*, E-article: https://www.omicsonline.org/open-access/marine-processing-waste--in-search-of-bioactive-molecules-2329-6836   1000e118.php?aid= 82242 Accessed on November 11, 2016.

44. Suleria, H. A. R., Hines, B., Addepalli, R., Chen, W., Masci, P., Gobe, G., & Osborne, S., (2017). *In vitro* anti-thrombotic activity of extracts from blacklip abalone (*Haliotisrubra*) processing waste. *Marine Drugs, 15*(1), 8–12.

45. Suleria, H. A. R., Masci, P., Zhao, K. N., Addepalli, R., Chen, W., Osborne, S., & Gobe, G., (2017). Anti-coagulant and anti-thrombotic properties of Blacklip Abalone (*Haliotisrubra*): *In vitro* and animal studies. *Marine Drugs, 15*(8), 240.

46. Vlieghe, P., Clerc, T., Pannecouque, C., Witvrouw, M., De Clercq, E., Salles, J. P., & Kraus, J. L., (2002). Synthesis of new covalently bound kappa-carrageenan-AZT conjugates with improved anti-HIV activities. *J. Med. Chem., 45*, 1275–1283.

47. Vo, T. S., & Kim, S. K., (2010). Potential anti-HIV agents from marine resources: An overview. *Mar. Drugs, 8*, 2871–2892.

48. Voultsiadou, E., (2010). Therapeutic properties and uses of marine invertebrates in the ancient Greek world and early Byzantum. *J. Ethnopharmacol., 130*, 237–247.

49. Wang, J. H., Liu, Y. L., Ning, J. H., Li, X. H., & Wang, F. X., (2013). Is the structural diversity of tripeptides sufficient for developing functional food additives with satisfactory multiple bioactivities? *J. Mol. Struct., 1040*, 164–170.

50. Whittle, K., & Gallacher, S., (2000). Marine toxins. *British Medical Bulletin, 56*, 236–253.

51. Wijesekara, I., & Kim, S. K., (2010). Angiotensin-i-converting enzyme (ACE) inhibitors from marine resources: Prospects in the pharmaceutical industry. *Mar. Drugs., 8*, 1080–1093.

CHAPTER 4

CHITIN AND CHITOSAN: APPLICATIONS IN MARINE FOODS

P. ANAND BABU, S. PERIYAR SELVAM, RESHMA B. NAMBIAR,
M. MAHESH KUMAR, and EMMANUEL ROTIMI SADIKU

ABSTRACT

In recent years, there has been an increasing interest in the application of
chitin and chitosan in the food industry due to its biocompatibility, biodegradability, bioadhesion, and nontoxicity. Chitin is the second most abundant natural biopolymer found in the shell of crustaceans, fungi, insect, and
arthropods. Each year, around 6 to 8 million tons of crustacean wastes are
produced globally. Yet these waste shells comprise useful chemicals like
nitrogen containing chitin, which could be sustainably used in food applications. Chitin a polysaccharide and its deacetylated derivative, chitosan,
are being used for the preservation of seafood products (fish, oyster),
thereby reducing the waste material from the seafood processing industry.
Marine-based products are highly perishable, mainly due to the high level
of polyunsaturated fatty acids (PUFAs), increased water activity, abundant
free amino acids, and the presence of autolytic enzymes. The antimicrobial, gelling, antioxidant, and film-forming property of chitosan makes
it a potential source of food preservative or coating material of natural
origin to improve the shelf life of seafood products. In addition, chitosan
nanoparticles are used as a carrier for vaccine delivery to protect the fish
from bacterial and viral infections. It acts as an effective and safe gene
delivery vehicle for treating genetic disorders. Furthermore, use of chitin
and chitosan as a diet supplement, promotes the growth and good health
in fish. The chapter focuses on the applications of chitin and chitosan on
marine-based food products that are inexpensive, renewable, and abundant
marine waste.

4.1 INTRODUCTION

Globally, 18–30 million tons of fish waste is being dumped as waste every year. Fishery wastes are very hazardous because of their high biological oxygen demand (BOD), chemical oxygen demand (COD), fat-oil-grease (FOG), total suspended solids (TSS), pathogenic microbes, organic matters, other nutrients, etc., [59]. Recently, marine wastes have attracted global attention due to their usefulness in the preparation of chitin and chitosan (CS).

Chitin is a nitrogenous polymer, comprising of 2-acetamido–2-deoxy-β-D-glucose with a linkage of β (1→4) and it is commonly present in invertebrates-exoskeleton of insect and arthropod (e.g., lobsters, shrimps, and crabs) and cell walls of yeast, and fungi [16]. Chitin is one of the most commonly available polysaccharides existing in the environment. Chitin and CS (derivative of chitin) are valuable in agriculture, biomedical, food, cosmetics, biotechnology, and these are used as chelating agents in textile companies for removing impurities from its effluents [42]. Chitin is water insoluble due to its intermolecular hydrogen bonds, and it occurs chiefly in the form of polymorphic crystalline structures, such as alpha and beta forms. The α-form is majorly obtained from crab and shrimp shells, and its chain is aligned in anti-parallel fashion; whereas the β-form is commonly obtained from mollusks like squid, which is organized in the parallel strand. The third form of chitin is γ-form, which consists of two parallel and one anti-parallel strand.

CS is the deacetylated form of chitin derivative, which is a useful bioactive polymer [74]. The deacetylation degree (DD) denotes the sum of the molar ratio of N-acetyl glucosamine (NAG) and D-glucosamine to the molar ratio of D-glucosamine units [17]. CS can be obtained by partial deacetylation (DA) of chitin, and it is a linear, semi-crystalline polymer consisting of 2-amino–2-deoxy-D-glucose glucosamine (GlcNH2) with β-D-(1 → 4) glycoside linkages [30]. Unlike chitin, CS is soluble in dilute aqueous acid solutions, based on the content of glucosamine, the degree of DA (DD, higher than 50%), which limits its applications. CS shows a range of interesting biological and physicochemical characteristics. The cationic amino-polysaccharide, which is non-toxic, biocompatible, biologically active and biodegradable properties, make CS more ideal candidate for imply in many fields, such as: supplementary foods, food ingredients and postharvest preservation of food products, cosmetics, water treatment and medicine [10,24].

Chitooligosaccharides (COS) isoligomers, having a molecular weight (MW) of approximately 10 kDa or less and they are derived from CS or chitin, by either chemical or enzymatic methods [42]. COS can be prepared from CS by several methods such as: acid hydrolysis or enzymatic hydrolysis with various non-specific enzymes (such as cellulases, lipases, proteases, chitosanases, and chitinases) [40]. Enzymatic hydrolysis methods have gained great attention owing to safety and ease of control. Currently, COS has been the topic of great interest because of their medicinal and pharmaceutical applications, due to its non-toxicity, high solubility and positive physiological effects [42].

In this chapter, authors have summarized the use of chitin and CS for the enhancement of the quality and storage life of marine products.

4.2 CHITIN

Sodium hydroxide (NaOH) treated chitin is combined with ice, leading to the formation of an alkaline solution. In this suspension, N-DA continues efficiently, and 50% of the deacetylated product is water soluble [58]. Currently, water-soluble CS is preferred as it can be readily introduced in the field of foods due to its non-toxicity and chemical similarity to chitin/chitosan.

4.3 CHITIN AND CHITOSAN DERIVATIVES

In recent years, there is an increasing attention in the alteration of CS by the chemical method to increase its solubility and also to extend its use [5]. It is a well-known fact that the CS may dissolve in acid solutions alone. Therefore, uses in food systems are limited because of alteration in taste, texture, aroma, and color. Different CS based water-soluble compounds include: carboxymethyl chitosan (CMC), iodoandtert-amino chitosan. On the other hand, the dependability of the safety of food products turns into the vital issue as the compounds are synthesized by following the chemical method. Phongying et al. [51] prepared nano-sized water-based CS by further deacetylating chitin whiskers.

4.3.1 *DEPOLYMERIZATION OF CHITIN AND CHITOSAN*

CS comprises of three categories of functional groups: An amino/acetamido in addition to the primary and secondary OH⁻groups at the positions of

C–2, C–3, and C–6. The amino groups are major reasons for changes in its physicochemical properties and structures, and they are interrelated with their chelation, flocculation, and biological functions [70]. The CS with high viscosity excludes its applications in numerous fields of biology [24].

Derivatization of CS by adding small functional groups, such as carboxymethyl or alkyl groups [5] may significantly improve the dissolution of CS at alkali and neutral media without disturbing its positive charge. The method of modification and graft copolymerization leads to the production of functional molecules by forming a covalent bond onto the backbone of CS. Previously, various scientists reported that the primary derivation and then graft modification of CS could significantly increase its solubility in water, antioxidant, and antimicrobial activities [71].

4.4   CHITOSAN AND ITS DERIVATIVES

The side chain alteration of CS provides a variety of compounds for several food applications. These derivatives show enhanced antibacterial activity against bacterial pathogens when compared to CS.

4.4.1   QUATERNIZED CHITOSAN

Quarternized CS (Trimethyl chitosan (TMC)) is commonly obtained by treating CS with iodomethane in the alkaline solution of N-methyl–2-pyrrolid-inone [7]. Other forms of modified CSs are: N-(2-hydroxyl) propyl–3-triethyl ammonium chitosan chloride (HTEC), N-(2-hydroxyl-phenyl)-N,N-dimethyl CS (NHPDCS) and N-(2-hydroxyl) propyl–3-trimethylammonium chitosan chloride (HTCC) [7]. CS and the water-soluble quaternized CS showed antibacterial property towards Gram-positive organisms as well as Gram-negative organisms.

4.4.2   CARBOXYALKYL CHITOSAN

Caboxyalkylation of CS is achieved by introducing carboxylic acids into CS backbone, which imparts unique physicochemical and biophysical properties like increased water solubility at broad pH range, antioxidant, antibacterial, antifungal activity, biodegradability, non-toxicity, and good biocompatibility. The CS subjected to carboxyalkylation yields: carboxymethyl, carboxyethyl chitosan, carboxybutyl chitosan, and other modified compounds through the carboxyalkylation reaction. Ahmed and Ikram [2] carried out carboxyalkyl CS grafting with succinic acid to improve water solubility and transfection efficiency.

4.4.2.1 CARBOXYMETHYL CHITOSAN

CMC occurs in the hydroxyl and amino side chains of CS. The O-carboxy-methyl chitosan (O-CMC)is mainly formed when the reaction is carried out at a low temperature or room temperature; and N-carboxymethyl CS (N-CMC) and N,O-carboxymethyl CS (N, O-CMC) are formed when the reaction is carried out at high temperature in an isopropanol/water suspension and in alkaline or monochloroacetic acid presence. The amphoteric nature of CMC improves the water solubility, biocompatibility, biodegradability viscosity, non-toxicity, and antimicrobial activity which support its candidature in food applications.

4.4.2.2 CARBOXYMETHYL CS BASED COMPOUNDS

CMC compounds can be replaced by amino and hydroxyl groups, resulting in the formation of N-CMC and O-CMC, respectively or the substitution of both -OH and -NH_2forms of N, O-CMC. These compounds are biocompatible, biodegradable, and they may exhibit antimicrobial activity than CS. Furthermore, N, O-CMC has an excellent gel-forming ability with outstandingH_2O retention behavior [28]. N-CMC is water soluble; however, it has distinctive physical, chemical, and biological activity, like huge hydrodynamic level and composites, high viscosity, less toxic and gel-forming property. Hence, these properties possibly make it an attractive choice to imply in food applications as well as cosmetics [14].

4.4.3 HYDROXYALKYL CHITOSAN

The hydroxyalkyl chitosan is synthesized by reacting the CS over epoxides at the -NH_2 group, which yields N-hydroxyalkyl at -OH groups by yielding O-hydroxyalkyl chitosan-based compounds. The ratio of O/N-substitution was estimated [45], based on the choice of catalyst (HCl or NaOH) and temperature used in the reaction. Hydroxyethyl (HE) CS is a multifunctional polymer, obtained from CS with high solubility in H_2O, retention of moisture content and gel-forming behavior [47]. Also, it has tremendous physicochemical behavior, which renders it suitable as an antioxidant agent [25]. Mainly, HE chitosan derivatives have been utilized as a potential carrier substrate for drug delivery. Hydroxypropyl chitosan (HPC) grafted with maleic acid showed superior inhibitory effects (99.9%) against both *S. aureus* and *E. coli*, within 30 mins [50].

4.4.4 PHOSPHORYLATION OF CHITOSAN

The preparation of phosphorylated CS includes the CS heating with ortho-phosphoric acid and urea in N,N-dimethyl formamide (DMF). Phosphory-lated CS showed improved ionic conductivity and swelling index. Although, it diminished the crystalline nature yet its tensile strength (TS) remained the same, which is same as CS. The phosphorylated CS showed considerable rough structure on the surface, dissimilar to CS [27]. CS reaction with phos-phorous pentoxide (P_2O_5) produced phosphorylated CS, which is soluble in water with a great level of replacement and high antimicrobial activity [64].

4.4.5 SULFATED CHITOSAN

CS is modified by utilizing sulphonic acid salts or sulphuric acid to yield a sulfated CS [53]. CS sulfates are water-soluble anionic derivative and have several biological activities, such as: antisclerotic, antiviral, antibacterial, antioxidant, and inhibition of enzyme activities [45]. During the sulfation of CS, few $-NH_2$groups are changed into negatively charged ions, and the polysaccharide improved its polyelectrolyte behavior. Hence, sulfated CS might be a suitable candidate for the development as an ideal drug carrier in the form of micelles or microcapsules [9].

4.4.6 COPOLYMER OF CHITOSAN

Copolymerization of CS is a method for altering the physical and chemical behavior of chitin and CS for broadening their commercial use [45]. Grafted copolymers enhance the physical and chemical behavior of both synthetic and natural polymers. Grafted copolymer like CS-g-PVCL (chitosan-graft-poly(N-vinylcaprolactam) is formed by grafting CS with a change in the lengths of poly (N-vinylcaprolactam) chain, *via* the activation of the terminal group of carboxylic acid through amidation reaction. The copolymers are thermo-sensitive, soluble in water at very less temperature and melting properties were based on the graft chain length. Grafted-CS compounds (CMCTS-g-MAAS and CMCTS-g-AAS) were synthesized by the methacrylic acid sodium (MAAS) and acrylic acid sodium (AAS) graft copolymerization on the etherification product of CS carboxy methyl CS (CMCTS) [20]. Furthermore, these grafted CS derivatives have better superoxide anion scavenging abilities.

4.5   CHITIN, CS AND COS: APPLICATIONS OF IN SEAFOODS

Seafood products are highly sensitive to quality deterioration during storage, caused by enzymatic, chemical, and microbial spoilage. Besides inhibiting the microbes, it is necessary to sustain the sensory attributes and various quality attributes, such as: texture, structure, colors, water holding capacity (WHC) as well as yield and shrinkage during cooking [48].

Nowadays, there is an increasing interest in the utilization of chitin and CS as coating materials, food supplements, flavor ingredients and preservative compounds against food contamination, for the removal of suspended solids and effluents, during the processing of seafood products, etc., [11]. Food and Drug Administration (FDA) has permitted the utilization of CS in specific food applications, like a packaging film to preserve the foods. Primex® of Norway produced the CS, which has the GRAS status [52] and is considered as a functional food.

4.5.1   CHITOSAN-BASED EDIBLE COATINGS AS SEAFOOD ADDITIVE

Generally, polysaccharides, proteins, and lipids can be used as edible coatings, which can prolong the storage life of food products by working as gas, solute, and vapor barriers. CS based polymer can be utilized as edible films or coatings owing to its distinctive characteristics of improved thickness during hydration. Moreover, CS films are hard, lifelong, and elastic and they have reasonable water vapor behavior and might prolong the shelf life of food products with more moisture content [12, 32]. It is extensively reported that using CS as an edible film might improve the storage quality of seafood products [29, 63]. CS or other forms of CS and storage in chilled conditions have combined influence on prolonging the storage quality of white shrimp [26]. Postharvest preservation of shrimp using 1.0% O-CMC and 1.5% CS decreased the psychrophilic bacterial growth throughout storage, lowered melanosis and maintained the freshness of shrimp even at $0 \pm 1°C$ for storage up to 10 days. Following treatment, water-soluble chitosan (CMC) and CS were diffused into the shrimp. Conversely, CMC is not as much effective on melanosis, since it slowly dissolves under high-humidity storage in the fridge.

CS exerts good antimicrobial properties by destroying the anionic outer layer of microorganisms and alterations in the CS configuration could additionally improve its antibacterial activity, which is an important characteristic necessary for the preservation of food (Table 4.1). Ye et al. [73] prepared a CS-coated active film, which was incorporated with five antibacterial agents, viz: nisin, sodium benzoate, potassium sorbate (PS), sodium diacetate and sodium lactate (SL) for preserving the cold-smoked salmon against *Listeria monocytogenes*. CS-coated films comprising4.5 mg/cm² SL–0.6 mg/cm² PS, 4.5 mg/cm² SL, and 2.3 mg/cm² SL–500 IU/cm²nisinare most active treatments against *Listeria monocyte genes* at room temperature. The treated group exhibited anti-listerial property during the cold storage. Recently, CS films or coatings incorporated with essential oils have attracted wide attention as it not only enhances the antibacterial, antifungal, and antioxidant activity of the film, it also reduces the water vapor behavior and oxidation of the lipid [72]. The combined effect of CS + CO (CS and cinnamon oil) coating on rainbow trout fillet, for the suppression of lipid oxidation and microbes in refrigerated condition, was studied by Ojagh et al. [49]. It was also observed that they maintained the sensory characteristics within acceptable limits, improved the shelf-life throughout the storage, and there was no significant change in texture, aroma, color as well as overall quality with limited microbes. Thus, cinnamon oil incorporated CS coating results in the antimicrobial coating, which could be employed as a better protection for fish at refrigerated storage conditions.

Tsai et al. [67] prepared CH-chitin (chemical method) and MO-chitin (microbial method) by the DA of CS products and tested it for its antimicrobial activity (Table 4.1). CS was mixed with salmon (*Oncorhynchus nerka*) fillets for bacteria analysis. The antimicrobial property was improved with an enhancement in the CS deacetylation degree, and this showed stronger activity against bacteria. The lethal concentration of CS (50–500 ppm) had a high level of DA, which effectively inhibited the microbes. The results indicate that CS, with a high degree of DA, preserved the fish fillets against numerous bacteria and extended their storage life.

TABLE 4.1 Antimicrobial Activity of Chitosan against Various Pathogens

Microorganisms		Seafoods	Reference
Bacteria Gram-positive	*Bacillus cereus* CCRC 10250	Seafoods Fish fillets	[3, 67]
	Bacillus sp.	Seafoods oysters	[13]
	Staphylococcus aureus CCRC 12652	Seafoods Fish fillets	[67]
	Staphylococcus sp.	Seafoods oysters	[13]
	Corynebacterium sp.	Seafoods oysters	[13]
	Micrococcus sp.	Seafoods oysters	[13]
	Lactic acid bacteria sp.	Seafoods oysters	[3,13]
Gram-negative Yeast	*Escherichia coli* CCRC 10674	Seafoods Fish fillets	[13,67]
	Aeromonas hydrophilia CCRC 13881	Seafoods Fish fillets	[67]
	Pseudomonas aeruginosa CCRC 10944	Seafoods Fish fillets	[67]
	Pseudomonas sp.	Seafoods oysters	[3, 13, 43]
	Salmonella typhimurium CCRC 10746	Seafoods Fish fillets	[67]
	Shigelladysenteriae CCRC13983	Seafoods Fish fillets	[67]
	Vibrio cholera CCRC 13860	Seafoods Fish fillets	[67]
	Vibrio parahaemolyticus CCRC10806	Seafoods Fish fillets	[67]
	Vibrionaceae sp.	Seafoods oysters	[67]
	Shewanella sp.	Seafoods oysters	[67]
	Alcaligenes sp.	Seafoods oysters	[67]
	Enterobacteriaceae sp.	Seafoods oysters	[3, 13]
	Moraxella sp.	Seafoods oysters	[13]
	Acinetobacter sp.	Seafoods oysters	[13]
	Flavobacterium sp.	Seafoods oysters	[13]
	Candida albicans CCRC 20511	Seafoods Fish fillets	[67]
Mold	*Aspergillus fumigatus* CCRC 30502	Seafoods Fish fillets	[67]
	Aspergillus parasiticus CCRC 30117	Seafoods Fish fillets	[67]
	Fusarium oxysporum CCRC 32121	Seafoods Fish fillets	[67]

4.5.1.1 CHITOSAN NANOPARTICLES AS ACTIVE COATING

CS and CS nanoparticles are also used as an effective coating on seafood. The influence of various ratios of CS and CS nanoparticles, as an edible film on fish sticks throughout the cold storage (−18°C), was studied by Abdou et al. [1]. Results indicated that fish fingers coated with either CS or CS nanoparticles had a lower total bacterial count and it increased the shelf life for up to 6 months under −18°C.

4.5.2   FOOD SUPPLEMENTS

Fish meal is an ideal source of animal protein in feed, and there is a need to formulate feeds with ingredients that are cheap and rich in nutrients for the optimum growth of fish. Chitin and CS incorporated feed prepared from fish silage has proven to be effective as it improved the stability, durability, keeping quality and also minimized loss, during the stacking of the fish feed [75]. Chitin and Chitosan were prepared from Indian white prawn and 2% CS was incorporated into feed containing fish silage from the filleting waste of freshwater carp, rohu (*Labeorohita*). The feed was compared with the control feed. It was observed that the fat content of CS feed was the least, i.e., 4.69% when compared to chitin feed (5.67%) and control (6.69%). Also, the stability (organoleptic score) and durability of CS feed in water (% loss) was better than chitin and control feed.

Experiments were performed to assess the impact of CS, chitin, and cellulose supplement feed on the development of Japanese eel, yellowtail, and red sea bream [35]. The development percentage of each fish orally administered with 10% chitin incorporated feed recorded the maximum percentage, demonstrating the food superiority. The maximum percentage for feed effectiveness was observed in red sea bream and Japanese eel, when supplied with the 10% chitin incorporated feed. The fishes administered with the CS supplemented feed had decreased development rate and diet efficiency, demonstrating the fact that the addition of 10% CS in the feed should have prevented the process, which is responsible for the basal diet, absorption, digestion, and assimilation. The red sea bream and yellowtail orally administered with the 10% cellulose incorporated diet exhibited a slightly higher developmental percentage and diet effectiveness than the percentage observed for the control feed [35].

4.5.3   CHITOSAN-BASED ACTIVE PACKAGING FILM

The keeping quality of perishable food was improved by suppressing the growth of microbes using CS packaging films. The active CS film has extraordinary characteristics like gas barrier property, mechanical property, availability, cost-effectiveness, and sensory attributes which prolongs the storage life of the food products without jeopardizing the buyer's well-being making it suitable for use in food packaging applications. Barracuda fish steaks were covered with CS film (CS, 0.3%) incorporated with ginger (*Zingiber officinale*) essential oil (GEO) and stored at 2°C for 20 days [54].

The film showed higher antibacterial activity against *Staphylococcus aureus* and *Escherichia coli*. In sensory attributes, CS-GEO film wrapped sample showed the most acceptable range until the end of storage (20 days) when compared to the unpacked control samples and fish steak wrapped in ethylene vinyl alcohol film only for 12 days. The outcomes revealed the CS-ginger essential oil (GEO) film was effective in prolonging the shelf life of fish steak. The incorporation of GEO heightened the moisture and antibacterial activity of CS without negatively varying the physical and mechanical properties, and it possibly improves the shelf life of fish during refrigerated storage [54]. Biodegradable CS films and coatings are used for a range of fishes for decreasing the microorganisms and subsequently, enhancing the overall acceptability and to extend the shelf life [29].

Gunlu and Koyun [23] studied the potential of increasing the keeping quality of sea bass. The outcomes showed that the time span of usability of the untreated and vacuum-treated groups lasted for 5 days, though that of vacuum-treated sea bass packed with CS film extended its shelf life upto25–30 days. Thus, the samples covered with CS extended the shelf life by at least20 days. The combined effect of vacuum treatment and packaging of sea bass fillets using CS-based biodegradable film indicated a decreased trimethylamine-nitrogen (TMA-N), total volatile basic nitrogen (TVB-N) and also suppressed the growth of psychrotrophic and mesophilic bacteria throughout the period of cold storage (4°C).

4.5.4   GEL ENHANCER

Surimi is a protein-rich product from refined fish, which forms an elastic gel when solubilized with sodium chloride and heated. It is frequently used as a protein supplement, and with further processing, by the addition of ingredients, such as: colors, flavors, binders, etc., it can be used as a seafood analog. Fish with less marketable value was used for a tough and flexible gel preparation; inferior grade surimi is also manufactured on shore by the support of gel-forming biopolymers (for example; starch). Therefore, CS might be a better choice for integration into this product to expand the techno-functional characteristics [31, 38].

The quality of the gels prepared using cheap quality walleye Pollock were increased two folds due to the incorporation of CS (1.5%), once the salted surimi pastes were stored at less than 25°C. The myosin chain polymerization might be augmented with the addition of 1.5% CS [31]. In gel formation beside CS, endogenous transglutaminase (TGase) plays a major role. Benjakul et al. [8] suggested that the surimi gel of barred garfish, added with 1% CS indicated arise in the breaking force. Hence, the improving property of CS was probably facilitated by the endogenous TGase during the period of product processing, thereby leading to the production of protein-protein and protein-CS complexes. In the presence of CS, it might not considerably alter the microstructural and rheological characteristics of gels obtained from horse mackerels. However, a minor lessening of gel flexibility was attained by the addition of CS under high-pressure conditions [21]. Studies reported by Li and Xia [38] revealed that the DD and MW of CS have diverse effects on the gelling activity of meat protein derivatives of silver carp fish. The gel comprising of CS (Degree of DA- 77.3%) exhibited very high storage modulus and penetration force. The gels diffusion forces were improved with growing MW of CS incorporation in the gel. Salt-soluble meat proteins and CS interaction were stable due to the electrostatic behavior and hydrogen bonds.

4.5.5   ENCAPSULATING AGENT (CARRIER MATERIALS)

Several bioactive compounds are more susceptible to various ecological parameters, including oxygen, light, and temperature. In view of these concerns, the encapsulation of new functional foods and bioactive compounds might be a potential approach to defeat these difficulties. Encapsulation technique is used to entrap active components that are later discharged under controlled conditions [18]. Various materials, like minerals, antioxidants, vitamins, enzymes, colorants, and sweeteners, are encapsulated for use in the food industry [61]. CS can be employed as a carrier material for the encapsulation process because of its nontoxicity, mucus adhesiveness, biodegradability, and biocompatibility [4]. Lately, an increase in the time span of usability and vitamin C release in rainbow trout was achieved successfully, by using nanoparticles (NP) made of CS/vitamin C, during storage for 20 days [4]. This investigation confirmed the storage period of vitamin C was prolonged in the feed during the storage at room conditions. Moreover, in rainbow trout, the CS nanoparticles were used to protect vitamin C from severe enzymatic and acidic conditions of the gastrointestinal tract.

CS nanoparticles are used to encapsulate DNA, after that it can be successfully incorporated into shrimp and Asian sea bass feed, to preserve them from an infection caused by white spot syndrome virus (WSSV) and *Vibrio anguillarum*, respectively. The outcome revealed [37] that the NP improved the viable percentage of WSS-infected shrimp for the period of

30 days after the experiment and the sea bass immunized with CS-DNA (pVAOMP38) complex orally which exhibited reasonable prevention of the infection of trial *Vibrio anguillarum* [36].

Likewise, Tian et al. [65] stated the CS microparticles incorporated with plasmid vaccine administrated orally to Japanese flounder prolonged the DNA liberate from CS microparticles in pH 7.4 PBS solution up to 42 days of period post intestinal diffusion. Klinkesorn and McClements [34] confirmed through an *in-vitro* absorption experiment, with pancreatic lipase enzyme that the encapsulation of tuna oil along with CS, influenced its digestibility and physical strength. The increased concentration of CS reduced the quantity of fatty acids released by the emulsions. The reason behind this might be the protection provided by the formation of CS coating around the lipid molecules, binding of fatty acid by CS or the interaction of CS with lipase directly. Results also suggested that the use of CS as coating materials for the effective release of omega-3 fatty acids and encapsulating with CS may possibly protect the emulsified poly-unsaturated lipids from oxidation during the storage period.

Encapsulation of tuna oil with CS by using the ultrasonic atomizer method was proven to be a potential method for use in the immediate future [33]. For the medical purpose, salmon calcitonin was synthesized by microencapsulating with CS beads, and the outcomes concluded that the CS beads incorporated with salmon calcitonin could be produced by gelling, the positively charged CS with a negatively charged counterpart, thereby resulting in a controlled releasing property [6].

4.5.6   EFFLUENT TREATMENT

Various researchers previously reported the use of CS (84% - DD with a viscosity of 2,400 cP) as a thickening agent for the total suspended waste removal from various processing plants, such as: sewage, poultry processing units and seafoods [60, 62]. Shrimp wastewater processing by using CS (10 mg/L) reduced the suspended solids by ~98% [11]. Biopolymer like CS might be utilized for production of several polyelectrolyte complexes with anionic polysaccharide like alginate. The CS-alginate complex stability could be affected due to various environmental factors, such as pH and ionic strength [44]. Experimental CS with an amount of 20 mg/L (surimi wash water protein, SWWP) improved the recovery of protein when compared to the commercial example [68].

The effluent of the fish-meal factory was treated by adding 10 mg/L CS at pH 7 showed ~85% reduction in the TSS [22]. The effectiveness of CS as a coagulating agent for the seafood products sewage treatment were primarily ascribed to its cationic charge and its interrelations with anionic compounds like proteins etc., present in the effluents. Moreover, the hydroxyl groups present in the CS molecule enhanced the proteins precipitation and TSS from the effluents [60, 69], while the coagulated by-products can use as a protein source in animal feed formulation.

4.5.7 GENE DELIVERY USING CHITOSAN NANOPARTICLES

For the encapsulation of genes, mostly viral vectors have been used. Due to its protection, strength, and capability to produce in huge amount, non-viral liberation systems, such as CS biopolymers are becoming increasingly popular as an alternative to viral vectors [66]. There are few reports on the successful formulation of CS and DNA complexes [57]. CS itself increases the transformation efficiency, more efficient gene delivery, via receptor-mediated endocytosis, was achieved by the addition of appropriate ligands to DNA-CS complex [46]. The DD and MW of the CS are the most essential criteria for the formation of effective CS-nucleic acid complex, which may enhance the stability and gene uptake by the cells.

The protective efficacy of CS nanoparticles encapsulated with DNA assemble holding the VP28 genetic material of WSSV for oral administration, was studied in black tiger shrimp. The study determined the changes in the immunological test, such as: prophenoloxidase, SOD, and hyperoxide compound in the hemolymph of CS-microencapsulated VP28 gene-treated black tiger shrimp. The report concluded that the oral delivery of CS/DNA assemble (pVP28) nanocarrier offer considerable preservation to black tiger towards WSSV and moreover enhanced the immune system of black tiger shrimp [37].

4.5.8 DRUG DELIVERY

Use of drugs and veterinary chemicals may prevent disorders in the fish farming industry, and various infected sewages; however it involves with water environments. External tissues of fishes are a good source of mucus [19]. Nevertheless, mucoadhesion is the unfamiliar approach in fish farming, as they might be followed to lessen the harmful effect of parasiticides and antibiotics.

Costa et al. [15]reported the use of fluorescent CS nanoparticles, to evaluate their adsorption in gills, skin, and digestive system, which are infection targets of tambaqui fish (*Colossomamacropomum*). In fish holding tanks, CS nanocarrier acts as a chemotherapeutic agent in immersion systems, which could favor the residual impact of the drug, and these NP remain adhered to the fish mucosal surface post-treatment. The study reported that the CS nanoparticles could keep on attached to the mucosal outer layer of the fish. CS nanoparticle showed enhanced microencapsulation of several medicines and improving their properties. Hence, these nanocarriers could be used for the preparation of pharmaceutical formulations for their lease of drugs to the fish to reduce the aquatic disorders.

4.5.9 DNA NANO-VACCINES

Vaccination is one of the numerous approaches that attempted to solve disease problem in aquaculture. The frequent use of adjuvants in aquaculture vaccination is harmful to fish, as it leads to pseudomelanosis, granulomas, and causing injury to the kidney. Administration of drug by injection causes severe ache and strain to fish. Rivas-Aravena et al. [56] loaded CS NP with inactivated ISAV (V) and DNA coding as adjuvant (Ad) for the replicas of alphavirus and vaccination against ISAV. The vaccine was orally administered to Atlantic salmon, which induced the expression of an immune molecule. However, it may not induce the humoral immunity. NP-V combined with NP-Ad administration protected the salmon by ~77% from infection whereas, immunization with NP-V resulted in a reasonable safety of fish towards Infectious salmon anemia disease (ISAV).

The potential use of CS as a cationic genetic carrier for administration of DNA vaccine orally against *Vibrio anguillarum* has been reported by Kumar et al. [36]. Eukaryotic cell expression vector pcDNA 3.1 and porin gene of *Vibrioanguillarum* were used to make a DNA vaccine named pVAOMP38. The result demonstrates that the CS nanoparticles can be used successfully to deliver DNA vaccine, orally into fish. After immunization, Asian sea bass was stimulated with *V. anguillarum* by injection into the muscle. The 46%of survivability was noted, which indicates that the Seabass vaccinated orally using DNA (pVAOMP38)-CS mixture exhibited modest defense against the infection. Therefore, the study confirmed that oral DNA vaccination (CS-pVAOMP38) of OMP38 gene triggers an antibody immune response against *Vibrio* in fish. Hence, CS could denote a possibility of the safe and as effective carrier for oral vaccination in fish towards the pathogens.

Li et al. [39] developed an effective CS nanoparticle loaded DNA vaccine, named pEGFP-N2-OMPK (pDNA) against *Vibrio parahaemo-lyticus* by using a eukaryotic expression vector pEGFP-N2 and *Vibrio parahaemolyticus* strain (OS4) surface protein K (ompK) genetic material. The NP showed encapsulation efficiency of 91.5% and loading percentage of 2.08%. Three weeks post-vaccination with CS /pDNA, the survival rate of seabream (black) was 72.3% preserved from *V. parahaemolyticus* (OS4). The experiment suggested that the administration of CS/pDNA orally stimulated an antibody immune defense in fish against OS4. Henceforth, the reports revealed that the CS nanoparticles might be used as potential carrier agents for an oral pDNA vaccine.

4.6   SUMMARY

CS gathered a great interest in its commercial applications in seafood companies, because of its biocompatibility, biodegradable property, nontoxicity, and mucus adhesion behavior. CS and its derivatives possess numerous useful characteristics, like antibacterial, gel enhancement, antioxidant properties, coagulating agent and encapsulating capacity. In addition, it has an excellent film-forming ability and barrier properties, hence proving it a possible candidate for preparing biodegradable films in order to improve the shelf-life of perishable marine foods. Considering these factors, CS might effectively be integrated into seafood for the improvement of its quality and the enhancement of person nutrition. Owing to these outstanding characteristics, CS may be useful as a functional component in sea-based foodstuffs as well as these attributes virtue for future studies.

KEYWORDS

- antibacterial
- antimicrobial
- antioxidant
- atlantic salmon
- barracuda
- biodegradability

- carboxyalkyl chitosan
- carboxymethyl chitosan
- chitin
- chitin
- chitooligosaccharides
- chitosan
- chitosan nanoparticles
- chitosanases
- coagulation
- copolymerization
- deacetylation
- depolymerization
- D-glucosamine
- DNA vaccine
- immunological parameters
- *Listeria monocytogenes*
- marine food
- melanosis
- *Paralichthys olivaceus*
- polysaccharide
- prophenoloxidase
- shrimp
- vector
- *Vibrio anguillarum*

REFERENCES

1. Abdou, E. S., Osheba, A. S., & Sorour, M. A., (2012). Effect of chitosan and chitosan-nanoparticles as an active coating on microbiological characteristics of fish fingers. *International Journal of Applied Science and Technology, 2*(7), 158–169.
2. Ahmed, S., & Ikram, S., (2015). Chitosan & its derivatives: A review in recent innovations. *International Journal of Pharmaceutical Sciences and Research, 6*(1), 14–27.
3. Alak, G., (2012). The effect of chitosan prepared in different solvents on the quality parameters of brown trout fillets (*Salmo truttafario*). *Food and Nutrition Sciences, 3*(9), 1303–1306.

4. Alishahi, A., Mirvaghefi, A., Rafie-Tehrani, M., Farahmand, H., Shojaosadati, S. A., Dorkoosh, F. A., & Elsabee, M. Z., (2011). Shelf life and delivery enhancement of vitamin C using chitosan nanoparticles. *Food Chemistry, 126*(3), 935–940.

5. Alves, N. M., & Mano, J. F., (2008). Chitosan derivatives obtained by chemical modifications for biomedical and environmental applications. *International Journal of Biological Macromolecules, 43*(5), 401–414.

6. Aydin, Z., & Akbuga, J., (1996). Chitosan beads for delivery of salmon calcitonin: Preparation and release characteristics. *International Journal of Pharmaceutics, 131*(1), 101–103.

7. Benediktsdottir, B. E., Gudjonsson, T., Baldursson, O., & Masson, M., (2014). N-alkylation of highly quaternized chitosan derivatives affects the paracellular permeation enhancement in bronchial epithelia *in vitro*. *European Journal of Pharmaceutics and Biopharmaceutics, 86*(1), 55–63.

8. Benjakul, S., Visessanguan, W., Phatchrat, S., & Tanaka, M., (2003). Chitosan affects transglutaminase-induced surimi gelation. *Journal of Food Biochemistry, 27*(1), 53–66.

9. Berth, G., Voigt, A., Dautzenberg, H., Donath, E., & Mohwald, H., (2002). Polyelectrolyte complexes and layer-by-layer capsules from chitosan/chitosan sulfate. *Biomacromolecules, 3*(3), 579–590.

10. Bhatnagar, A., & Sillanpää, M., (2009). Applications of chitin- and chitosan-derivatives for the detoxification of water and wastewater - A short review. *Advances in Colloid and Interface Science, 152*(1–2), 26–38.

11. Bough, W. A., (1976). Chitosan a polymer from seafood waste, for use in the treatment of food processing wastes and activated sludge. *Process Biochemistry, 11*, 13–16.

12. Butler, B. L., Vergano, P. J., Testin, R. F., Bunn, J. M., & Wiles, J. L., (1996). Mechanical and barrier properties of edible chitosan films as affected by composition and storage. *Journal of Food Science, 61*(5), 953–955.

13. Cao, R., Xue, C. H., & Liu, Q., (2009). Changes in microbial flora of Pacific oysters (*Crassostrea Gigas*) during refrigerated storage and its shelf-life extension by chitosan. *International Journal of Food Microbiology, 131*(2–3), 272–276.

14. Chen, L., Du, Y., & Zeng, X., (2003). Relationships between the molecular structure and moisture-absorption and moisture retention abilities of carboxymethyl chitosan. *Carbohydrate Research, 338*(4), 333–340.

15. Costa, A. C. da. S., Brandao, H. M., Silva, S. R. da, Bentes-Sousa, A. R., Diniz Jr, J. A. P., Pinheiro, J. de J. V., et al., (2016). Mucoadhesive nanoparticles: A new perspective for fish drug application. *Journal of Fish Diseases, 39*, 503–506.

16. Crini, G., (2006). Non-conventional low-cost adsorbents for dye removal: A review. *Bioresource Technology, 97*(9), 1061–1085.

17. Croisier, F., & Jerome, C., (2013). Chitosan-based biomaterials for tissue engineering. *European Polymer Journal, 49*(4), 780–792.

18. Deladino, L., Anbinder, P. S., Navarro, A. S., & Martino, M. N., (2008). Encapsulation of natural antioxidants extracted from *Ilex paraguariensis*. *Carbohydrate Polymer, 71*(1), 126–134.

19. Dickerson, H. W., (2006). *Ichthyophthirius multifiliis* and *Cryptocaryon irritans* (Phylum Ciliophora). Chapter 4. In: Woo, P. T. K., (ed.), *Fish Diseases and Disorders* (pp. 116–153). CAB International, Cambridge, MA.

20. Fernandez-Quiroz, D., Gonzalez-Gomez, A., Lizardi-Mendoza, J., Vazquez-Lasa, B., Goycoolea, F. M., San Roman, J., & Argüelles-Monal, W. M., (2015). Effect of

the molecular architecture on the thermosensitive properties of chitosan-g-poly (N-vinylcaprolactam). *Carbohydrate Polymers*, *134*, 92–101.

21. Gomez-Guillen, M. C., Montero, P., Solas, M. T., & Perez-Mateos, M., (2005). Effect of chitosan and microbial transglutaminase on the gel-forming ability of horse mackerel (*Trachurus spp.*) muscle under high pressure. *Food Research International*, *38*(1), 103–110.

22. Guerrero, L., Omil, F., Mendez, R., & Lema, J. M., (1998). Protein recovery during the overall treatment of wastewaters from fishmeal factories. *Bioresource Technology*, *63*(3), 221–229.

23. Günlü, A., & Koyun, E., (2013). Effects of vacuum packaging and wrapping with the chitosan-based edible film on the extension of the shelf life of sea bass (*Dicentrarchuslabrax*) fillets in cold storage (4°C). *Food and Bioprocess Technology*, *6*(7), 1713–1719.

24. Harish, P. K. V., & Tharanathan, R. N., (2007). Chitin/chitosan: Modifications and their unlimited application potential: an overview. *Trends in Food Science & Technology*, *18*, 117–131.

25. Huang, R., Mendis, E., & Kim, S. K., (2005). Factors affecting the free radical scavenging behavior of chitosan sulfate. *International Journal of Biological Macromolecules*, *36*(1–2), 120–127.

26. Huang, J., Chen, Q., Qiu, M., & Li, S., (2012). Chitosan-based edible coatings for quality preservation of postharvest white leg shrimp (*Litopenaeus vannamei*). *Journal of Food Science*, *77*(4), C491–C496.

27. Jayakumar, R., Nagahama, H., Furuike, T., & Tamura, H., (2008). Synthesis of phosphorylated chitosan by novel method and its characterization. *International Journal of Biological Macromolecules*, *42*(4), 335–339.

28. Jayakumar, R., Chennazhi, K. P., Muzzarelli, R. A. A., Tamura, H., Nair, S. V., & Selvamurugan, N. V., (2010). Chitosan conjugated DNA nanoparticles in gene therapy. *Carbohydrate Polymers*, *79*(1), 1–8.

29. Jeon, Y. J., Kamil, J. Y., & Shahidi, F., (2002). Chitosan as an edible invisible film for quality preservation of herring and Atlantic cod. *Journal of Agricultural and Food Chemistry*, *50*(18), 5167–5178.

30. Jolanta, K., Małgorzata, C., Zbigniew, K., Anna, B., Krzysztof, B., Jorg, T., & Piotr, S., (2010). Application of spectroscopic methods for structural analysis of chitin and chitosan. *Marine Drugs*, *8*(5), 1567–1636

31. Kataoka, J., Ishizaki, S., & Tanaka, M., (1998). Effects of chitosan on gelling properties of low-quality surimi. *Journal of Muscle Foods*, *9*(3), 209–220.

32. Kittur, F. S., Kumar, K. R., & Tharanathan, R. N., (1998). Functional packaging properties of chitosan films. *Z. Lebensm. Unters. Forsch.*, *A206*, pp. 44–47.

33. Klaypradit, W., & Huang, Y. W., (2008). Fish oil encapsulation with chitosan using ultrasonic atomizer. *LWT-Food Science and Technology*, *41*(6), 1133–1139.

34. Klinkesorn, U., & McClement, D. J., (2009). Influence of chitosan on stability and lipase digestibility of lecithin-stabilized tuna oil-in-water emulsions. *Food Chemistry*, *114*(4), 1308–1315.

35. Kono, M., Matsui, T., & Shimizu, C., (1987). Effect of chitin, chitosan, and cellulose as diet supplements on the growth of cultured fish. *Nippon Suisan Gakkaishi*, *53*(1), 125–129.

36. Kumar, S. R., Ahmed, V. P. I., Parameswaran, V., Sudhakaran, R., Babu, V. S., & Hameed, A. S. S., (2008). Potential use of chitosan nanoparticles for oral delivery of

DNA vaccine in Asian sea bass (*Latescalcarifer*) to protect from *Vibrio (Listonella) anguillarum*. *Fish & Shellfish Immunology*, *25*(1/2), 47–56.

37. Kumar, S. R., Venkatesan, C., Sarathi, M., Sarathbabu, V., Thomas, J., Basha, K. A., & Hameed, A. S. S., (2009). Oral delivery of DNA construct using chitosan nanoparticles to protect the shrimp from white spot syndrome virus (WSSV). *Fish & Shellfish Immunology*, *26*(3), 429–437.

38. Li, X., & Xia, W., (2010). Effect of chitosan on the gel properties of salt-soluble meat proteins from silver carp. *Carbohydrate Polymers*, *82*(3), 958–964.

39. Li, L., Lin, S. L., Deng, L., & Liu, Z. G., (2013). Potential use of chitosan nanoparticles for oral delivery of DNA vaccine in black seabream *Acanthopagrus schlegelii* Bleeker to protect from *Vibrio parahaemolyticus*. *Journal of Fish Diseases*, *36*(12), 987–995.

40. Lin, S. B., Lin, Y. C., & Chen, H. H., (2009). Low molecular weight chitosan prepared with the aid of cellulase, lysozyme, and chitinase: Characterization and antibacterial activity. *Food Chemistry*, *116*(1), 47–53.

41. Liu, X. F., Guan, Y. L., Yang, D. Z., Li, Z., & De Yao, K., (2001). Antibacterial action of chitosan and carboxymethylated chitosan. *Journal of Applied Polymer Science*, *79*(7), 1324–1335.

42. Lodhi, G., Kim, Y. S., Hwang, J. W., Kim, S. K., Jeon, Y. J., Je, J. Y., et al., (2014). Chitooligosaccharide and its derivatives: Preparation and biological applications. *BioMed. Research International*, 1–13.

43. Lopez-Caballero, M. E., Gomez-Guillen, M. C., Perez-Mateos, M., & Montero, P., (2005). A chitosan-gelatin blend as a coating for fish patties. *Food Hydrocolloids*, *19*(2), 303–311.

44. Mi, F. L., Sung, H. W., & Shyu, S. S., (2002). Drug release from chitosan-alginate complex beads reinforced by a naturally occurring crosslinking agent. *Carbohydrate Polymers*, *48*(1), 61–72.

45. Mourya, V. K., & Inamdar, N. N., (2008). Chitosan-modifications and applications: Opportunities galore. *Reactive and Functional Polymers*, *68*(6), 1013–1051.

46. Murata, J., Ohya, Y., & Ouchi, T., (1998). Design of quaternary chitosan conjugate having antennary galactose residues as a gene delivery tool. *Carbohydrate Polymers*, *32*(2), 105–109.

47. Nanaki, S. G., Koutsidis, I. A., Koutri, I., Karavas, E., & Bikiaris, D., (2012). Miscibility study of chitosan/2-hydroxyethyl starch blends and evaluation of their effectiveness as drug sustained release hydrogels. *Carbohydrate Polymers*, *87*(2), 1286–1294.

48. No, H. K., Meyers, S. P., Prinyawiwatkul, W., & Xu, Z., (2007). Applications of chitosan for improvement of quality and shelf life of foods: A review. *Journal of Food Science*, *72*(5), 87–100.

49. Ojagh, S. M., Rezaei, M., Razavi, S. H., & Hosseini, S. M. H., (2010). Effect of chitosan coatings enriched with cinnamon oil on the quality of refrigerated rainbow trout. *Food Chemistry*, *120*(1), 193–198.

50. Peng, Y., Han, B., Liu, W., & Xu, X., (2005). Preparation and antimicrobial activity of hydroxypropyl-chitosan. *Carbohydrate Research*, *340*(11), 1846–1851.

51. Phongying, S., Aiba, S. I., & Chirachanchai, S., (2007). Direct chitosan nanoscaffold formation via chitin whiskers. *Polymer*, *48*(1), 393–400.

52. Preuss, H. G., & Kaats, G. R., (2006). Chitosan as a dietary supplement for weight loss: A review. *Current Nutrition and Food Science*, *2*(3), 297–311.

53. Rajasree, R., & Rahate, K. P., (2013). An overview on various modifications of chitosan and its applications. *International Journal of Pharmaceutical Sciences and Research*, *4*(11), 4175–4193.

54. Remya, S., Mohan, C. O., Bindu, J., Sivaraman, G. K., Venkateshwarlu, G., & Ravishankar, C. N., (2016). Effect of chitosan-based active packaging film on the keeping quality of chilled stored barracuda fish. *Journal of Food Science and Technology, 53*(1), 685–693.

55. Riva, R., Ragelle, H., Rieux, A. D., Duhem, N., Jerome, C., & Preat, V., (2011). Chitosan and chitosan derivatives in drug delivery and tissue engineering. *Advances in Polymer Science, 244*, 19–44.

56. Rivas-Aravena, A., Fuentes, Y., Cartagena, J., Brito, T., Poggio, V., Torre, J. L., et al., (2015). Development of a nanoparticle-based oral vaccine for Atlantic salmon against ISAV using an alphavirus replicon as adjuvant. *Fish & Shellfish Immunology, 45*(1), 1–10.

57. Roy, K., Mao, H. Q., & Leong, K. W., (1997). DNA-chitosan nanospheres: Transfection efficiency and cellular uptake. *Proceedings of the Controlled Release Society, 24*, 673–674.

58. Sannan, T., Kurita, K., & Iwakura, Y., (1976). Studies on chitin. 2. Effect of deacetylation on solubility. *Macromolecular Chemistry and Physics, 177*(12), 3589–3600.

59. Sapkota, A., Sapkota, A. R., Kucharski, M., Burke, J., McKenzie, S., Walker, P., & Lawrence, R., (2008). Aquaculture practices and potential human health risks: Current knowledge and future priorities. *Environment International, 34*(8), 1215–1226.

60. Savant, V. D., (2001). Protein absorption on chitosan-polyanion complexes: Application to aqueous food processing wastes. *PhD Thesis, Food Science and Technology Department* (p. 123), Oregon State University, Oregon.

61. Shahidi, F., & Han, X., (1993). Encapsulation of food ingredients. *Critical Reviews in Food Science and Nutrition, 33*(6), 501–547.

62. Shahidi, F., Kamil, J. Y. V. A., & Jeon, Y. J., (1999). Food applications of chitin and chitosan. *Trends in Food Science and Technology, 10*(2), 37–51.

63. Souza, B. W., Cerqueira, M. A., Ruiz, H. A., Martins, J. T., Casariego, A., Teixeira, J. A., & Vicente, A. A., (2010). Effect of chitosan-based coatings on the shelf life of salmon (*Salmo salar*). *Journal of Agricultural and Food Chemistry, 58*(21), 11456–11462.

64. Tachaboonyakiat, W., Netswasdi, N., Srakaew, V., & Opaprakasit, M., (2010). Elimination of inter-and intramolecular crosslinks of phosphorylated chitosan by sodium salt formation. *Polymer Journal, 42*(2), 148–156.

65. Tian, J., Yu, J., & Sun, X., (2008). Chitosan microspheres as candidate plasmid vaccine carrier for oral immunization of Japanese flounder (*Paralichthysolivaceus*). *Veterinary Immunology and Immunopathology, 126*(3/4), 220–229.

66. Tomlinson, E., & Rolland, A. P., (1996). Controllable gene therapy pharmaceutics of non-viral gene delivery systems. *Journal of Controlled Release, 39*(2/3), 357–372.

67. Tsai, G. J., Su, W. H., Chen, H. C., & Pan, C. L., (2002). Antimicrobial activity of shrimp chitin and chitosan from different treatments and applications of fish preservation. *Fisheries Science, 68*(1), 170–177.

68. Wibowo, S., (2003). Effect of the molecular weight and degree of deacetylation of chitosan and nutritional evaluation of solid recovered from surimi processing plant. *PhD Thesis, Food Science and Technology Department* (p. 138). Oregon State University, Oregon.

69. Wibowo, S., Savant, V., Cherian, G., Savange, T. F., Velaquez, G., & Torres, J. A., (2007). A feeding study to assess nutritional quality and safety of surimi wash water proteins recovered by a chitosan-alginate complex. *Journal of Food Science, 72*(3), 179–184.

70. Xia, W. S., (2003). Physiological activities of chitosan and its application in functional foods. *Journal of Chinese Institute of Food Science and Technology, 3*(1), 77–81.
71. Xie, W. M., Xu, P. X., Wang, W., & Lu, Q., (2002). Preparation and antibacterial activity of water-soluble chitosan derivative. *Carbohydrate Polymers, 50*(1), 35–40.
72. Yanishlieva, N. V., Marinova, E. M., Gordon, M. H., & Raneva, V. G., (1999). Antioxidant activity and mechanism of action of thymol and carvacrol in two lipid systems. *Food Chemistry, 64*(1), 59–66.
73. Ye, M., Neetao, H., & Chen, H., (2008). Effectiveness of chitosan-coated plastic films incorporating antimicrobials in inhibition of *Listeria monocytogens* on cold-smoked salmon. *International Journal of Food Microbiology, 127*(3), 235–240.
74. Younes, I., & Rinaudo, M., (2015). Chitin and chitosan preparation from marine sources. structure, properties, and applications, *Marine Drugs, 13*(3), 1133–1174.
75. Zynudheen, A. A., George Ninan., & Mannodi, S. B., (2011). Effect of chitin and chitosan on the physicochemical quality of silage based fish feed. *Fishery Technology, 48*(2), 149–154.

PART II
Novel Processing Techniques: Fish and Fish Products

CHAPTER 5

FISH FREEZING: PRINCIPLES, METHODS, AND SCOPE

VIJAY SINGH SHARANAGAT, VIDUSHI KANSAL, and
LOCHAN SINGH

ABSTRACT

Fish freezing is a preservation process, where the temperature of the fish is reduced to a specific level or range. The process accompanies the formation of ice crystals and a reduction in water activity, which consequently enhances the shelf life of the product. At the same time, the process also proves economically feasible in reducing the rates of reactions responsible for deteriorating the quality of food product, especially in case of perishable foods such as meat, fruits, vegetables, and fishes. Recently, different advanced techniques like impingement freezing, high pressure assisted freezing, and cryogenic freezing has been developed, which supersede the traditional methods in terms of freezing time and end products quality. Besides these techniques, considerations must also be given to different quality governing attributes like pre-treatments, storage environment, place, and system of freezing, thermodynamics of freezing, time, load, the size of desired ice particles, etc. The present chapter discusses different freezing techniques used for the fishes and its products. It also focuses on the role of different attributes in determining the product quality during and after freezing process and will provide a deep insight into the engineering aspects of the fish freezing.

5.1 INTRODUCTION

5.1.1 FREEZING: A PRESERVATION METHOD

Although there are many less expensive preservation methods available, freezing is still the most preferred method owing to its better quality of the product when compared to its cost. In this preservation process, the temperature is reduced to the point that permits ice crystal formation and solute concentration. Theoretically, freezing should be done below the eutectic point of food freezing system to arrest all deteriorative reactions, but it is not economically viable. Therefore, foods are frozen to temperature, where microbial activity arrests to a point to obtain the desired shelf life. Formation of ice leads to reduced water activity and increased solute concentration in the unfrozen portion, which affects the rate of chemical reactions. This consequently enhances the shelf life of the product. However, the final influence of temperature on chemical reactions may be grouped as [18]:

- **Normal stability**: Reduction in the rate of deterioration reaction with a decrease in temperature thus ensuring better stability on storage;
- **Neutral stability**: Reaction rate is not influenced by temperature thus no significant effect of storage;
- **Reversed stability**: Increased reaction rate with a decrease in temperature thus decreasing the stability of food during the storage.

Considering engineering aspects of the freezing process, one has to deal with computations required to calculate refrigeration loads and time required to obtain desired low-temperature storage of the product. Also, one should be aware of changes taking place during freezing and storage.

Freezing starts at a temperature when the ice crystals formed after nucleation and the aqueous phase are in equilibrium. Thereon, ice crystals might grow larger or might nucleate more depending on the rate of freezing, i.e., slow freezing or fast freezing procedure, respectively. An engineer should be well acquainted with freezing temperature range for a specific food product to obtain high-quality product at the end. Freezing temperature is defined by the equilibrium condition of the system, which is further defined by thermodynamic and kinetic factors. Under equilibrium conditions, the characteristics of the system are defined by thermodynamic factors whereas kinetic factors define the rates at which equilibrium condition can be obtained [33].

Freezing consists of two stages: Nucleation (formation of seed ice crystals) and crystal growth. The initial burst of crystallization (nucleation) depends on the extent of super-cooling. Thereon, the rate of heat removal determines the size of the crystal. The graph for the process of freezing with time and temperature is described later in this chapter.

5.1.2 FISH FREEZING

Common preservation methods for fishes include: freezing, drying, curing, pickling, smoking, irradiation, and fermentation. However, the freezing dominates owing to its higher quality and yield product with no significant change in organoleptic and nutritional properties.

Fish can be frozen either at sea or on land. Frozen fish may either be consumed directly after thawing or may have to go under another reprocessing action. Not only the end product's use determines freezing parameters but also the source of fish. For example, the tropical fish has more preservative effect on freezing whereas North Antarctica fish resist freezing due to the presence of antifreeze proteins. Thus, North Antarctica fishes are commonly cured or smoked or dried. Also, muscles of raw fish have a more profound effect on freezing when compared to that of cooked fish; because cooked fish has already altered the tissues thus freezing and thawing do not affect it. Fermented fish products show better microbial quality during freezing as acids reduce tolerance of microbes to freezing and frozen storage. Several other pre-treatments can also be done to increase the stability of fish while freezing and during frozen storage.

If the frozen fish must be used for reprocessing, then the loss of soluble proteins should be as minimum as possible that is ensured by following of proper procedures. This soluble protein helps in the formation of pellicle (firm, glossy surface) while processing. Poor frozen and stored fish show less extraction of proteins leading to no pellicle formation giving dull matte surface.

Freezing may also be used for caviar products. However, if caviar is processed first and then frozen, quality is preserved. But if the caviars are frozen first, then the products formed would be of the thin membrane and sticky. Small eggs with a firmer outer membrane like herring and whitefish freeze well, whereas sturgeon eggs with thin membrane are being damaged by freezing.

The present chapter discusses different freezing techniques used for the fishes and its products. It also focuses on the role of different attributes in determining the product quality during and after freezing process and provides a deep insight into the engineering aspects of the fish freezing.

5.2 ATTRIBUTES GOVERNING FISH FREEZING

5.2.1 PRE- FREEZING PARAMETERS

5.2.1.1 TYPES OF FISH

Fishes can be classified according to their physical and chemical composition. Chemically, fish differs in oil and protein content. Therefore, five classifications of fish are: (i) low oil, high protein, e.g., Cod Fish; (ii) medium oil, high protein, e.g., Mackerel; (iii) high oil, low protein, e.g., Lake trout oil; (iv) low oil, very high protein, e.g., Skip Jack Tuna; (v) low oil, low protein, e.g., Clams. Most of the fishes lie in the first two groups. Basic classification of fishes based on skeleton structure is shown in Figure 5.1.

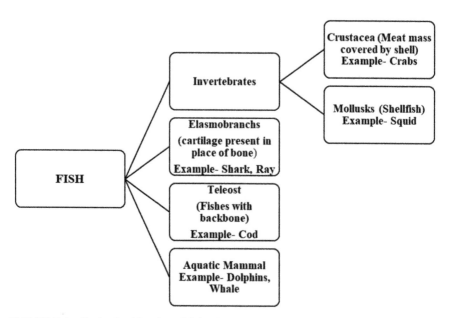

FIGURE 5.1   Basic classification of fishes based on the skeleton.

Generally, fishes with high oil content or enzyme activity are more preferred for freezing as the preservation process slows down the physicochemical and biochemical reactions, which might have caused spoilage. However, the effectiveness of this method depends on the initial quality of fish being frozen, storage temperature, duration of storage, the rate of freezing, fluctuation in storage temperature and thawing procedures.

Fishes may either belong to tropical zones or non-temperate zones. Fishes in tropical regions live in warm water thus all its biochemical process takes place at this temperature. Upon freezing, the rate of retardation of these processes is much higher than that observed in fishes belonging to cold water. Therefore, keeping the quality of tropical fish is much better than North Pacific fishes. Also, the onset of rigor in cold-water fish takes place at a higher temperature thus increasing the chances of thaw rigor. However, in warm water, the fish onset of rigor is faster at 0°C, thus making it more susceptible to freeze rigor, i.e., contraction during freezing, which leads to drip loss when thawed. It has also been observed that Antarctic and tropical species of fishes are less susceptible to texture deterioration when frozen, compared with species of North Temperate region [29].

Types of fishes with their influence on freezing are shown in Table 5.1. Table 5.2 indicates that the economic importance of fishes is commonly used for freezing. Generally small fish spoils faster than larger fish. Flatfishes are more stable than round fishes. Bony fishes are edible for a longer period than cartilaginous fishes [24].

5.2.1.2   TIME OF FREEZING

The time of freezing does not refer to the time required to carry out the freezing process. Rather it refers to the time at which fish has to be frozen after being caught. Fish may be frozen in sea vessel or first stored in ice before bringing to land and then frozen. Some fishes are even frozen after processing like smoking, curing, cooking, etc. Fish might be refrozen when cooked after being thawed.

5.2.1.2.1   Freezing at Sea

When the large sea vessel is employed for catching fish that is found in deep-ocean, then freezing at sea is the best option. The reason for this is simple that these vessels cannot leave until they are full to their capacity. Secondly, these vessels take days to reach on the shore; thus if the fishes are stored in ice and not in proper storage, then the post-harvest losses would be very high. Vertical plate freezers are generally used at sea for processing of bulk products. It is used in spaces, where headroom is compact such as on the ship. Also, this method does not require any type of packaging thus reducing the cost. They can also be used for freezing squids and octopus.

TABLE 5.1 Types of Fishes With Their Influence on Freezing

Type of fish/fish products	Description	Issues related to freezing/ storage	Form in which frozen
Clams	• Bivalve mollusks	• Rancidity, toughening, and thaw drip • If meat, then degradation is faster and leaving spongy texture	• In shell or as meat
Crustaceans	• Two types of muscles: fast acting (Tonic) and Slow acting (Phasic) **Example:** Crabs, Lobsters, Shrimps	• Cooked Lobster show poor frozen storage • Toughening of crustacean meat due to non-enzymatic respiration at low temperature • Poor freezing or storage leads to rapid loss in flavor, spongy or stringy texture and darkening	-
Crustaceans	• Loss of quality due to enzymes present on the surface **Example:** Shrimps and prawns	• Has to be frozen as soon as possible after being caught • To be frozen below −28.88°C and to be stored below −17.77°C	• Raw or cooked • IQF cooked whole • Battered/Breaded uncooked • IQF and glazed
Fatty Fish	• Lipid Content 20–30% (5% fatty) • Lipid content may vary with time **Example:** Herring, Sablefish, Mackerel, Eel, Dogfish, Chinook salmon, Rainbow trout	• Development of rancid flavor • Lower freezing requirement • To be frozen within 12–18 hours of capturing	-
Lean Fish	• High moisture content • Lipid Content < 2% (phospholipids) **Example:** Code, Pullock, Plounder, Pacific Ocean perch, Halibut, Squid	• High freezing load due to more water to freeze • Toughening of muscle while storage	Drawn, Dressed, fillet, or Whole

TABLE 5.1 (*Continued*)

Type of fish/fish products	Description	Issues related to freezing/ storage	Form in which frozen
Mussels	• Sold live or on half shell, smoked or cooked meat • Bivalve mollusks	• Freezing rate should be high • Core temperature to reduce to −25°C in 24hrs	• In the shell, half shell or as IQF meat
Oysters (mollusk)	• *Lamellibranchs* (Bivalves) can be frozen with shell or as shucked meat	• Changes in flavor, color, drip loss • Slow freezing rate, high storage temp • Worsens darkening of storage	• Glazed or properly packed
Sea Cucumbers	• Before boiling viscera to be removed or may give the bitter product • Upto 12 months shelf life on a freezing	• Thaw drip major in boiled and then frozen	• Raw or boiled
Smoked fish product	• Either hot smoked or cold smoked • Better keeping quality in cold storage of cold smoked fish products	• Rapid freezing (1–2 hr) to avoid big crystals • Storage at −26 to −21 °C, high cost	• Require good barrier packaging (mostly vacuum packed)
Surimi and Surimi Analogs	• Fishes showing poor keeping quality in frozen storage could be turned to surimi products for Freeze stability **Example**: Pacific Pollock	• Freeze denaturation which can be avoided by use of cryoprotectant like sorbitol	• Properly packaged

TABLE 5.2 Types of Fishes With Economic Importance of Freezing

Cod fishes	Flat fishes	Herrings and sardines
Atlantic Cod	Halibut	Pacific Herring
Atlantic Red fishes	Rainbow smelt	Rainbow Trout
	Eulachon	Atlantic salmon
Haddock	Pacific halibut	—
Hake	American Plaice	Tuna and Swordfish
Ocean perch, Rockfish, and redfish	English sole	Chum salmon
Pacific Cod	Atlantic halibut	Atlantic herring
Pacific Ocean perch		Pink salmon
Pollock	Greenland halibut	Manhaden
Red and striped mullet	Dover sole	Coho salmon
Walleye Pollock	Yellowtail Flounder	
Whiting	Petrel sole	Salmon
	Rock sole	Pacific salmon
Yellow eye rockfish	Smelt	Sockeye salmon

However, freezing at sea has its own drawbacks, which must be considered. Rigor mortis plays an important role in deciding the operations of the freezing process at sea. If the whole fishes are to be frozen, then it should be made sure that they are not bent before rigor sets in and then frozen. Because this may cause gaping of outer expanded portion and compact appearance of inner contract portion. Although frozen storage will maintain this appearance without worsening the condition, yet when the fish is thawed then fillets from inner portion will differ from fillets from the outer portion. Also, if the bent fish is then straightened before freezing then gaping of inner portion occurs. Therefore, it should also be avoided by chilling the fishes immediately after the catch to nullify the effect of rigor mortis.

If the fishes are filleted in the sea itself, then it is important to maintain chilling conditions throughout the line to avoid any shrinkage due to rigor mortis, which ultimately results in corrugated appearance and distorted shape. Effect of rigor mortis is more profound on fillets than in the whole body due to no restrictions posed to shrinkage by a skeleton. Chilling delays the onset of rigor mortis as high temperature acts as a catalyst in the processing.

5.2.1.2.2   Freezing on Land

When the fishes are caught near the sea and can be transported to the processing center within 5–6 hours of catch, then the fishes are frozen according to the demand. High demand for the fish, less amount frozen and more are directed towards the market and vice versa. Fishes are transported in baskets filled with ice and brought to a processing center, where they are filleted and then frozen.

5.2.1.2.3   Re-Freezing

When fish is frozen at sea or in a remote area where full processing is not possible, it is brought to the processing plant for subsequent final processing. Fishes like salmon are frozen and processed only when the market offers high prices. After processing, these are frozen again to ease out the distribution without any alteration to quality.

5.2.1.3   ROLE OF FORMATION OF ICE CRYSTALS

It is an important factor, which influences the quality of end-product. The freezing process should be designed according to that. Ice crystal formation influences fish products in the following ways: (i) Size of ice crystals determines the extent of thaw drip loss; (ii) Distribution of ice crystals in the fish affects frozen food properties like thermal conductivity, density, thermal diffusivity, etc.; (iii) Ice content affects the stability of fish while storage, (iv) Uncontrolled ice formation affects texture and appearance. Therefore, to control the size of ice crystals, it is important to know about the ice forming process. In fish muscles, the rate of freezing and stage of rigor influence the shape, size, and place of ice crystals.

In the fish, water is present in two forms: either in the cells or between the cells that are intracellular and intercellular. When the freezing process starts, ice crystals begin to form in intercellular water due to its higher freezing point. Intracellular water has dissolved solutes, which depress its freezing point. When freezing is done at a slow rate, then the intercellular water is translocated out of the cell to maintain osmolality. This results in the growth of ice crystal which eventually disrupts the cell membrane. However, when freezing is done at high speed then no translocation can take place due to lack of time; and small ice crystals are formed within and outside the cells.

Therefore, there is no disruption of cells, which result in lower thawing drip unlike in slow freezing.

Ice is formed in two steps: nucleation and propagation. Nucleation is the formation of seed ice crystals, which can support the further growth ice. It can be homogenous or heterogeneous. Propagation refers to the further growth or progress of ice crystals. To control ice crystal formation during freezing, one may do either of the followings: (i) Limit nucleation by addition of cryoprotectants; (ii) Limited rate of heat transfer; or (iii) Lowering the temperature below the glass transition temperature.

If heat removal is rapid, the propagation of a few crystals is insufficient to match the heat flux. Hence incipient super-cooling occurs, which results in an increased probability of nucleation. Therefore, a large number of ice crystals are formed [45]. Furthermore, the glass transition temperature plays an important role in the formation of ice crystal during freezing. Glass transition temperature refers to the temperature at which rubber state of food turns to glassy state when the temperature is reduced below this temperature. An aqueous solution of protein or carbohydrate on cooling forms an unfrozen matrix with a discontinuous phase of ice crystals as suspension [17]. Below the glass transition temperature, nucleation or propagation of existing ice crystals is not possible. However, controlling the freezing is not enough to control ice crystal formation. Number and size of ice crystals change during freeze storage by the phenomenon of Ostwald ripening and accretion. Ostwald ripening refers to the growth of large ice crystals at the expense of smaller ones, due to fluctuations in temperature while storage and is closely correlated with the drip loss. Accretion, also called sintering, refers to attaching of adjacent ice crystals, particularly small size, that is in contact. Thus, it is important to maintain a constant temperature for freezer storage. Different mathematical models are now available for calculating ice crystal distributions in frozen foods [57].

5.2.1.4   PRE-TREATMENT OF FISHES

During storage, rancidity, and protein denaturation are major chemical changes, which lead to value degradation. To prevent and control these changes, several treatments can be permitted. Storage stability of fish may be increased by few pre-freezing treatments. Summary of these treatments with their significance is listed in Table 5.3.

TABLE 5.3   Pre-Treatments with Their Significance

Pretreatment	Significance
Brining	Reduces drip and free water loss
Cryoprotectants	Stabilize proteins during storage
Dehydrins	Suppress Ostwald ripening
Glazing	Prevents Rancidity, desiccation, flavor, and color loss (for whole fish)
Heating	Increases lipid oxidation stability
Phosphate Dips	Increase water holding capacity
Skinning	Storage stability increased
Sulfating agents	Prevents melanosis
Wrapping	Prevents desiccation, weight loss, rancidity (for fillets)

Heating the mince prior to freezing affects the oxidative stability through altered pro-oxidative enzymes like lipoxygenases and lipoxidases [26, 58] and microsomal enzymes [48]. It also changes the pro-oxidative properties of myoglobin and other haemo-proteins [23] and enhances the production of antioxidants with aqueous and lipid soluble nature [63]. Skinning is done to remove unstable tissues, which increase storage stability. In fishes like herrings, under skin layer have dark muscles that are high in fat content with the low level of tocopherols (an antioxidant). These tissues also show abundance in hemoprotein and enzymes, which under skin layer are highly susceptible to oxidative changes. Therefore, to stabilize, either tissue is removed or is protected [11].

Cryoprotectants are added to stabilize the fish proteins, which are more susceptible to denaturation during the freezing compared to beef, pork or poultry. Denaturation of protein results in a change in myofibrillar proteins forms insoluble protein complexes and reduces the enzymatic activity of sarcoplasmic proteins. Cryoprotectants work by cryoprotection and cryo-stabilization. In cryoprotection, low molecular weight favors native protein state thermodynamically, whereas in cryostabilization high molecular weight polymers help to raise the glass transition temperature. Cryostabilization aims at reducing the number of ice crystals, which eventually disrupt muscle proteins. Cryoprotectants (like sugars, polyalcohol, carboxylic acids, amino acids, and polyphosphates) are used out of which sucrose, sorbitol, and polyphosphate mixture are common.

Carbohydrates act as anti-denaturants. Lactose, sucrose, galactose, fructose, glucose, and maltose are most effective. Carboxylic acids denature themselves in order to protect proteins. Glutaric, maleic, malonic, glyceric, methylmalonic, L-malic, gluconic, tartaric, citric, and 2-aminobutyric are most effective. Amino acids like cysteine, aspartic acids, glutamic acid, and glutathione protect proteins from freeze damage. Polyphosphates in combination with sugars improve water retention. Modified starch, maltodextrin, ethylene di-amine, creatine, tetraacetic acid, monosodium glutamate, and pyrophosphoric acid provides protection to proteins. Absorption on Na^+ and K^+ ions shifts the isoelectric points and leads to hydration of highly charged protein micelles. Thus, increases their water holding capacity. Polyphosphate dips are done to hydrate proteins. They also act as a germicide and reduce the microbial softening reactions.

5.2.1.5   HANDLING OF FISHES

Freezing does not improve the quality of fish rather it helps to preserve the quality. If high-quality fresh fish is soon frozen after harvesting, then it would be far superior in terms of organoleptic, biological, and nutritional quality than fresh fish available in the market.

Care must start as soon as fishes are being caught. Minimum bruises should be ensured on the body, which is possible only when suitable catching method is used. Trawling, dredging, spearing or harpooning, and gillnetting differ in the extent of damage done to fish while catching. Trapping is said to be the best way as fish is alive until killed prior to processing. Thus, there is no release of digestive enzymes from the gut of fishes, which might cause spoilage of other fishes also. In trawling, the most practiced method leads to scale abrasion and rupturing of the skin of the fishes at the end. Fatty and lean fishes are handled in a similar way after being caught. But the only difference lies in the form; they are stored. Fatty fishes do not generally undergo size reduction procedure or any other restructuring. Therefore, fatty fishes are only stored as whole or fillets, whereas lean fishes are stored in the form of mince, blocks as well.

After the fishes have been caught, they should be cooled down as soon as possible. Thus, instead of leaving the catch lying on deck exposed to the sun, it should be moved to the shade. After the fish is caught, the procedure stated in Figure 5.2 should be followed.

Washing cools down the temperature of fishes stored on deck. It also removes any extraneous material adhering to the surface of the fish and reduces the microbial load on the surface. Since water acts as a cleansing agent; it should be free from bacteria and pathogens. If the handling is done on the deck of the ship, then water should be taken offshore. Water near shore is contaminated with industrial effluent. But, if handling is done in pre-processing plants, potable grade water having chlorine concentration maximum of 5 ppm should be used.

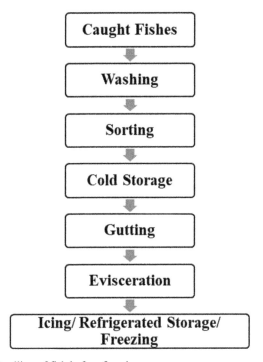

FIGURE 5.2   Handling of fish before freezing.

Sorting followed washing and done based on the composition of the catch. The trawl catch would consist of various kinds of fishes when compared with the pelagic shoal catch. Generally, fishes are sorted based on species and size. During sorting, they are inspected for any microbial contamination, enzymatic degradation, physical, and chemical damage. Fishes found unfit for consumption are either thrown back to sea or processed into fish meal or manure. If there is any time lag between gutting and sorting, then fishes are immediately cold-stored to avoid any further spoilage. Onboard, fish could be stored on ice, or could be preserved in refrigerated seawater (RSW) or chilled sea water (CSW).

Gutting is an important step in determining the quality of the final product. Gutting is the process of gut removal from the fish. While removing gut, liver should also be removed as it contains some lipids, which are perishable in nature even at low temperatures. While evisceration, it should be taken care that no bruises or cuts are made on soft belly portion to avoid easy penetration of bacteria. Complete viscera should be removed followed by blood. Since blood is the major source for microbial growth, it should be washed away with chilled water as soon as possible. Thus, gutting should be done as soon as possible after the fish has been caught. Complete bleeding should also be ensured as traces of blood might lead to permanent pink or red discoloration of flesh during storage. Bleeding operation takes 15–30 minutes thus delays the operation of freezing. Bleeding could be done in two stages: First, dip the fish in the water tank and then rinse it in the cold water spray. Consequently, fish is preserved.

However, some markets demand ungutted fish like herring rather they prefer whole fish. In that scenario, handling becomes more difficult. Any mishandling can trigger microbiological as well as enzymatic action in fish. Thus, any rough handling should be avoided.

Although handling is a necessary part of fish processing, yet some of these steps have proven to have negative effects on fish storage. The first step is washing. Here, washing does not refer to the step which is done after when fish is caught. But it refers to the step, which is done when there is any size reduction activity like mincing. Washing reduces the stability of herring mince by removing pro-oxidative enzymes from cooking, reducing the antioxidants and relatively increasing the phospholipids and free fatty acids in fat [16].

5.2.1.6   PRINCIPLE OF THERMODYNAMICS

Thermodynamics properties provide means for studying heat transfer, internal energy, conversion of heat energy to different forms of energy. During the fish freezing process, these properties help in estimation of equilibrium freezing point of fish, fish's equilibrium unfrozen water content as a function of temperature during freezing, heat removal requirements for freezing, the effect of pressure and ice crystal size on freezing points. Freezing point depression, enthalpy, specific heat, latent heat, and thermal conductivity are discussed in this chapter.

5.2.1.6.1   Freezing Point Depression

Freezing point depression is a property of a solution. According to this property, the presence of solutes in water depresses the freezing point of pure water, which is below 0°C. The extent of depression depends on the molar concentration of solutes. Food systems like fishes could be considered a mixture of a solid and aqueous solution. While freezing of food systems, water freezes to ice whereas the solutes get dissolved in remaining water leading to increased concentration. This further leads to depression in freezing point. Therefore, freezing of food system is observed over a range of freezing points. This phenomenon is explained through the phase diagram (Figure 5.3) that demonstrates a typical freezing curve for a food material [61].

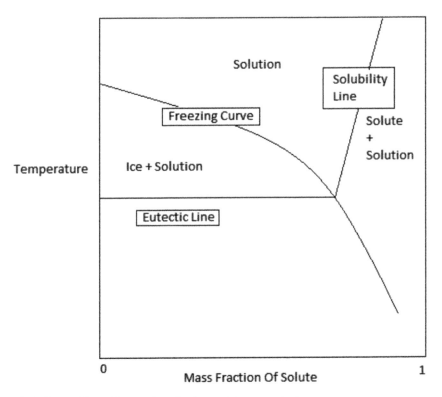

FIGURE 5.3   Phase diagram for a food material at a constant pressure.

The freezing curve is made by joining freezing points of food material. This curve shows equilibrium points between ice content and concentrated solution. If the solution contains a single solute, cooling induced ice formation causes the temperature to drop until it reaches eutectic temperature, where it remains while the remaining water and the solute solidify. After both completely solidify, continued cooling will cause the temperature to drop further and gradually approach to the temperature of coolant [52]. For a non-ideal solution, the freezing point depression (ΔT_F) can be calculated as follows [61]:

$$\Delta T_F = \frac{RT_w^2}{M_w \lambda_w} \left[ln \left(\frac{\sum_{j=1}^{n}(\tau_j X_{j0} / M_j)}{\frac{X_{w0} - X_{aw}}{M_w} + \sum_{j=1}^{n}\left(\frac{\tau_j X_{j0}}{M_j}\right)} \right) \right] \tag{1}$$

Since it is very difficult to know the molecular weight of each component present in food material, some empirical equations have been established. Freezing point depression is written as a function of water content, an easily measurable quantity. Also, the equilibrium freezing point of aqueous solutions can be calculated by the following equation [57]:

$$ln(a_w) = ln(\gamma_w X_w) = \frac{-18.02 \Delta H_{av}(T_0 - T)}{RT_0 T} \tag{2}$$

Where, a_w is the thermodynamic activity of water, γ_w is the activity coefficient of water, and Xw the total mole fraction of water in the solution T_0 and T are in Kelvin. Water's molecular weight is 18.02. The ΔH_{av} is the average latent heat of fusion between T_0, and T. The R is ideal gas law constant in J (kg mole)$^{-1}$K^{-1}, ΔT_F is freezing point depression, k is thermal conductivity (W/m°C), M is molecular weight (g/mole), m_x is weight fraction of x-component in food material, R is universal gas constant (=8.314 kJ/kg mol K), T is temperature (K or °C), λ_w is latent heat of fusion (J/g), τ_j is molecular dissociation of compound j in food material, γ_w is the activity coefficient of water, τ is molecular dissociation, m is mass (kg), M is molecular weight (g/mol), n is number of component in a food, X is mass fraction, j is the jth component of the food material, j_0 is the jth component of the food material at the initial freezing point, w is water, w_0 is water at the initial freezing point of a food material.

5.2.1.6.2 Ice Content

Various thermodynamic properties are dependent on the ice content and liquid water present in food systems. For example, ice has a higher thermal conductivity coefficient when compared to water. Therefore, as the freezing proceeds, it becomes easier to extract heat from inner core or food due to increased heat transfer. Therefore, change in properties is governed by ice content, which can be calculated with the following equation [52]:

$$n_i = n_{wo} - n_w \approx \left(n_{wo} - Bn_s \right) \left[\frac{T_i - T}{T_0 - T} \right]$$ (3)

Where, n_i is weight fraction of ice in food, n_s is combined weight fraction of solutes and solids, n_{wo} is total weight fraction of water in the food prior to freezing. B is the ratio of the mass of un-freezable water to mass of total dry solids in a food material. For fish, B ranges from $0.14 - 0.32$ [26]. T_i is the initial freezing point.

5.2.1.6.3   Volume Changes

When fishes freeze, they tend to expand due to different density of water (1000 Kg/m³) and ice (967 Kg/m³). Any space in the fish structure is occupied by these ice crystals. If ice crystals are small, they tend to fit compactly without disturbing cell membrane unlike in case of large crystals.

5.2.1.6.4   Enthalpy

At a temperature below the initial freezing point (T_i), enthalpy of food is equal to the sum of individual enthalpy(H) contribution by each component:

$$H = n_w H_w + n_1 H_1 + n_s H_s$$ (4)

Where H, H_w, H_1, H_s is the enthalpy per unit mass for the food, water, ice, and the combined solids and solutes, respectively. H is measured with respect to 233.16K where enthalpy is zero for all components. By combining equations (3) and (4), and rearranging we get:

$$H = (T - T_R) \left[C_f + \frac{(n_{wo} - Bn_s) \Delta H_0}{T_0 - T_R} \left[\frac{T_0 - T_i}{T_0 - T} \right] \right]$$ (5)

5.2.1.6.5 Effective Heat Capacity

The heat capacity of food continuously changes with the progress of the freezing process. Therefore, effective heat capacity is calculated by differentiating equation of ice content with respect to T:

$$C_e = C_f + \frac{\Delta H_0 \left(n_{wo} - 3n_s\right)\left(T_0 - T_i\right)}{\left(T_0 - T\right)^2} \tag{6}$$

Equation (6) is only valid for $T < T_i$.

5.2.1.6.6 Thermal Conductivity

Thermal conductivities of ice and water are 2.25W/m/k and 0.569W/m/k, respectively at 0°C. Therefore, the thermal conductivity of food material increases considerably on freezing. Although thermal conductivity is a kinetic property, yet it can still be related to temperature as follows:

$$k = k_f + \left(k_0 - k_f\right)\left[\frac{T_0 - T_i}{T_0 - T}\right] \tag{7}$$

Where k_0 and k_f are thermal conductivity of fully thawed food and fully frozen food.

5.2.1.6.7 Effects of Crystal Size

The diameter of crystal size affects the equilibrium temperature and depression in freezing point. Small crystal sizes have slightly lower equilibrium temperature than that for large ice crystals owing to their high surface energy compared to weight. Also, smaller is the diameter; higher is the effect on freezing point depression. For example, 1μm crystal depresses the point by 0.1K, whereas 10 μm crystal depresses the point by 0.01K.

5.2.2 ROLE OF PHYSICAL PROPERTIES OF FISH

Physical properties of frozen food include: water content, freezing point, ice content, latent heat, enthalpy, specific heat, thermal conductivity, and thermal diffusivity. Table 5.4 summarizes the physical properties of fish.

TABLE 5.4 Physical Properties of Seafood*

Fish	Water Content % (mass)	Highest freezing point, °C	Specific heat		Latent heat of fusion kJ/ kg
			Above freezing/kg kJ °C	Below freezing kJ/ kg°C	
Whole Fish					
Haddock, cod	78	−2.2	3.63	1.82	261
Halibut	75	−2.2	3.55	1.79	251
Herring, kippered	70	−2.2	3.43	1.72	235
Herring, smoked	64	−2.2	3.28	1.65	214
Menhaden	62	−2.2	3.23	1.62	208
Salmon	64	−2.2	3.28	1.65	214
Tuna	70	−2.2	3.43	1.72	235
Fish Fillets or Steaks					
Haddock, cod, perch	80	−2.2	3.68	1.85	268
Hake, Whiting	82	−2.2	3.73	1.87	275
Mackerel	57	−2.2	3.1	1.56	191
Pollock	79	−2.2	3.65	1.84	265

*Source: Wiley Encyclopedia of Food Science and Technology, Second Edition, Editor: Frederick J. Francis.

a. **Specific Heat:** Specific heat refers to the quantity of heat required or removed to raise or drop the temperature of a unit mass of food material by a unit degree. The SI unit for it is J/kg K. It should be noted that specific heat of food material is independent of its mass density. Specific heat for a food mixture is calculated by summing up the specific heat of individual components and then taking the mass average. Specific heat of food is changed while freezing due to phase and temperature change. The apparent specific heat of frozen food increments with increasing temperature until it reaches the initial freezing temperature. Apparent specific heat of frozen food is maximum at its initial temperature and then starts to decrease when the temperature is increased. During freezing, rapid decrease in the apparent specific heat is observed because temperature just passes the initial freezing temperature. Initial freezing temperature is passed rapidly because latent heat of fusion is removed near this point.

b. **Latent Heat:** Latent heat refers to the amount of heat required or removed to transform the food material from one state to another.

It is a hidden heat as no temperature change is observed while it absorption or release. The SI unit for latent heat is J/g. Reduction in the latent heat of fusion of ice occurs with the decrease in the temperature of freezing. The temperature dependence can be expressed as [40]:

$$\lambda_w = 334.2 + 2.12T + 0.0042T^2 \tag{8}$$

c. **Thermal Diffusivity:** This property helps to determine the transfer rate of heat in a solid food material having any shape. Theoretically, it relates to the ability of food to conduct and store heat. SI unit is m²/s. Thermal diffusivity for frozen food shows sharp decrease near the initial freezing point because of a sharp increase in apparent specific heat. However, for unfrozen food, thermal diffusivity does not vary with temperature due to compensation of changes in specific heat, thermal conductivity, and density.

5.2.3 FREEZING CURVE FOR FISH

The freezing curve for water and fish is shown in Figure 5.4, respectively. When the temperature is cooled down, the water temperature first drops to below 0°C. This is known as super-cooling. At this stage, nucleation takes place. Due to crystallization, the heat of crystallization is released raising the temperature again to 0°C. This temperature remains constant until all water has frozen to ice that is until all latent heat of fusion has been extracted out. On further cooling, the temperature drops steadily. However, when we freeze the fish, the process is the same during freezing, but the curve differs greatly.

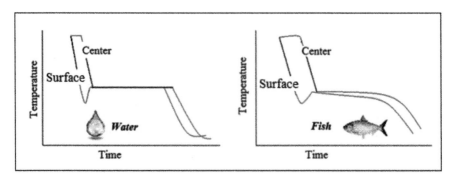

FIGURE 5.4   Freezing curve for water and fish.

This is due to the phenomenon called freezing point depression. Due to an increase in solute concentration in residual water, the freezing point of fish decreases. Thus, there is no sharp freezing point but a range of freezing point. Thus, during removal of latent heat of fusion of fish, no constant temperature is observed. This region is also called as freezing plateau. The pace with which the operator passes this phase determines the amount and size of crystals in the frozen product. Therefore, this region is also called as a critical freezing zone. Faster freezing supports more nucleation and small crystal size. This region is also called as thermal arrest region, as there is a slight change in temperature when compared to the amount of heat being extracted.

5.3   FREEZING METHODS

5.3.1   CONVENTIONAL FREEZING METHODS

Conventional freezing methods can be classified based on the mode of heat transfer and refrigerant used. Three broad classifications are:

- Through conduction: Contact freezing.
- Through convection: Air Blast Freezing, Cryogenic freezing.
- Through conduction and Convection: Immersion freezing.

Most efficient method among these for fish freezing is immersion freezing. However, its high cost has not made it the most desirable commercial method. Use of Freon as a refrigerant has proven to be costly; thus brine solutions are used frequently.

5.3.2   EMERGING FREEZING METHODS

These methods have not been developed yet to qualify as commercial methods. More developments are going on to reduce the cost and to increase the effectiveness of these methods. Emerging freezing techniques for fish include pressure shift freezing (PSF) and ultrasonic assisted freezing (USAF).

5.4   FREEZING SYSTEMS

5.4.1   CONTACT FREEZING

In contact freezing, fish is in intimate contact with metallic plates, which ensure high thermal conductivity. However, one disadvantage to this method is that regular shaped products of uniform size can only be used at a time.

Plate Freezers are made up of metals like extruded aluminum, aluminum-zinc, etc., which ensures high thermal conductivity. Although these freezers tend to distort surface appearance, yet these are highly preferred due to high heat transfer rate, lower evaporative loss due to the absence of circulating air, easy to clean, operates at room temperature, low footprint, etc. It is used when fish is in uniform pieces like block or slabs of up to 10 cm thickness. Uniform pieces ensure better contact with freezers thus allowing better heat transfer. However, if whole fishes are frozen, then one side is in contact with plate whereas the other side is exposed to cold air. This provides a final product with one side as flat and other in the rounded shape of fish making it unacceptable by consumers. These freezers are used for bulk freezing and distribution, not for retail sale. Refrigerants used to cool the plates are Freon, ammonia or CO_2. Operating temperatures are $-30°C$ to $-40°C$ with high heat transfer coefficients of 550 to 600 W/ (m^2K). These freezers may have either horizontal or vertical double contact plates. These freezers provide shortest freezing time and lowest operating cost.

Specialized contact freezers (Surface hardeners) are used when soft, humid fish fillets tend to lose their shape on the belts of continuous air blast freezers. To avoid stickiness, deformation or dehydration, the product is first crust freeze. This is done by using high heat transfer rate of metallic belts over which fish is passed. The one mm thick layer of crust is formed within a minute. And then these crust freeze fishes are transferred to continuous belt freezers.

5.4.2   AIR BLAST FREEZING

Air blast freezers use cold air, which is gushed over fishes. It is less efficient than plate freezing but can be used for large volumes but also leaves surface dehydrated. Proper packaging barrier eliminates this problem but adds to the barrier of heat transfer and thermal arrest time. These freezers can be sub-classified as: natural convection air blast freezers or forced convection air blast freezers. These freezers may be batch or continuous. According to plant capacity, any of them may be chosen. In natural convection, the flow of cold air is not regulated; therefore, it results in non-uniform freezing and freezer burn. Thus forced convection freezers are in practice. Continuous air blast systems include: carton/box freezer, tunnel/belt freezer, fluidized bed freezer, Spiral freezer, Impingement freezer.

Fluidized bed freezers are used for small granular fishes, mince, and prawns. Products pass on a mesh belt through a tunnel, and cooled air is blown from the bottom (underneath) of the belt to the products. Sufficient air velocities are maintained to fluidize the products partially, thus avoids clumping and ensuring individual freezing of the product. These systems are compact with high heat transfer coefficients and short freezing time.

Impingement freezing aims at increasing heat transfer coefficients by breaking up the static surface boundary layer of gas surrounding the fish by impinging very high velocity (20–30 m/s) gas jets. This method is best suited for fillets not thicker than 20mm.

Air blast freezers and plate freezers have the capacity to reduce the temperature of round fish from a chilled temperature to −18°C within 24 hours at the warmest point. In terms of thickness, portion 2-inch-thick freezes in 2.5 hours, whereas portion 6-inch-thick freeze in about 72 hours.

5.4.3   CRYOGENIC FREEZING

Cryogenic freezing employs either CO_2 or N_2 as a refrigerant in freezing systems like immersion freezers, tunnel, and spiral freezers and mechanical freezers. Liquid CO_2 is fed into the nozzle, which sprays into equal parts of solids and vapors. Solid CO_2, which settles on the fish surface, sublimes instantly taking about 85% of latent heat. Rest of the heat is taken by currents created due to CO_2 snow. Similarly, when N_2 is sprayed, it is a mixture of vapors and droplets. When droplets touch the surface, it evaporates taking 50% latent heat and rest is extracted by gaseous N_2 flowing along with freezers.

Immersion freezing: Product is dipped into an aqueous medium having a temperature lower than the freezing point of the product. The aqueous medium can be water, water, and salt (NaCl or $CaCl_2$) or aqueous solution of propylene glycol. However, these products have their own shortcomings. First, food could absorb the solutes from the refrigerant: For example, salty taste of fishes frozen in a brine solution. Secondly, food could dilute the aqueous solution, thus altering the freezing process. Third, it allows cross-contamination. These problems could be overcome with the help of proper packaging. Immersion freezers can either operate in batch mode or in continuous mode. The batch mode has high labor cost but is randomly used now. In continuous mode, fishes wrapped in plastic films are placed on the conveyer belt, which immerses in the aqueous liquid. Length of the belt in liquid equals to the holding time of the product. The conveyor belt is chosen according to the buoyancy of product in freezing medium: For example, a product which floats in the medium is carried by the auger conveyer system. After immersion, an excess of liquid is either drained off or blown off.

Fish frozen in the combined blast-plate freezer is much superior in quality to those frozen in blast freezer alone. Furthermore, this rapid freezing reduces drip loss considerably.

5.4.4 PRESSURE SHIFT FREEZING

Pressure shift refers to the phase change caused by pressure change at a theoretically constant temperature. In PSF, the pressure is applied over food in the refrigerated atmosphere, which ensures uniform nucleation all over the food. This method exploits the property of water where increasing pressures (above atmospheric pressure) results in a steady descent of freezing point. This can be observed in the pressure-temperature equilibrium diagram for the water-ice system. In PSF, when the pressure is released ice crystals start to form along with which heat of crystallization is released which increases the temperature. Thus, this heat is removed by the refrigeration system. Faster pressure release leads to uniform super-cooling and more ice will be formed with shorter thermal arrest time.

5.4.5 ULTRASONIC ASSISTED FREEZING

In this method, ultrasound does not play any role in freezing. Rather it assists in freezing. Ultrasound, especially low-intensity ultrasound, can be used to determine the ice content that further helps in realizing the position of the frozen/unfrozen boundary.

5.5 SELECTION OF FREEZING METHODS AND SYSTEMS

Freezing method and the system can be chosen either in combination or alone depending on cost and end product quality. Freezers ability is defined by freezing capacity and freezing time. Several factors influence the choice of freezing technology, which can be broadly classified according to the quality of the product and yield, environmental factors, factory management requirements, and financial considerations.

5.5.1   QUALITY OF PRODUCT

a.   *Moisture Loss:* Every freezing method causes some moisture loss either in the form of evaporative loss or drip loss. These losses affect the weight, texture, and appearance of fish, which consequently decreases its value. Evaporative losses can be minimized by a rapid lowering of temperature which leads to lower water vapor pressure at the surface thus discouraging evaporation. Also, rapid freezing leads to crust formation at the surface, which inhibits water migration from the body to the surface and then to the atmosphere. Drip loss can be controlled by crystal size as discussed earlier.

b.   *Color:* Color is affected by freezing rate. Slower freezing rate encourages more loss of pigment in drip and more intense browning reactions. Cryo-mechanically frozen fishes have shown more resemblance to the original color.

5.5.2   ENVIRONMENTAL FACTORS

Due to increasing legislation about carbon footprint, the organization must choose its freezers accordingly. Suppose a freezer with low cost, but high carbon footprint can land the organization into the legal issues.

5.5.3   FACTORY MANAGEMENT

a.   *Space Available:* Cryogenic freezers have less space requirements when compared to other counterparts. Thus, it makes it more flexible.

b.   *Turnover:* High cost of the freezer can be paid back only when expected turnover is high. Also, freezer with the ability to expand is preferable due to low-cost investment. Cryogenic and impingement freezers can be easily adapted to demand by changing tunnel length.

c.   *Product Changes:* Produce can change product line only when the previous one has exited fully so that operator can make changes in holding time of the product in freezers. Thus, if one requires changing the line frequently, he must go for freezers with low residence time.

d. Cleanability and Maintenance: Freezers offering easy cleanability and low maintenance are more popular as they reduce the opportunity cost of standing time of these freezers.

5.5.4 COST CONSIDERATIONS

- **Capital investments and variable cost:** Mechanical freezers require higher initial investment when compared to cryogenic freezers but have low variable cost on the other hand. Thus, one has to a trade-off between the two costs and should choose the most suitable freezer.
- **Labor cost:** Mechanical freezers require more labor and more maintenance. But, the skill requirement for the labor to operate mechanical freezer is lower than that of cryogenic freezers.

All the freezing options are summarized in Table 5.5.

5.5.5 FREEZING LOAD CALCULATION

The capacity of the freezer is defined as the ability of freezer to extract the defined maximum amount of heat in each unit time. If the freezer is loaded with more than its capacity, then it might overheat leading to increased cost of cooling the atmosphere. Thus, the operator should be aware of the load that will be subjected to the freezer. Freezing load is equal to the sum of sensible heat required to lower down the temperature and latent heat required to cross the region of thermal arrest. Calculations for the freezing loads are follows:

$$\text{Sensible heat} = H_1 = S_{uf}. W. (T_i - T) \qquad (9)$$

$$\text{Latent Heat} = H_2 = W. H_f \qquad (10)$$

$$\text{Sensible heat} = H_3 = S_f. W. (T_f - T_s) \qquad (11)$$

$$\text{Freezing load, } H_{fl} = H_1 + H_2 + H_3 \qquad (12)$$

where, S_{uf} is specific heat of unfrozen fish (kJ), W is weight of unfrozen fish, T_i is the initial temperature of fish, T is the initial freezing point; H_f is the latent heat (kJ) of fusion of fish; S_f is specific heat of frozen fish (kJ), W is weight of unfrozen fish, T_f is final freezing point, and T_s is storage temperature.

TABLE 5.5 Summary of Various Freezing Techniques

Criteria	Batch freezer	Tunnel or spiral freezer	Fluidized bed freezer	Cryogenic	Plate freezer	Impingement freezer	High pressure assisted freezer
Cleanability	Cabinets are easy to clean	Easy cleanability due to flexible lines	Requires efforts	Difficult	Easy	—	—
Cost Structure	Low capital cost, High Labor Cost	Economical, low maintenance	High Initial investment	Low capital investment and low maintenance but high operating cost	Low cost	High capital Investment, but lower variable cost when compared to cryogenic	High cost
Location in factory	Fixed, siting limited to space available	Fixed	Fixed	Flexible due to less footprint	Fixed	More flexible than mechanical	Still in the development stage
Range products	Wide range	Continuous in-line freezing	Small granular diced fish, uniform IQF product	Wide product range	Regular shaped product with <50 mm thickness	Flat product, not thicker than 20 mm	Large volume product
Speed of freezing	Slow	Reasonable	Fast	Variable	Reasonable	Faster	

Thus, the freezing load is the sum of H_1, H_2 and H_3 However, this sum is sub estimate as it does not include heat resistances offered by packaging material and glaze.

5.5.6 CALCULATION OF FREEZING TIME

Freezing time refers to the time required to lower down the temperature of the thermal center from an initial temperature to the desired value when placed in a freezer. The time required to freeze the product is calculated to ensure: (i) Rapid freezing; (ii) Complete freezing; and (iii) Optimized freezing process.

Freezing time determines the duration of stay of a product in the freezer. As freezer is expensive and is a bottleneck process, the time required for freezing determines the cost and speed of other upstream and downstream processes. Freezing time is influenced by all those factors, which affect the overall heat transfer coefficient. Some of these factors are: freezer type, airspeed in a blast freezer, initial productinitial temperature, thickness, shape, density, and packaging, etc. Freezing time for different fish products [9] are mentioned in Table 5.6. Freezing time can be calculated by either measuring it or by calculating it using mathematical models.

5.5.6.1 MEASURING FREEZING TIME

This method is usually used, when predicting time through models is difficult or verification of freezing time has to be done. Methods simulating the actual conditions as closely as possible must be designed in order to provide a measurement of temperature history for minimum one location as the freezing process undergoes completion [54]. Thus, a temperature sensor can be either a probe or thermocouple, which is placed at the center of fish. Method to be chosen for measurement must compromise among accuracy, precision, cost, and ease of application.

5.5.6.2 MATHEMATICAL MODELS TO ESTIMATE FREEZING TIME

These models are used when food characteristics, thermal properties, and process conditions are known, and heat conduction is uni-directional. Also, these models are applicable only for the range of food shapes and operative

TABLE 5.6 Freezing Time for Different Fish Products

Fish product	Freezing method	Initial temperature (°C)	Operating temperature (°C)	Freezing Time
Cod fillets laminated block 57mm thick in a waxed carton	Horizontal Plate	6	–40	1h, 20 min
Haddock fillets	Air Blast (4m/s)	5	–35	2h, 5min
Packaged fillets 50mm thick	Sharp freezer	8	–12 to –30	15h
Packaged fillets 50mm thick	Air Blast (2.5 to 5m/s)	5	–35	5h, 15 min
Scampi meat 18mm thick	Air Blast (3m/s)	5	–35	26 min
Shrimp meat	Liquid nitrogen Spray	6	–80/variable	5 min
Single haddock fillets	Air Blast	5	–35	13 min
Single tuna 50 kg	NaCl Immersion	20 to 18 at center	–12 to –15	72h
Single tuna 90 kg	Air Blast	20 to 45 at center	–50 to –60	26h
Whole Cod Block 100mm thick	Vertical Plates	5	–40	3h, 20 min
Whole lobster 500g	Liquid nitrogen Spray	8	–80/ variable	12 min
Whole Round fish 125mm, e.g., cod, salmon	Air Blast (5m/s)	5	–35	5h

conditions for which they were developed. They are some assumptions to these models:

- Fish has been pre-chilled ensuring all heat extraction at initial freezing temperature.
- Physical properties are constant throughout the body.
- Boundary conditions are symmetric, i.e., no variable operating conditions.

a. **Plank's Model:** This model is based on the phase change period of the freezing process. This equation equates overall heat transfer to the sum of convective heat transfer and latent heat required to convert water into ice in the food material. This model has some additional assumptions:

- Constant phase change temperature.
- Enthalpy involves only latent heat and does not account sensible heat at pre- and post-phase change.
- In frozen phase pseudo-stationary, the state is obtained with a linear temperature profile.
- Frozen and unfrozen phase have the same physical properties.

Thus, freezing time according to planks equation is as follows:

$$t_f = \frac{\rho_f \Delta}{T_{if} - T_a} \left(\frac{PD}{h} + \frac{RD^2}{k_f} \right) \tag{13}$$

where, t_f is freezing time; ρ_f is density of food; T_{if} is initial temperature of food; T_a is temperature of refrigerant or freezing atmosphere; D is the cross-sectional diameter if food material; h is convective heat transfer coefficient; k_f is conductive heat transfer coefficient of food material; P and R are shape factors.

P for the slab, infinite cylinder and sphere are 0.5, 0.25 and 0.166, respectively; and R is P/4. Plank's equation under-estimates freezing time by 40%. Although the relationship between P and R is under the same operating conditions yet for different shapes, it correlates similarly in experimental values also.

b. **Cleland and Earle's Method:** Cleland and Earle suggested some modifications to Planck's equation after conducting a series of experiments. They proposed new expressions for the constants P and R. This equation is valid only for slabs.

$$t_f = \frac{\Delta H^*}{T_{if} - T_a}\left(\frac{PD}{h} + \frac{RD^2}{k_f}\right)\left(1 - \frac{1.65 Ste}{Kf}\ln\left(\frac{T_c - T_a}{T_{ref} - T_a}\right)\right) \qquad (14)$$

where, ΔH^* is the difference in enthalpy between the temperature a beginning of the phase change and the reference temperature ($-10°C$), and D is the thickness or diameter of the food product; and

$$P = 0.5[1.026+0.5808 \; PK + Ste \; (0.2296 \; PK + 0.105)] \qquad (15)$$

$$R = 0.125 \; [1.202 + Ste \; (3.41PK + 0.7336)] \qquad (16)$$

where, Ste refers to Stefan number. Advantage of this method over other methods is that it allows the calculation of different freezing times with respect to different final center temperature. There are several other methods but are not discussed here.

To reduce freezing time, one can do either of the following: (i) reduce the thickness of product; (ii) Increase surface area by means of size reduction; (iii) pre-chilling the product; (iv) increase air velocity. Sometimes, by increasing air velocity, it increases the load of the system leading to heating, which is not desirable.

5.6 STORAGE FACILITIES FOR FROZEN FISH

Rapidly frozen and minimally dehydrated fish from the freezer is transferred to cold storage to maintain its high quality. Fish is stored when the core temperature reaches $-18°C$. This is obtained after equilibration of the temperature of fish to that of ambient temperature. The time required for it is called equilibration time that is the duration of subsequent handling of fish after freezing. Storage temperature should be as low as possible as an unfrozen fraction of water increases with temperature. But cold storage with constant temperature is preferred over one with a lower temperature of storage but higher fluctuations. When temperature fluctuates, some fractions of water present in fish thaw, spread over and get refrozen. With the increase in temperature, enzymes present in water also get activated. These enzymes digest proteins. Thus, upon thawing, soft flesh fish is obtained. Storage should be done below $-30°C$ as it is below the cryohydratic point of NaCl. This ensures arrested denaturation process for actomyosin while storage but may induce textural changes. Table 5.7 indicates states about average storage life of seafood [54].

TABLE 5.7   Practical Storage Life of Frozen Seafood at Several Temperatures [54]

Product	Storage life (in months) at temp. of		
	–12°C	–18°C	–24C
Clams and Oysters	4	6	>9
Cooked Shellfish		3	
Fatty Fish, glazed	3	5	>9
Lean Fish Cod, sole, etc.)	4	6 to 9	>12
Lobster, crab, shrimps in Shell	4	6	>12
Shrimps (cooked/ peeled)	2	5	>9

5.6.1 STORAGE PARAMETERS

Storage parameters influence the final quality of the product. Even if fish is handled properly while and before freezing, improper storage results in considerable reduction in final yield. Selected parameters are discussed here.

  a. **Storage Temperature:** Physical and chemical changes slow down with decreasing storage temperature. Storage temperature should be chosen according to the species, product type and desired time of storage. Generally, fishes are stored at –30°C, as microbial actions and enzymatic actions are reduced greatly. However, if fish is intended to be stored for a shorter duration of time, the higher temperature could be chosen to save energy cost.
  The International Institute of Refrigeration (IIR) recommends a storage temperature of –18°C for lean fish such as cod and –24°C for fatty fishes such as herring. The code also recommends that for lean fish intended to be kept in cold storage for over a year, the storage temperature should be –30°C. If the storage has different types of fishes, the lowest temperature required for freezing is used.
  b. **Shelf Life-Limiting Factor:** Several chemical changes like protein denaturation, rancidity, discoloration, desiccation take place while storage. These changes are more profound at higher storage temperature, e.g., –18°C. But at –30°C, all these changes practically are arrested. Thus, storage temperature should also be decided based on the allowed forgone quality of the product during storage.

c. ***Construction Material and Insulation:*** Construction material should allow minimum heat transmission from the ambient atmosphere to cold storage. Generally, bricks are only used. But to add on the resistance to heat transfer, an additional insulation layer is attached to the wall. This layer should have a minimum heat transfer coefficient. It should be odor free, antiriot, fire resistant and impermeable to water vapor. Commonly used insulation materials are: polystyrene, Styrofoam, and polyurethane.

d. ***Cold Air Distribution:*** A very low air movement is required to avoid excessive desiccation, water loss, and microbial contamination due to condensation of vapor on the surface. High humid air is used to reduce weight loss from the product. Generally, 1m/s velocity of air is used. Desired speed is generated using fans.

e. ***Room Construction and Size:*** Total internal space required is determined by racking, stacking, and loading system. Frozen stores may vary from small room with manual racking to a large room with automatic racking. The height of the store should be kept in accordance with rules and regulations of planning authorities.

f. ***Floor of Storage:*** Floors for cold stores are divided into various layers serving purposes. Seven layers make up the floor. Sub-base provides flexible but firm support. Base slab provides the main support for the whole structure. Heaters prevent freezing below insulation. Then comes the vapor barrier, insulation layer, slip layer, and wear floor. Wear floor provides strong, hygienic, and wear resistant working surface.

g. ***Doors of Storage Rooms:*** Doors size, position, duration of not being locked and fitting of any infiltration protection determine the amount of heat it has will permit to enter during loading or unloading of product. Ingress of warm, moist air can cause many problems like increased refrigeration load and ice build-up on pipes. It is advisable to avoid positioning of the door on the opposite walls. Also, doors can have vestibules (air locks) to lessen the quantity of air ingression. They are highly effective but suffer from the disadvantage of heavy maintenance, large footprint, and restriction for trucks. Thus, persons still use PVC strip curtain although they are unhygienic and need considerable maintenance. Air curtains fitted with the fan are also an alternative infiltration protection technique.

h. ***Refrigeration Load:*** Operator of cold storage should be aware of the maximum load; which plant can handle and should be ready for the contingency. It is important to maintain a constant temperature of storage as was discussed earlier. Refrigeration load for cold storage includes heat from frozen food, people, light, machinery, doors, evaporative chamber and infiltration through walls. Thus, the operator should develop the load profile of the day and should try to maintain temperature accordingly.

5.6.2   CHANGES DURING STORAGE

5.6.2.1   PHYSICAL CHANGES

a. **Unfrozen Water Fraction:** Bound water present in food does not freeze even at very low temperatures. This fraction of water is high in solute concentration; thus, its freezing point is suppressed to the point that it is not practically feasible to freeze the food product to that temperature. Thus, food is stored with some unfrozen water fraction, which is highly reactive. It might lead to protein denaturation over the period. Thus, even properly frozen fish also shows a minimum of 5% of drip loss.

b. **Weight Loss:** Generally, fish is frozen up to a temperature of $-40°C$ but is stored at a temperature of $-30°C$ or above. Thus, a heat and moisture gradient develops between the surface and surrounding environment [5]. Globally, 1.67 to 6.15% of weight loss in the fish product is observed.

c. **Recrystallization:** Whenever there is any fluctuation in temperature during storage, frozen water melts, spreads over and then is crystallized again. Due to the migration of water towards the surface, crystals now formed are of bigger size. This further leads to cell disruption and consequently more drip loss. Migration of water could be controlled with the help of cryo-protectants.

d. **Protein Denaturation:** Protein solubility and extractability decrease with storage, which further leads to a diminished nutritional value of fish [1]. The water-holding capacity of the fish and the biochemical properties of actomyosin such as enzymatic activity, viscosity, and surface hydrophobicity, are affected by freezing [53]. Denaturation of proteins is confirmed by the presence of nucleic acids in drip exudate after thawing. The possible reason behind denaturation is that the formation of ice crystals would have disrupted the organized H-bonding system in protein structure resulting in exposure of hydrophobic and hydrophilic regions of proteins to the environment.

Thus, intermolecular cross-linkages between protein molecules [49] are allowed. This further leads to protein aggregation, which results in chewy, rubbery texture and more thaw drip. Pollock and cod have more sensitive proteins when compared to other fishes.

e. ***Freezer Burn:*** Due to evaporation of water from the surface, white patches might be observed which might be mistaken with mold growth. These patches disappear on rehydration with water. But if severe, dehydration might denature proteins irreversibly. Fish can be protected by either packaging or by glazing.

f. ***Functional Properties:*** Rancid flavor, progressive toughening and cold store flavor are principle sensory changes observed in fishes [50]. Also, fishes show faster deterioration in texture and functional properties than animal tissues [1]. They also lose their juiciness on frozen storage. In soft-finned fishes belonging to Gadoid species, following chemical reactions are observed during frozen storage (Figure 5.5).

FIGURE 5.5 Chemical reactions during freeze storage.

Thus, on the first bite of this spongy texture fish, all the water exudate thus later leads to chewing of the very dry product [49]. Whereas non-gadoid species show toughening of muscles and dry texture.

5.6.2.2 CHEMICAL CHANGES

a. ***Rancidity:*** In fishes, lipids become rancid due to hydrolytic oxidation. Amount of lipid present in the fish is not directly proportional to the risk of rancid deterioration. It is the stability of lipids, which decides it. For example, Chinook salmon with higher lipid content is more stable than lower lipid content pink salmon. Free fatty acids, triacylglycerides, and phospholipids were responsible for oxidation in mackerel mince [19]. Since mincing disrupts the cells as well increases surface area for exposure to air, it destabilizes the fish. Thus, to avoid oxidation, either components susceptible to oxidation can be removed, or antioxidant or components like cryoprotectants could be added which protect lipids [60]. Pre-heating and pre-washing can help to control of oxidation as stated earlier in this chapter.

b. **Flavor Loss:** Decomposition of TMAO into fatty acids leads to protein structure changes. This effect could be observed in the form of flavor changes, which occur in three distinct phases. First the original flavor concentration decreases, then taste becomes bland, and consequently, off-flavors develop due to carbonyl and acids compounds formed during lipid oxidation.

c. **Release of Enzymes:** Latent malic enzymes present in fish are solubilized by the disruption of the tissue caused by freezing [15].

d. **Acetaldehyde Formation:** Fluctuation in storage temperature affects the DMA and FA formation in frozen fish.

5.6.3 GUIDELINES FOR STORAGE

a. **Structure:** Although cold storage has a long life, yet regular maintenance of insulating material is required. Any small crack can increase the refrigeration and defrosting load considerably. Thus, immediate repairing and restructuring of these cracks are required, or ice will form in these crevices. Formation of ice hampers insulating property of the material. The operator should regularly check heater circuits fitted in doors and floors. This monitoring will reveal any damage to insulating material before any serious ice has built up.

b. **Refrigeration system:** High efficiency of cold storage can only be defined by its refrigeration system. Although refrigeration systems are fully automatic and do not need constant attention these days, yet the system still needs constant cleaning and maintenance operation. One should ensure full flow through heat exchangers, cleaned finned coils on condenser and evaporator, routine cleaning of filters and strainers, appropriate oil and lubricant level. Also, the operator should keep a check on ice build-up in the refrigeration system. Ice tends to reduce the life of construction material.

c. **Monitoring:** Monitoring is the most crucial step, one should never avoid. Continuous monitoring can only ensure high performance of devices and problem identification at initial stages. Monitoring can be performed by comparing the desired temperature profile of the cold storage with an actual temperature profile. Any deviation detected by sensors should be corrected by taking appropriate control actions.

5.7 THAWING

Frozen fish has to be thawed before it can be used any further. Thawed fish can be either reprocessed then frozen or can be consumed after cooking. Thawing is a more complicated process than freezing:

a. It takes much longer time than freezing. This is because the unfrozen layer is less thermally conducting than frozen layer leading to reduced heat transfer with increasing unfrozen layers in the thawing process. Also, the temperature difference between thawing medium and food is comparatively less than freezing.

b. Risk of microbial contamination is higher in case of thawing due to high-temperature exposure of surface for a long time, which is an optimal condition for microbial growth. Thus, while thawing, one should maintain the maximum surface temperature of 7.2°C and an ambient temperature of 18°C [4, 6, 8].

c. Result of mishandling of fish while freezing can be seen in the form of drip loss while thawing. This may reduce the yield of the product.

d. If the fish is kept at glass transition temperature for long while thawing, it may lead to increased recrystallization rate.

e. Textural changes in fillets which are more when compared to whole fish due to the presence of the backbone providing structural support.

In the case of fish, the thawing process is complete only when all the ice has melted, and the coldest point has reached the temperature of –1°C. Commercially, fishes are tempered, i.e., they are thawed just below their freezing point yielding firmer texture product suitable for further processing.

Thawed fish are as susceptible to spoilage as wet fish. Thus, they should be utilized immediately or should be stored at a low temperature until required. They may also be left partially frozen so that its own temperature is enough to preserve it until used.

5.7.1 THAWING CURVE

During thawing, surface ice commonly known as glaze melts quickly due to its higher thermal conductivity compared to water. Thus, there is a sharp increase in the temperature of the fish initially. But, gradually rate of increase in temperature arrest due to the attainment of latent heat peak. Now, the inner core of fish remains at a constant temperature (which is lower than the initial freezing point of fish), until all ice melts. This period of latent heat absorption is called a freezing plateau. This is the time when maximum drip loss is observed [12]. Figure 5.6 summarizes the temperature changes during thawing.

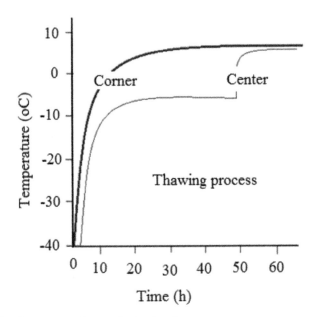

FIGURE 5.6   Temperature changes during thawing.

The surface of the fish heats up rapidly towards to ambient temperature, thus developing non-uniform temperature distribution throughout the body. This increases the risk of local microbial contamination. The extent of non-uniformity of treatment increases with size and speed of thawing. Thus, a balance between speed of thawing and desired uniformity of treatment should be stroked according to the size of fish. Thawing rate also depends on the type of thawing method used, fish species, the thickness of the product, thawing time. Thawing should be done at a quicker rate to avoid any biochemical degradation. However, it should not to be so rapid to avoid drip loss. It is believed that in rapid thawing, melted water does not get enough time to get reabsorbed into the muscles. If the time required by fish to reach −1°C from −5°C is kept to higher than 50 min, then the surface drip loss could be minimized. Drip loss is influenced by other factors, which include: postmortem treatment and pH, the orientation of the cut surface, the rate of freezing, cut-surface-to-volume ratio, fluctuation during frozen storage.

5.7.2 THAWING METHODS

a. *Conventional Surface Heating Technologies:* Conventional surface heating technologies are based upon heat transfer through the surface. The first surface is heated, and then consequently heat travels inside. This process is little long due to increasing resistance of unfrozen layers with the process. This technique includes air thawing, water thawing, steam vacuum thawing, direct contact thawing, frying in frozen state emerging volumetric effect technologies (ohmic thawing, dielectric thawing, power ultrasound thawing, low-frequency acoustics thawing, pressure- assisted thawing). In this technique, internal heating is done. Commercially, air thawing and water thawing are being used. Other methods are still struggling with the high-cost issue. Table 5.8 summarizes the methods, which are being used or can be used for fish freezing.

TABLE 5.8 Common Methods of Thawing of Fishes

Thawing method	Fish type	Specifications
Air thawing	Fish fillets, whole shellfish, shucked shellfish, salmon	100 mm thick – 8 to 10 hours –max 20°C
Dielectric thawing	Herring, white fish, fish fillets	100 mm thick – 1h–5Kv, 80 MHz
Direct contact thawing	Block or Slabs of fish	Data not available
Low-frequency acoustics thawing	Pacific cod [22]	91 mm thick –1500 Hz
Ohmic heating	Shrimp Blocks	50 mm thick – 15 to 20 min after immersion
Power ultrasound thawing	Cod [61]	7.6 cm thick – 2.5h-500 kHz
Pressure-assisted thawing	Pollock whiting fillets, whiting fillets [4]	max 150 MPa
Steam vacuum thawing	Headed and gutted whiting, shrimp	100 mm thick–80 min
Water thawing	Whole round Fish, headed and gutted fish, squid, fish trunks, rock lobster tails	100 mm thick – 4.5 h – max 18°C and 22 m/s

b. *Air Thawing:* Air thawing can be either done in still air or moving air. Still, air thawing is time and space consuming but requires a minimal cost. The product is kept for overnight in air at a maximum of 18°C. A 100mm thick cod may take up to 20 hours which cause desiccation. Time can be reduced to 8–10 hours by increasing exposed surface area. Whereas, air blast can complete thawing 2 hours with a minimum air velocity of 6m/s at 20°C. Fishes are laid on mesh trays which are stacked in height double to fish height. Air circulated should not exceed 20C as it may damage the surface. Also, air should be saturated to improve heat transfer to fish as dry air has less thermal conductivity. It also prevents drying of the surface. Thawing time depends on the temperature of the air, the velocity of air, shape, and size of the block, exposed surface area. Thus, it is common practice to place fish distantly in air blast equipment. Air thawing process may be accelerated by dividing the process into two stages:

- First: drastic thawing is done at 20°C combined with high air velocity to increase heat input. As initially thawing is faster due to less resistance by unfrozen layer.
- Second: follows cold thawing at 5°C. This reduces surface damage due to prolonged hours of thawing at high temperature.

Air has low thermal conductivity and low volumetric heat capacity resulting in low heat transfer coefficient and long process time. Thus, water thawing was introduced. However, when uniform final product temperature is required, air thawing is preferred. Also, air thawing efficiency can be increased by additional energy like radiations or HVEF. High voltage electrostatic field (HVEF) was applied in addition to air blast thawing of tuna. It was observed that HVEF application reduced thawing time as well as specific energy consumption [30].

c. *Water Thawing:* Water thawing is ten times faster than air thawing owing to high thermal conductivity and volumetric heat capacity of water. Higher heat transfer coefficient is obtained by circulating water with a pump, spraying it, or using jet impingement [56]. However, due to the high heat transfer coefficient, the temperature of water and surface is almost the same unlike in air thawing where the surface is colder than air. Thus, the temperature of the water should be kept either cold enough (<5°C) or hot enough (>40°C) that no microorganism proliferates on the surface. Water thawing may be a cost-effective method but recirculation and cleaning of water to avoid cross-contamination increase its cost. Also, it reduces market acceptability of fishes as it results in leaching out of nutrients and bleaching of the surface. Water thawing requires a minimum of

5mm/s velocity of water and maximum of 18°C temperature to thaw 100mm thick whole fish.

d. **Steam Vacuum Thawing:** It is an expensive method but gives less water-logged fish with better appearance when compared to water thawing. Steam is introduced in a vacuum chamber where it condenses rapidly on the surface of fish giving high heat transfer rate. Vacuum provides two-fold advantages: (1) lowers the condensation temperature of the steam and (2) enhances heat transfer by eliminating air molecules which protect the surface of fish. Thawing for fish products can be completed within one hour.

e. **Direct Contact Thawing:** In this method, only uniform block and slab of fishes could be used. This technique is similar to plate freezing; the only difference lies in the circulation of hot fluid instead of cold fluid. Since the product is sandwiched between two metal plates, direct contact is established ensuring high heat transfer from highly conducting plates. However, the advantage of high heat transfer is not enough to camouflage the disadvantage of high equipment cost. Thus, it is not a popular commercial method.

f. **Ohmic Thawing:** Here, current flows through fish and due to the resistance of fish, it starts to heat up. The only problem is that fish in the frozen state is a poor conductor of electricity thus does not heat up. However, it is a good conductor of heat which helps in solving this problem. Fish is first dipped in water, which raises its temperature enough to conduct electricity but fish is still unthawed. Thus, this method is a hybrid one. Current is passed through to generate heat, which is sufficient to thaw fish completely. In practice, fillets or small whole fish in blocks up to 50 mm thick are first immersed in tap water for 1530 minutes, depending on temperature and type of block; and are then heated electrically for a further 1520 minutes until they are thawed [20]. This method is suitable for large or medium scale caterers. However, there is no commercial equipment available.

g. **Dielectric Thawing:** Dielectric thawing uses radio frequency (RF) and microwave (MV). The rapid reorientation of dipolar molecules (water and other) under electromagnetic field generated heat inside the products [56] RF operates at a range of 3 kHz to 300 MHz (longer wavelength), while MV operates at 300 MHz to 300 GHz (Shorter wavelength). The longer the wavelength higher the penetrating power though the RF penetrating power is more compare to MV and therefore it may be more suitable for large foods, while MV gives better energy absorption and hence higher heating rates [31].

Blocks of fishes are placed between two parallel metal plates. These metal plates are not in contact with the fishes, but the high-frequency alternating current is applied across; the plate is large enough to generate heat, which can thaw blocks of fish. Prior to thawing, fish can be immersed in water so that uniform electrical conditions are generated during thawing. This method is fastest when compared to those above. It can thaw whole 100mm thick cod in 40 minutes whereas an 80mm thick herring in 13 minutes only. Although this method provides greater flexibility and short thawing time, yet its capital cost and operating cost are much higher. Therefore, commercial application is limited right now.

5.7.3 SELECTION CRITERIA FOR THAWING SYSTEM

There are many factors that affect the type thawing system used. Comparison of various systems depending upon these factors [58] is given in Table 5.9. Some of the factors are discussed here.

a. **Operational Considerations:**
- Mode of Operation: According to space available, batch or continuous operation is decided. If one has ample amount of space and capital, continuous operation is preferred. But for small-scale industries, batch operated water thawing or air thawing systems are preferred.
- Capital and Operating Cost: Water used for thawing have salts, blood, fats, and oils which may cross-contaminate the product; thus there is a need to be discharged or clarified which increases its cost of operation. Thus, air thawing seems to be the most inexpensive thawing system.
- Labor Requirements: As per the availability of labor in nearby areas, machines should be employed. Batch system requires more labor than a continuous one. Also, except for the water thawing system, the rest of the systems require the same number of working personnel.
- Water Availability: Water is important for any plant. Thus, the cost of drawing fresh water from the resources should be considered. Low water availability might increase water cleaning cost.
- Flexibility and Reliability: Batch type air blast thawer is commonly used due to its flexible nature. It could handle various kinds of fishes simultaneously. Thus, there is a saving of energy

TABLE 5.9 Comparisons Among Various Thawing Methods [58, 59]

Parameter	Method						
	Air Blast	Water	Vacuum dielectric	Dielectric	Electrical resistance	Microwave	Tempering
Capital	1	1.8	1.4	4.3	0.5–1.0	5–10	
Cleaning	Difficult	Difficult	Easy	Easy	Difficult		Easy
Fuel	1	0.7	1	3.8	2	4	0.8
Labor	1	0.7	1	1	-		
Maintenance	1	1.1	0.1	1.4	0	2	Low
Odor	High	Low	Low	Low	Low		Low
Relative speed	1	0.9	1.1	4	2	High	Low
Temperature distribution	Uneven	Uneven	Uneven	Uneven	Uniform	Uneven	Uniform
Typical output (t/h)	1	1	1	1	0.1	0.01	1
Versatility	High	High	High	High	Low	Low	Low

required to freeze fishes in different batches classified according to their size and running the thawer under capacity.

- Cleanability and Hygiene: Surfaces which are in direct contact of the product should be easily cleanable. Sharp edges and dent surface might accumulate the product which later becomes a host for microbial spoilage. Also, hygiene should be maintained while cleaning the equipment surfaces. Clean water should be employed to avoid any cross-contamination.

- Speed of Operation: Methods like dielectric thawing, direct contact thawing offers high speed. But this advantage is usually offset by the initial capital cost and operating cost. However, where speed matters, this disadvantage might be overlooked. For example, caterers are needed to present the dish in little time of ordering. If they use conventional method, it might take too long, and they might end up losing customers. However, the use of dielectric method helps them not only in serving on time but also save wastage.

b. *Input Considerations*

- Species of fish: Usually lean fishes require more energy for thawing due to high ice content present when compared to fatty fishes. For example, 1 kg of lean white lean fish takes 300 kJ to thaw completely, whereas 1kg fatty herring fish just took 240 kJ of energy.

- Type of product: Seafood which is stored at −30°C would require more heat as compared to that stored at −15°C. Smoked fishes are stored at lower temperatures when compared to fish fillets. Also, fish fillets, which are pre-rigor frozen, are needed to be thawed slowly so that shrinkage which would occur due to completion of rigor mortis after thawing, could be controlled. Pre-rigor frozen whole fish does not show considerable shrinkage.

- Extent of thawing: When the fish has to be utilized immediately, complete thawing is done. But if there is a time lag, then fish are left partially unfrozen so that their own cold reserve is enough to maintain low temperature and avoid further spoilage.

- Variety of fishes handled: If a thawing unit handles one type of fish, then the equipment could be optimized according to the specifications. But if a unit handles a variety of fishes, then it has to employ methods, which would fitto all types which may lead to certain compromises. Warm water thawing can handle all types of whole fishes at a time.

c. *Output Considerations*

- Output of plant: Depending on the output required, the system should be chosen. The continuous system will offer better output than the batch system.
- Flexibility of output: Amount of fishes to be thawed depends on the demand for the same. Thus, the system should be flexible enough to work out a wide range of output without energy wastage. For example, if a unit has three batch thawing systems, they can use all three systems or less according to the requirement.

5.8 SUMMARY

Although many new techniques have been introduced in the field of fish preservation, yet freezing has never lost its significance due to the high quality of end product. Fish freezing has been a very old practice but has been improved continuously with the advent of emerging technology. The fish freezing technique not only requires mechanical and technological knowledge of freezing process and fish respectively, but it also demands optimization of both these understandings to reduce the overall cost of a plant. For example, a phenomenon like the desirability of rapid freezing could be explained in terms of heat of crystallization as well as percolation of water through cell membranes. Thus, the rate of freezing could be determined according to both processes, i.e., freezing rate could be chosen in a way that it is enough to stop percolation of water and process of nucleation could be increased by several pre-treatments instead of doing it with the rate of freezing. The high freezing rate increases the cost of the plant.

KEYWORDS

- cryogenic freezer
- dielectric thawing
- eutectic point
- fillets
- fish freezing
- freezing curve

- glazing
- gutting
- Ostwald ripening
- phosphate dips
- plank's model
- re-crystallization
- rigor mortis

REFERENCES

1. Alvarez, C., Huidobro, A., Tejada, M., & Vazquez, I. E., (1999). Consequences of frozen storage for nutritional value. *Food Sci. Tech. Int., 6*, 493–498.
2. Borgstrom, G., (1965). *Fish as Food* (Vol. IV, p. 323). Academic Press, New York.
3. Burgess, G. H., Cutting, C. L., Lovern, J. A., & Waterman, J. J., (1967). *Fish Handling and Processing* (p. 232). Chemical Publishing Co. Inc. New York.
4. Calvello, A., (1981). Recent studies on meat freezing. Chapter 5, In: Lawrie, R., (ed.), *Developments in Meat Science* (Vol. 2, pp. 125–158). Applied Science Publishers, UK.
5. Campanone, L. A., Roche, L. A., Salvadori, V., & Masheroni, R. H., (2002). Monitoring of weight losses in meat products during freezing and frozen storage. *Food Science Tech. Int., 8*, 229.
6. CFR (Code of Federal Regulations) by FDA (Food and Drug Administration) (2001). *Title 21. Part 110. Current Good Manufacturing Practice in Manufacturing, Packing, or Holding Human* Food (p. 121). Food and Drug Administration, US Government, Washington, D.C.
7. Cho, S. Y., Endo, Y., Fujimoto, K., & Kaneda, T., (1989). Autoxidation of ethyl eicosapentanoate in a defatted model system. *Nippon Suisan Gakkai Shi, 53*(3), 545–545.
8. Chourat, J. M., (1997). Contribution an l'etude de la decongelation par haute presion. *PhD Thesis* (p. 125). Universite de Nantes, France.
9. FAO Corporate Document Repository 5: Freezing Time, Freezing and Refrigerated storage in Fisheries, (2002). *Fisheries Aquaculture Department of FAO* (p. 201). Rome – Italy.
10. FAO, (2000). *Code of Practice for Fish and Fishery Products* (pp. 81, 82). Agenda item 4, CX/FFP, *Codex Alimentarious* Commission, United Nations, Rome.
11. FAO, (1981). *Refrigerated Storage in Fisheries* (p. 112). FAO Fisheries Technical Paper 214, Fisheries Aquaculture Department of FAO, Rome – Italy.
12. Fellows, P., (2000). *Food Processing Technology, Principles and Practices* (p. 201). CRC Press, Baton Raton–FL.
13. Fenemma, O. R., Powrie, W. D., & Marth, E. H., (1973). *Low-Temperature Preservation of Foods and Living Matter* (p. 207). Marcel Dekker, New York.
14. Fujita, Y. T., Oshima, K., & Koizumi, C., (1994). Increase in the oxidative stability of sardine lipids through heat treatment. *Fish. Sci., 60*(3), 289–293.

15. Gould, E., (1971). An objective test for determining whether fresh fish have been frozen and thawed. In: Kruezer, R., (ed.), *Fish Inspection and Quality Control* (p. 72). Fishing News (Books) Ltd. London.

16. Hall, G. M., & Ahmad, N. H., (1997). Surimi and fish-mince products. In: Hall, G. M., (ed.), *Fish processing technology* (2nd edn., pp. 74–92). Chapman and Hall, London.

17. Hartel, R., (2001). *Crystallization in Foods* (pp. 192–231). Aspen Publishers Inc., Maryland.

18. Heldman, D. R., & Lund, D. B., (2007). *Handbook of Food Engineering* (2nd edn., p. 525), CRC Press, Taylor and Francis Group, Boca Raton – FL.

19. Hwang, K. T., & Regenstein, J. M., (1996). Lipid hydrolysis and oxidation of mackerel (*Scomber Scombrus*) mince. *J. Aquatic Food Product Technol.*, *5*, 17–21.

20. Jittinandana, S., Kenney, P. B., & Slider, S. D., (2005). Cryoprotectants affect physical properties of trout during frozen storage. *J. Food Sci.*, *70*(1), C35–C42.

21. Johari, G. P., Hallbrucker, A., & Mayer, E., (1987). The glass-liquid transition of hyper-quenched water. *Nature*, *330*, 552–553.

22. Jul, M., (1984). *The Quality of Frozen Foods* (p. 223). Academic Press, London.

23. Kasapis, S., (2006). Glass transitions in frozen foods and biomaterials. Chapter 2, In: *Handbook of Frozen food Processing and Packaging* (pp. 110–115). CRC Press, Taylor and Francis Group, Boca Raton – FL.

24. Kennedy, C. J., (2002). The selection and pretreatment of fish. *Managing Frozen Food*, 103–110.

25. Kissam, A. D., Nelson, R. W., Ngao, R., & Hunter, P., (1981). Water thawing of fish using low-frequency acoustics. *Journal of Food Science*, *47*(1), 751–755.

26. Khayat, A., & Schwall, D., (1983). Lipid oxidation in seafood. *Food Technol.*, *37*(7), 130–133.

27. Kolbe, E., & Kramer, D., (2007). *Planning for Seafood* (p. 85). Alaska sea grant college program, University of Alaska, Fairbanks.

28. Merritt, J., (2010). *Guidelines for Industrial Thawing of Ground Fish in Air and in Water* (p. 112). Report IDD 109, Nova Scotia Department of Fisheries, Halifax.

29. Miles, C. A., Morley, M. J., & Rendell, M., (1999). High power ultrasonic thawing of frozen foods. *Journal of Food Engineering*, *39*(2), 151–159.

30. Mousakhani-Ganjeh, A., Hamdami, N., & Soltanizadeh, N., (1999). Thawing of frozen fish (*Thunnus albacares*) using still air method combined with a high voltage electrostatic field. *Journal of Food Engineering*, *39*(2), 110–115.

31. Myres, M., (1981). *Planning and Engineering Data, I: Fresh Fish Handling* (p. 83). FAO, Fisheries Circular 735, FAO, Rome.

32. Murakami, E. G., & Okos, M. R., (1989). Measurement and prediction of thermal properties of foods. In: Singh, R. P., & Medina, A. G., (eds.), *Food Properties and Computer Aided Engineering of Food Processing Systems* (pp. 3–48). Kluwer Academic Publishers, Amsterdam.

33. Ohlsson, T., Bengtsson, N. E., & Risman, P. O., (1974). The frequency and temperature dependence of dielectric food data is determined by a cavity perturbation technique. *Journal of Microwave Power and Electromagnetic Energy*, *9*(2), 129–145.

34. Park, J. W., (1994). Cryoprotection of muscle proteins by carbohydrates and polyalcohols: A review. *J. Aquat. Food Prod. Technol.*, *3*(3), 23–41.

35. Porsdal, K., & Lindelov, F., (1981). Acceleration of chemical reactions due to freezing. In: Rockland, L. B., & Stewart, G. F., (eds.), *Water Activity: Influences on Food Quality* (pp. 100–120). Academic Press, New York.

36. Pham, Q. T., (1987). Calculation of bound water in frozen food. *Journal of Food Science, 52,* 210–212.

37. Pham, Q. T., (2001). *Thawing, Operation in Food Refrigeration* (pp. 332). CRC Press, Boca Raton- FL.

38. Rahman, M. S., (1995). *Food Properties Handbook* (pp. 87–177). CRC Press, Boca Raton-FL.

39. Rahman, M. S., & Ruiz, J. F. V., (2007). *Handbook of Food Preservation* (p. 635). Second Edition, CRC Press, Boca Raton- FL.

40. Rahman, M. S., (2001). Thermo-physical properties of foods. In: Sun, D. W., (ed.), *Advances in Food Refrigeration* (pp. 70–109). Leatherhead Publishing, Leatherhead.

41. Regenstein, J. M., Schlosser, M. A., Samson, A., & Fey, M., (1982). Chemical changes in trimethylamine oxide during fresh and frozen storage of fish. In: Martin, R. E., Flick, G. J., Hebard, C. E., & Ward, D. R., (eds.), *Chemistry and Biochemistry of Marine Food Products* (pp. 137–148). AVI Publishing Co., Westport, Connecticut.

42. Reid, D. S., Doong, N. F., Sinder, M., & Foin, A., (1986). Changes in the quality and microstructure of frozen rockfish. In: Kramer, D. E., & Liston, J., (eds.), *Seafood Quality Determination* (pp. 1–15). Elsevier Science Publishers, Amsterdam.

43. Reid, D. S., (1993). Basic physical phenomenon in the freezing and thawing of plant and animal tissues. In: Mallett, C. P., (ed.), *Frozen Food Technology* (pp. 20–28). Chapman & Hall, London.

44. Reid, D. S., (1996). Fruit freezing and processing of fruits. In: Somogyi, L. P., Ramaswamy, H. S., & Hui, Y. H., (eds.), *Science and Technology, Volume 1. Biology, Principles and Applications* (pp. 169–178). Technomic Publishing, Lancaster, CA.

45. Reid, D. S., (1990). Optimizing the quality of frozen foods. *Food Technology, 44*(7), 78–82.

46. Reid, D. S., (1987). The Freezing of food tissues. In: Grout, B. W. W., & Morris, G. J., (eds.), *The Effects of Low Temperatures on Biological Systems* (pp. 478–489). Arnold, London, UK.

47. Reid, D. S., (1987). The significance of the glassy state in the storage of frozen foods. In: Grout, B. W. W., & Morris, G. J., (eds.), *The Effects of Low Temperatures on Biological Systems* (pp. 468–478). Arnold, London, UK.

48. Rudolph, A. S., & Crowe, J. H., (1985). Membrane stabilization during freezing: The role of two natural cryoprotectants, trehalose and proline. *Cryobiology, 22,* 367–377.

49. Santos-Yap, E. E. M., (1996). Fish and seafood. In: Jeremiah, L. E., (ed.), *Freezing Effects on Food Quality* (pp. 109–115). Marcel Dekker, New York.

50. Saymons, H., (1994). Frozen foods. In: Man, C. M. D., & Jones, A. A., (eds.), *Shelf Life Evaluation of Foods* (pp. 298–305). Blackie Academic & Professional, London.

51. Schwartzberg, H. G., (1990). Food freeze concentration. In: Schwartzberg, H. G., & Rao, M. A., (eds.), *Biotechnology and Freeze Concentration* (pp. 127–202). IFT Basic Symposium Series. Marcel Dekker, Inc., New York.

52. Schwartzberg, H. G., (2002). Food freezing: Thermodynamics. In: *Encyclopedia of Agriculture Food, and Biological Engineering* (2nd edn., pp. 537–538).

53. Sebranek, J. G., (1996). Poultry and poultry products. In: Jeremiah, L. E., (ed.), *Freezing Effects on Food Quality* (pp. 85–90). Marcel Dekker, New York.

54. Singh, R. P., & Heldman, D. R., (2009). *Food Freezing, Introduction to Food Engineering* (4th edn., p. 528). Fourth edition, Academic Press, New York.

55. Slabyj, B. M., & Hultin, H. O., (1982). Lipid peroxidation by microsomal fractions isolation from light and dark muscle of herring (*Clupea harengus*). *J. Food Sci., 47*, 1385–1395.

56. Smith, P. G., (2011). *Introduction to Food Process Engineering* (2nd edn., p. 233). Springer, New York.

57. Sutton, R. L., Evans, I. D., & Crilly, J. F., (1994). Modeling ice crystal coarsening in concentrated disperse food systems. *J. Food Science, 59*(6), 1227–1233.

58. Tanaka, T., Nagasake, T., & Takahashi, K., (1984). Thawing of frozen tuna meat. *Aspect of Meat Color and Contraction, 1*(2), 183–187.

59. Torrey, S., (2017). *Thawing Frozen Food*. FAO Corporate Resource Document. http://www.fao.org/wairdocs/tan/x5904e/x5904e00.htm Accessed on July 31.

60. Undeland, I., Stading, M., & Lingnert, H., (1998). Influence of skinning on lipid oxidation in different horizontal layers of herring (*Clupea harengus*) during frozen storage. *J. Sci. Food Agric., 78*, 441–449.

61. Undeland, I., Ekstrand, B., & Lingert, H., (1998). Lipid oxidation in minced herring (*Clupea harengus*) during frozen storage: Effect of washing and pre-cooking. *J. Agriculture Food Chemistry, 46*, 2319–2325.

62. Wang, L., & Weller, C. L., (2002). Thermo-physical properties of frozen foods. In: *Handbook of Frozen Food Processing and Packaging* (pp. 105–106). CRC Press, Taylor and Francis, Boca Raton – FL.

63. Wang, Y. J., Miller, L. A., & Addis, P. B., (1991). Effect of heat inactivation of lipoxygenase on lipid oxidation in lake herring (*Coregonus Artedit*). *J. Am. Oil Chem. Soc., 68*(10), 752–758.

64. Waterman, J. J., (1964). *Bulking, Shelving or Boxing* (p. 12). Torry Advisory Note 15, Central Science Laboratory, Aberdeen.

65. Zaritzky, N., (2006). Physical-chemical principles: Freezing. In: *Handbook of Frozen Food Processing and Packaging* (pp. 111–118). CRC Press, Taylor and Francis Group, Boca Raton, FL.

CHAPTER 6

FISH FREEZING: PRINCIPLES AND PRACTICES

JUHI SAXENA and HILAL A. MAKROO

ABSTRACT

Generally, the loss of quality in fish is caused by autolytic degradation followed by microbial spoilage, both of which are functions of storage temperature of fish. Since the deteriorative changes can be minimized at reduced temperatures, the purpose of freezing is to lower the temperature to the extent that slows down spoilage as well as a result in a product which when thawed is virtually indistinguishable from its fresh counterpart. Every fish species has its own spoilage rate, and calculation of relative spoilage rate allows a fish technologist to determine the equivalent length of time of storage at 0°C. The process of freezing results in numerous structural and biochemical changes in the fish (partial dehydration of proteins, the interaction of lipids, the action of enzymes etc.), which is largely dependent on the type of freezing systems employed such as: air blast, plate, immersion, spray, CO_2 and liquid nitrogen freezing. This chapter discusses the prediction of freezing times by numerical methods as suggested by various investigators along with solved numerical examples. The chapter concludes with the application of freezing systems in practice including onshore processing, freezing on board as well as changes reported in the quality of frozen fish and advanced measures to reduce the adverse effects of freezing to a minimum.

6.1 INTRODUCTION

Fish is among the most perishable foodstuffs, and therefore fish preservation is one of the most important aspects of the fishery industry. Freezing is the most widely accepted preservation method of choice primarily because it does not alter the product appreciably when compared to salting, drying, smoking, and canning; and thawing yields a seafood product, which is virtually indistinguishable from the fresh product.

The loss of quality in fish is initiated by the action of enzymes present in the gut and the flesh causing autolytic degradation, followed by the growth of microorganisms on the fish surface developing a slimy appearance. Subsequently, the bacteria invade the flesh of the fish resulting in tissue breakdown and a general deterioration of the product. The rates at which the autolytic and microbial degradation proceed are a function of the storage temperature of fish. Freezing functions as a preservative by: (a) *Reducing temperature*, which lowers the molecular activity; and (b) *Lowering water activity*, which reduces microbial growth. Normally, freezing step often gets the major emphasis; however, it is imperative that thawing must receive equal consideration in reducing quality loss. Generally, fish is packed in ice in order to avoid the possibility of a huge temperature drop so that slow freezing sets in and the resultant loss in quality during defrosting.

The storage life of chilled fish varies from species to species and is affected by breeding conditions, time of catch and the area of catch. For instance, tropical water fish will last longer when ice packed than temperate water fish; herring caught in summers (when it is feeding, and the fat content is high) may last if 4 days on ice, as opposed to it, caught in winters when it can last for approximately 12 days on ice [23]

6.1.1 RELATIVE SPOILAGE RATES

Bacterial spoilage of fish and muscle food at a temperature range of –2 to 20°C can be accurately calculated by a formula that resulted from tremendous research conducted by CSIRO, Division of Food Research, Australia [14, 19, 20]. The formula is applicable to various kinds of fish species, and many studies have applied the same finding to calculate fish spoilage [3, 4]. The equation has been simplified to Eq. (1), which links temperature and spoilage for calculation of the rate of spoilage in relation to that of the rate at a reference temperature (0°C).

$$R = (0.1T + 1)^2 \qquad (1)$$

where, R = Relative rate of spoilage; and T = Temperature (°C).

The relative rate of spoilage is minimum (i.e., = 1), when fish is properly chilled and kept at 0°C temperature. If the fish is at 4°C, then $R = [(0.1\text{x}4)+1]^2 = 1.96 = 2$ (approx.) indicating that the rate of spoilage is twice as fast at 4°C than at 0°C. Therefore, if the temperature history from the time of catching is known, the quality or spoilage history can be calculated by integrating relative fish spoilage with time to acquire the quality of fish and storage time equivalent to that of stored at 0°C.

This chapter discusses the basic phenomena of freezing, the associated biochemical changes in the fish on prolonged storage at very low temperatures and the actors that affect such changes (quality loss) in the product. The chapter also highlights the freezing systems generally employed for fish freezing and the prediction of freezing times from empirical equations by earlier researchers. The practices, which are followed on board as well as onshore, have also been discussed briefly.

6.2   THE FREEZING PROCESS

About 60–80% of fish is composed of water, which some dissolved and colloidal substances, resulting in freezing point depression to less than 0°C. The freezing process primarily involves freezing water to ice, and this process requires the removal of heat from the fish. The freezing process is accomplished in three stages (Figure 6.1).

- **Cooling stage**: As soon as the fish is subjected to a freezer; it cools rapidly to sub-freezing temperature (below 0°C). When the temperature of a biological system is reduced below 0°C, the solution first 'supercools' and then the solutes start crystallizing out of the solution, leading to the beginning of ice crystal formation. Ice crystal formation begins after 'nucleation.' Nucleation is the random aggregation of water molecules to a critical size. Nucleation is followed by crystal growth, where the water molecules attach to the existing crystals as freezing proceeds.
- **During the stage of "thermal arrest,"** a large amount of heat is removed to change the maximum of water into ice, and there is only a slight fall in temperature. More than 50% of water is frozen at this point.
- Beyond, the thermal arrest, the temperature begins to fall rapidly and the remaining water freezes (third stage).

Slow removal of heat results in slow freezing producing a small number of large ice crystals, which cause textural damage on defrosting while rapid removal of heat results in quick freezing producing many small ice crystals causing negligible shrinkage and rupture.

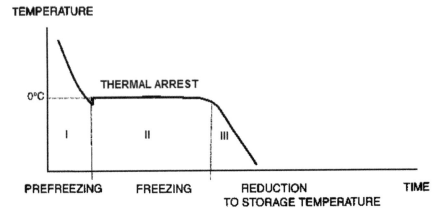

FIGURE 6.1   Fish freezing curve.

Fish generally freezes at −1°C to −2°C. As the water is frozen subsequently through the stages, the concentration of the dissolved salts keeps increasing, causing a depression in the freezing point. Thus, even at low temperatures of −25°C, only 90–95% water is frozen. However, a major portion of free water (75–80%) is frozen at −1°C to −5°C. This temperature range is also known as the critical zone [10].

6.3   BIOCHEMICAL CHANGES IN FISH DURING FREEZING

6.3.1   EFFECT ON PROTEINS

The textural changes in the fish during frozen storage are a consequence of the damage to the structural proteins. The overall damage to the proteins is caused equally by both freezing and thawing. Prolonged storage at −18°C plays a vital role in the deterioration of the fish protein. Although the rate of change is slow at such low temperatures, yet with time it becomes significant to cause adverse effects in the sensorial attributes of the fish. The peptide groups, side chains, and functional groups retain their native form in the protein if they are not subjected to a stressful environment. Exposure to freezing temperature for a considerable time causes denaturation of proteins.

Ultracentrifugation and Electronphotomicrographic studies have shown that dimers, trimers, and high molecular weight polymers are formed by side-to-side attachment of a molecule of myosin in frozen solution [1]. It has also been concluded that disulfide bonds are formed in denatured fish actomyosin during frozen storage [2]. This evidence was further proven by the technique of polyacrylamide-dic-gel electrophoresis, where myofibrillar proteins isolated from carp (*Labeorohita*) were shown to undergo two stages of denaturation: accelerated (during the first week) and slower rate (during the rest of the 7 weeks) during frozen storage at –20°C [18].

Sarcoplasmic protein fraction undergoes minimum damage due to freezing and frozen storage. It is the myofibrillar proteins that undergo maximum deterioration, and the factors that are responsible for the damage include the formation of formaldehyde and free fatty acids (FFA) [8], lipid oxidation [9, 21] and concentration effect of inorganic salts. Increase in FFA has been reported in tropical fish of India such as milkfish [26], mackerel [27], oily sardine [24] among others during storage at –18°C or –20°C.

Some of the degradation products of oxidizing lipids such as malonaldehyde and hydroxynonenal may react with lysine, aspartic acid or thiol groups of protein causing denaturation. Also, there are speculations that the different reactive groups of inorganic salts react with the proteins to degrade their quality. However, no significant conclusions could be made experimentally [7].

6.3.2   ENZYME ACTIVITY

Enzymes are mostly located in water-soluble sarcoplasmic proteins, and their activity is greatly reduced as a result of freezing. Reduction in the activity of aldolase, dehydrogenase has been reported during frozen storage. The activity of sarcoplasmic ATPase of cod gets reduced drastically. A gradual increase in FFA content is indicative of enzymatic activity. The response of different enzymes to freezing temperatures is different, and sometimes the products of enzyme activities themselves are causes of profound protein denaturation.

6.3.3   LOSS OF SOLUBILITY

Myosin-Actomyosin proteins of fish are the main proteins responsible for the loss of solubility during frozen storage. Studies reported that temperate and cold-water fishes do not undergo any significant change in their sarcoplasmic proteins or albumin fraction during freezing or storage. No changes were reported in the electrophoretic pattern of extracts of the water-soluble protein fraction from frozen cod stored for 7 months at −12°C [8]. However, many freshwater species like tilapia (*Oreochromismossambica)* [22] and marine species like oily sardines [17], Indian mackerel [19] and catfish showed loss of solubility of the sarcoplasmic fraction. The potential cause for the loss of solubility was assumed to be a loss of extractable proteins in the drip.

Loss of solubility of salt-soluble proteins or myofibrillar proteins have also been reported in temperate and cold water as well as tropical fishes like tilapia, oily sardine, Indian mackerel among others. In comparison, it was found that the loss of solubility of myofibrillar proteins was more severe than sarcoplasmic proteins [25].

6.3.4 CHANGES IN FATTY ACID PROFILE

The fatty acid profile of oily sardine frozen and stored at −18°C for a period of 150 days showed no change in the total saturated and monoenoic acid contents. However, the EPA and DHA contents were affected and showed a gradual decrease. The loss of EPA and DHA was more severe in samples with higher lipid content [27]. In another study of milkfish (*Chanos chanos*), the extracted lipids from the sample frozen at −18°C showed that during the first 3 weeks of storage there was a considerable loss of saturated fatty acids, which was followed by PUFA during the next 7 weeks and MUFA from 10th week onwards. The fatty acid profiles of marine catfish (*Tachysurusdussumieri)* stored at −20°C revealed an increase of the PUFA neutral lipid fraction and however reduction in PUFA has been observed in the phospholipid fraction [25]. A linear relationship was recognized between phospholipid decrease and FFA increase, which suggested the hydrolysis of phospholipid to yield FFA during frozen storage.

6.3.5 FUNCTIONAL PROPERTIES

The functional properties (water holding capacity, emulsifying ability) of fish muscle are the results of nature and quality of surface proteins present in them. Frozen fish stored for the long duration of time at −18°C have a deteriorated protein quality and thus exhibit poor texture. Some indices that are used to measure the quality of surface proteins and their functional properties are: hydration, fat absorption parameters, foaming stability and emulsifying capacity and stability.

6.3.6 QUALITY LOSS IN FROZEN FISH

Prolonged storage at –20°C or –18°C have shown to develop one or more of the undesirable attributes such as: rancid odor and flavor (oxidation of lipids); toughening (protein denaturation and aggregation); flesh discoloration (oxidation reactions) and desiccation (freezer burn). The factors affecting the quality loss in fish during frozen storage may be categorized as:

a. **Intrinsic factors:** Intrinsic factors involve the fish condition at the time of freezing (nutritional status or stage of spawning); the pH of the flesh post-mortem (which depends on the degree of struggling induced during fish capture) and the rigor mortis stage at the time of freezing (pre-rigor freezing may distort the product during cooking and thawing).

b. **Extrinsic factors** include:

- *The holding procedures prior to freezing*: It has been reported that a good quality storage life of 12 months at a storage temperature of –18°C can be achieved by holding whiting fish (headed and eviscerated) in ice (0°C) for 2 days during freezing, whereas those held for 4 days during freezing can be stored for 6 months [15].

- *Glazing*: Fish is a good source of PUFA (polyunsaturated fatty acids) which make it more prone to oxidation than saturated fats in other meats. To overcome this problem, *wrapping* or *glazing* of the product is generally employed. *Glazing* is the application of a layer of ice to the surface of a frozen product by spraying; brushing, dipping that confers protection to the product against dehydration and oxidation during cold storage. The main principle behind glazing is a sublimation of ice layer rather than the fish along with simultaneous surface air removal to reduce the oxidation rate. Although it is not easy to achieve completely uniform glazing but spray glazing methods are commonly employed. Ice formation due to abrupt freeing results in the fractured glaze (for example when the fish surface temperature is ≤ -70°C like during cryogenic freezing) and this is an improper glaze, which can be easily dislocated during handling. On the other hand, long, glazing time results in the thick and soft glaze which is also unsuitable and can be easily dislocated during handling. Heat added by the glazing process may be significant to require re-cooling in a freezer before the cold storage step.

- *Freezing Rate*: Slow freezing results in the formation of bigger ice crystals due to slow freeing damages the tissue cells hence are responsible for protein denaturation, water holding capacity loss, compromised cell membrane permeability, and enzyme release due to cell disruption. Collectively all these changes result in '*drip loss*' which ultimately increases the possibility of flavor changes in the fish.
- *Freezer temperatures*: During storage fish also undergo chemical changes similarly as in fruit and vegetables. Increase in temperature enhances the deteriorative changes while as such changes can be controlled by reducing the storage temperature.

6.4   FREEZING SYSTEMS

The freezing systems employed for fish freezing largely depend upon the financial, function, and feasibility considerations. In addition to the product damage and product water loss, both the capital and equipment running cost will be taken into financial consideration. The type of freezer (batch or continuous) will be determined by the functional requirements and the capability of the freezer to run in operation. For instance, for freezing large tuna (whole), a horizontal plate freezer would not be suitable. Feasibility is another deciding factor for the operation of freezers located in tile plants. For example, cryogenic freezing (although being high in costs) may be suitable in every respect for the operation. However, these freezing systems cannot be considered suitable in a plant area located, where there no guaranteed supply of liquid nitrogen.

6.4.1   AIR BLAST FREEZING (ABF)

This system involves vigorous circulation of cold air to cause rapid freezing of the product. Tunnel freezing is the most commonly used freezing system where a long, slow-moving mesh belt carries the product and the cold air is introduced in the opposite (counter-current flow) to the direction of the product entering the tunnel. The air temperature ranges from –18°C to –34°C and air velocities ranging from 100 to 3500 feet/min. These systems are suitable for irregular shaped, different sized and non-deformable foods such as fish fillets, breaded fish portions or fish sticks. Moreover, these freezers are available in batch or continuous operation.

There is a certain consideration to be met when designing the ABF system. The rate of air flow should be quite high to freeze the product in the desired time. Additionally, consistent airflow over every individual product or package would be required for the uniform rate of freezing. Continuous air blast freezers may be convenient in terms of airspeed. However, they are expensive and occupy relatively higher floor area. A required freezing capacity with a smaller freezer can be achieved by increasing the air velocity which consequently decreases the freezing time. Hence freezing cost could be reduced using increased airspeeds. In this way, continuous freezing could be economically justified by using the air velocity as high as 10–15 m/s.

Practically the air in contact with the product surface remains stagnant because of the friction in between two, and this stagnant air due to insulation effect becomes a hurdle for the heat transfer. The thickness of this stagnant air is determined by the velocity of the air, the amount of turbulence observed and other factors. Hence, the velocity of air quoted for air blast freezers are generally an average estimation of the velocities between the product or fish spaces. A simple calculation for average airspeed is shown in Problem1.

For the acceptable rise in product temperature, the rate of air flow must be taken into consideration. In case there is an unacceptable rise in temperature (greater than acceptable), it could be raised changes in product freezing time in between upstream and downstream placed products in the freezer space. Generally, in normal and well-designed air blast freezers up to 30% of the refrigeration load may be contributed by a fan. However, it can be worse if the design of the freezer is poor. The rise in air temperature depends on the heat load, and commonly it is expected to be in a range of 1 to 3°C (See Problems 1 and 2).

Hence, the fan should be positioned before the cooler while designing the freezer, which helps in uniformity in the air flow; as the cooler is responsible for giving a high resistance to air flow relatively.

There are continuous air blast freezers for freezing fish, where trucks or trolleys are used for continuous operation so that the coldest air can pass over the coldest fish. One disadvantage with this type is that a complete consignment (row) of trucks must be in motion at once with a fully loaded freezer. In this way, difficulty is being faced as the wheels require special bearings and lubricants and are mostly covered with frost at such low temperatures. Therefore, the use of conveyor belts is preferred for continuous operation.

A continuous ABF system is the Torry freezer, developed by the Ministry of Agriculture, Fisheries, and Food Research Station in Aberdeen. This uses a continuous flat belt made from stainless steel and offers the advantage of high heat transfer rates. It is particularly suitable for freezing individual fish fillets, where the bottom surface is flat, and the top surface is irregular. However, this type of freezer can be used only for products possible to freeze in a short time (approx. 30 min). The freezing time poses limitation especially since long freezing times may require freezer of great lengths which will be cumbersome. The dimensions of the freezer are determined by the freezing time, rate (kg/h) and the product loading density on the belt.

The space can be decreased when the double or triple belt or spiral freezers are used. Since partially frozen fish can easily stick to open metal mesh belts, double/ triple belt freezers may not be suitable for use. Such arrangements are more suitable for the battered and breaded type of products. Spiral belt freezers are widely used for Individual Quick Freezing (IQF) products. However, the type of mesh link belts used affects the appearance of the product. Direct loading of the fish on the belt gives it a crinkled or indented appearance. Open mesh belts can prove to be difficult for product removal after freezing, and may cause a loss in weight because of physical damage, particularly in the case of skinless fillets where the fish may stick to the belt. Problem #3 shows the solved example.

6.4.2   PLATE FREEZING (PF)

Plate freezers (PF) may not be as versatile as air blast freezers as they can only be used for freezing regular shaped blocks and packages. However, they are widely used systems for freezing fish. In PF, the plates may be arranged horizontally to make a series of shelves horizontally or vertically to make a series of bins.

6.4.3   HORIZONTAL PLATE FREEZERS (HPF)

Freezing of pre-packed cartons of fish and fish fillets in rectangular blocks are two major applications of HPF. The thickness of the package is 32 to 100 mm. Pre-packing of the fish inhibits any physical contact between the plate and the fish itself. The HPF will be suitable only when there is good contact on top as well as on the bottom of tray or package going in the freezer. When the fish is subjected to only one-sided freezing due to inadequate upper surface contact, the freezing time may be enhanced by 3–4 times than if there was adequate surface contact between the fish and the upper and lower plates. The plates are grouped together by means of a hydraulic piston to allow efficient heat transfer to the fish surface. The pressure applied to the product is variable and usually ranges in 70×10^6 to 280×10^6 bar depending on the type of product to be frozen.

6.4.4   VERTICAL PLATE FREEZER (VPF)

Vertical plate freezer (VPF) is specifically suitable for bulk freezing and does not require the packaging of the product. It is usually used for whole fish freezing rather than freezing of fillets. The maximum size of block suitable for freezing by VPF is usually 1070 mm × 535 mm. Other dimensions, however, are dependent on the physical characteristics of the fish to be frozen. The heat transfer rate can be improved by loading unwrapped and or unpacked fish in between the plates. Dense blocks with block density near 800 kg/m^3 can be produced by cod and haddock type of fishes. Fish that are high in fat-containing fishes like herring are generally wrapped before freezing, hence do not form blocks as dense as that of lean or fat less fish. Sometimes water is added inside the wrappers in case of fatty fish to strengthen the block, and to prevent dehydration and oxidation during frozen storage. Rigid block formation is particularly important when freezing at sea, as otherwise, the blocks are prone to breakage which subsequently poses difficulty in fish filleting or splitting.

Freezing time is quite high when the fish is frozen in wrappers especially since the wrappers act as insulation by preventing close contact with the plates. Usually, the VPF is designed to have unloading from the top, as in this way unloading blocks of the product is easier and simple for the operator. However, it is possible to make VPF with unloading from any side (top, bottom or sides).

6.4.5   AUTOMATIC PLATE FREEZER (APF)

APF is used for freezing fish packed in cartons. Simply put, APF is a form of HPF that operates in a continuous mode. The design of the freezers is catered to particularly the processing line purpose and based on that the usual capacity of APF is up to 2 tons/h. They have an advantage over other freezers in terms of labor requirement for loading and unloading of the products.

6.4.6   IMMERSION FREEZING

This freezing method ensures intimate contact between the refrigerant and fish surface enabling good heat transfer. Utilization of liquid for cooling allows effective control of the freezing rates particularly since the liquid is more suitable for removing a large amount of heat than a gas refrigerant. Gas cooling agents make a stagnant boundary layer, which retards the heat transfer rate; therefore, circulation is the key factor that allows effective freezing in this type. Despite having suitable refrigeration and heat transfer properties, various refrigerants are considered unsuitable for the freezing application due to textural, and taste changes resulted in the product due to direct contact in between the refrigerant and the product.

In general, the refrigerant used is sodium chloride having a eutectic point $-21.2°C$. Large and thick skin fishes have low salt uptake. However, many other fishes may not be suitable for sodium chloride brine due to adverse effects on taste and texture. Shrimps and other fishes have been frozen in syrup and brine, but a small degree of flavor change is usually observed.

6.4.7   SPRAY FREEZING (CRYOGENIC FREEZING)

Cryogenic freezing employs the use of a refrigerant with a very low boiling point to freeze unpacked or thinly packed fish products. In this method, the refrigerant is sprayed on to the product, and the heat removal is facilitated by phase change of the refrigerant. Generally, two refrigerants are most commonly used: liquid nitrogen and solid carbon dioxide.

6.4.8   LIQUID NITROGEN

In liquid nitrogen freezer, the liquefied gas is sprayed onto the product traveling through a tunnel on a moving conveyor belt as shown in Figure 6.2.

The nitrogen gas maintained at $-50°C$ is passed in a countercurrent fashion over the conveyor belt. As the fish moves through the pre-cooling stage, up to 50% of the heat is removed by liquid nitrogen, and the fish is frozen partially. In post-precooling stage, the completion of the freezing process is done by passing the product under a liquid spray. Maintenance of the coolant in a vacuum insulated pressure vessel is essential to prevent pressure , and regular venting is essential to maintain cooling of the contents. A major limitation in this method is the cost incurred during freezing, and is approximately four times more than the conventional air blast freezing. Other limitations include: storage requirements for the liquid nitrogen tank; high-cost liquid nitrogen delivery and uncertain supplies.

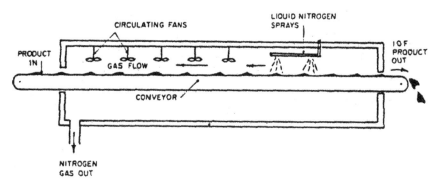

FIGURE 6.2 Spray freezing of fish fillets using liquid nitrogen as a refrigerant; Source: www.fao.org.

6.4.9 CARBON DIOXIDE

This type of freezer uses liquid carbon dioxide for freezing the product. In some systems, a layer of solid carbon dioxide is laid down on the conveyor belt, and the product is placed on top. Liquid carbon dioxide is then sprayed overhead; and sublimation of dry ice occurs at $-78°C$. It is possible to freeze to at least $-75°C$. Freezing is rapid, and drip losses are reduced to less than 1%.

It is possible to recover about 80% of the refrigerant used and can be liquefied again. Losses during storage can be minimized by the use of insulated vessels. Higher concentration of carbon dioxide in a factory atmosphere is unsafe, hence a freezer using carbon dioxide as refrigerant should have proper arrangements for the discharge of exhausted gas.

6.5 CALCULATION OF FREEZING TIME

Freezing time is the time required for reducing the product temperature from one point to the final point at its thermal center. The freezing time of the fish is inadvertently affected by the freezer type (immersion or air blast); operating temperatures; product temperature; thickness; shape, density, packaging, and fish species. The species of fish is of critical importance during freezing as the maximum of heat extracted will be required to convert liquid water into ice. Therefore, a lower amount of heat will be required, when the water content in fish is less (as in case of oily fish).

There are no precise models for predicting freezing process in any food system. However, it is critical to make accurate predictions of time of freezing to evaluate the quality of the end-product, processing considerations, and to know whether or not the process will be economical in a practical situation. Literature has proposed many models, which can be used for calculating and predicting the freezing time reasonably accurate for products uniform in shape such as blocks of the fillet. However, calculations are rough estimation for non-uniform or irregular products. The formulae for calculating freezing times have been simplified for practical applications, and it is commonly assumed that complete heat removal occurs at an initial temperature and the product is chilled prior to the freezing process.

The mathematical models given in this section indicate the complexity of the heat transfer during the freezing process. All the equations in this section give a close indication of the actual freezing time. However, practical confirmation of such predictions is generally important.

6.5.1 PLANK'S FORMULA

Planks formula assumes steady-state conduction in the frozen section with the product initially at its freezing point. It was derived based on the energy balance principle, and the general form of the equation is shown in Eq. (2).

$$t_f = \frac{L}{T_f - T_m}\left[\frac{P.a}{h_c} + \frac{R.a^2}{k}\right] \tag{2}$$

Where, t_f is the freezing time (s); k is the thermal conductivity of the frozen material (W/mK); h_c is the surface heat transfer coefficient (W/m²K); T_f is the freezing point of food (°C); T_m is the temperature of the freezing medium (°C); L is the latent heat of freezing; ρ is the density of the material; and a is the diameter of sphere or cylinder/ thickness of slab/ smallest dimension of a brick (m).

The values of P and R will change with the shape of the product. They represent the minimum possible distance from the surface to the center of the product for a given conformation.

Parameter	Slab	Cylinder	Sphere
P	1/2	1/4	1/6
R	1/8	1/16	1/24

In the presence of a packaging material, Eq. (2) can be modified to Eq. (3):

$$t_f = \frac{L}{T_f - T_m}\left[P.a\left(\frac{1}{h_c} + \frac{x}{k_1}\right) + \frac{R.a^2}{k}\right] \tag{3}$$

Where x is the thickness of the packaging material (m), and k_1 is the thermal conductivity of the packaging material (W/mK).

From Equations (2) and (3), it can be observed that the freezing time can be considered to have an inverse relationship with temperature change and at the same time has a direct proportionality to the square of product thickness. The solved example is shown in Problem #4. The Plank's formula is limited by the following assumptions:

• The material must be at freezing temperature uniformly throughout.
• The cooling medium is at a constant temperature.
• The material is homogeneous; and both the freezing point and latent heat of fusion can be defined accurately.

6.5.2 NAGAOKA FORMULA

Plank's formula does not take into consideration the time taken changing temperature from initial to the temperature of freezing medium or cooling temperature. Therefore in 1955, Nagaoka et al. [13] suggested a modification of the Plank's equation, based on data obtained from fish freezing. The modifications involved the time taken to decrease the temperature from an initial temperature T_i above the freezing point. The total enthalpy change ΔH replaces the latent heat of fusion in Eq.(2). Sensible heat is included in Latent heat, which must be removed while decreasing the temperature from T_i, and an additional empirical factor is included. Therefore, Nagaoka equation is as follows:

$$t_f = \frac{\Delta H}{T_f - T_m}\left[1 + 0.008\left(T_i\right)\right]\left[\frac{P.a}{h_c} + \frac{R.a^2}{k}\right] \tag{4}$$

where: ΔH is the total change in enthalpy or heat load, or the heat to be removed from the fish lowering it from initial temperature to the final temperature; and is given by Eq. (5). Problem #5 presents a solved example for Eq. (5).

$$\Delta H = c_{pu}\left(T_i - T_f\right) + L + c_{pf}\left(T_f - T_{final}\right) \tag{5}$$

where, T_{final} = final temperature of the fish; C_{pu} = heat capacities of the unfrozen fish; C_{pf} = heat capacities of the frozen fish.

6.5.3 LEVY EQUATION

Levy [11] considered the following definition of enthalpy to modify the Plank's equation.

$$t_f = \frac{\Delta H}{T_f - T_m}\left[1 + 0.008\left(T_i - T_f\right)\right]\left[\frac{P.a}{h_c} + \frac{R.a^2}{k}\right] \tag{6}$$

$$\Delta H = c_{pu}\left(T_i - T_f\right) + L + c_{pf}\left(T_f - T_{final}\right) \tag{7}$$

6.5.4 CLELAND AND EARLE EQUATION

Cleland and Earle [5] modified Plank's equation using dimensionless numbers such as: Stefan Number (N_{Ste}) and Plank's number (N_{PK}).

$$N_{Ste} = \frac{C_{Pf}\left(T_f - T_m\right)}{\Delta H_{ref}} \tag{7}$$

$$t_F = \frac{\Delta H_{ref}}{E(T_f - T_m)}\left[\frac{P*a}{h_c} + \frac{R*a^2}{k}\right]\left(1 - \frac{1.65 N_{Ste}}{k}\ln\left[\frac{T - T_m}{T_{ref} - T_m}\right]\right) \tag{8}$$

$$N_{PK} = \frac{C_{Pu}\left(T_f - T_m\right)}{\Delta H_{ref}} \tag{9}$$

$$P = 0.5\left[1.026 + 0.5808 N_{PK} + N_{Ste}\left(0.2296 N_{PK} + 0.105\right)\right] \tag{10}$$

$$R = 0.125\left[1.202 + N_{Ste}\left(3.410 N_{PK} + 0.7336\right)\right] \tag{11}$$

where, T_{ref} = reference temperature (°C) taken as -10°C; ΔH_{ref} = change in enthalpy from T_f to T_{ref}, and $0.15 < N_{Ste} < 0.35$; $0.2 < N_{Bi} < 20$ and $0 < N_{PK} < 0.55$, where N_{Bi} is the Biot number.

Use of equations (7) to (11) is shown in Problem #7.

6.5.5 METHOD BY CLELAND, CLELAND AND EARLE [6]

The method of Cleland et al. [6] is based on the relationship between the Biot number (N_{Bi}), N_{Ste}, and N_{PK}.

$$t_f = \frac{1.3179 C_{pi} a^2}{kE} \left[\frac{0.5}{N_{Bi} N_{Ste}} + \frac{0.125}{N_{Ste}} \right]^{0.9576} N_{Ste}^{0.0550} 10^{0.0017 N_{Bi} + 0.1727 N_{PK}}$$

$$* \left[1 - \frac{1.65 N_{Ste}}{k} \ln \frac{T - T_m}{T_{ref} - T_m} \right] \tag{12}$$

$$N_{Bi} = \frac{h_c a}{k} \tag{13}$$

where, $T_{ref} = -10°C$; $E = 1$ for infinite slab and 2 for an infinite cylinder and 3 for a sphere.

6.5.6 PHAM METHOD

The Pham method [16] calculates the freezing time as follows:

$$t_f = \frac{1}{E} \sum_{i=1}^{3} \Delta H_i a \frac{\left(1 + \frac{N_{Bi}}{a_i} \right)}{2 \Delta T_i h_c} \tag{14}$$

where,

$$\Delta H_1 = C_{pu} \left(T_i - T_{f,avg} \right) \tag{15}$$

$$\Delta T_1 = \frac{\left(T_i - T_m \right) - \left(T_{f,avg} - T_m \right)}{\ln \left(\frac{T_i - T_m}{T_{f,avg} - T_m} \right)}, a_1 = 6 \tag{16}$$

$$N_{Bi_1} = 0.5 \left(\frac{h_c a}{k_f} + \frac{h_c a}{k_u} \right) \tag{17}$$

$$\Delta H_3 = C_{Pf} \left(T_{f,avg} - T_{avg} \right), N_{Bi_3} = N_{Bi_2} \tag{18}$$

$$\Delta T_3 = \frac{\left(T_{f,avg} - T_m \right) - \left(T_{avg} - T_m \right)}{\ln \left(\frac{T_{f,avg} - T_m}{T_{avg} - T_m} \right)} \tag{19}$$

$$\Delta H_2 = \Delta H_f \tag{20}$$

$$\Delta T_2 = T_{f,avg} - T_m \tag{21}$$

$$N_{Bi2} = \frac{h_c a}{k}, a_2 = 4 \tag{22}$$

$$T_{avg} = T - \frac{T - T_m}{2 + \dfrac{4}{N_{Bi_3}}}, a_3 = 6, T_{f,avg} = T_f - 1.5 \tag{23}$$

where, k_u is the thermal conductivity of unfrozen food (W/mK).

6.6   FISH FREEZING PRACTICES

6.6.1   FREEZING ON BOARD

Fishes are usually frozen in block form, which requires plate freezing facilities. For freezing whole fish, VPF is required. Sometimes, larger fish such as Mackerel are frozen in bags. In such circumstances, water is placed in the bag to cushion the fall to avoid rupture of the base of the bag. If frozen without water, the frozen block may be placed in a polythene bag, and the outer area may be fiberboard after freezing to aid in handling [12].

A variety of blocks are produced onshore, ranging from laminated fillets to minced fish blocks. Laminated skin-less fillets are the most labor-intensive blocks to produce. The blocks are produced by placing the fillets on a polyethylene laminated board carton, lengthwise or breadthwise but parallel to each other. The corners of the carton are filled first to avoid any empty spaces as they produce non-uniform blocks resulting in wastage. After complete filling, the lid of the carton is folded-down, ensuring the lid locks with the outside of the liner wall and the carton is then placed on horizontal PF. It is essential to ensure that water droplet, if any, are removed from the plates as the water may freeze on the plates and reduce the heat transfer rates. Pelagic fish, like herring and mackerel, are frozen whole and the addition of water to fill the voids between fish is a normal procedure. The freezing of whole white fish (cod) is done after evisceration, and since this process is machine operated, the final yield of the fillet flesh may be lower than if it were done manually. With large catches, the fish may go into rigor much before freezing possibly due to the higher temperature of ambient air causing severe muscle contraction resulting in gaping. Thus, fish must never be frozen while in rigor. Air blast freezers (tunnel type) are generally fitted in factory ships for on-board freezing.

6.6.2 ON-SHORE PROCESSING

Onshore processing typically involves the production of value-added products such as fish-finger and fish-portion. These are portioned products that are produced from fish-fillet blocks that allow regular-portions to be cut causing minimum waste. The coating of fish with batter is a common practice and the batters used are a suspension of flour in water. The batter may be the adhesive type (which vary in their degree of viscosity) where improved adhesion is achieved by incorporation of gums, starch or breadcrumb; or tempura type where puffiness is produced in the product on frying due to the incorporation of sodium bicarbonate and leavening acid. The products coated with tempura batters are always flash fried prior to freezing as it sets the batter and provides mechanical strength to the coating, which reduces handling losses. Subsequent freezing of the fish is carried out immediately after flash frying as delaying would promote defrosting of the fish which will destabilize the coating as well.

A recent development in this aspect is a molding press, which takes tempered frozen portions placed in a mold and applies very high pressures causing the frozen block to take the shape of the mold. Such a technique has opened various other doorways for frozen products.

6.7 SUMMARY

This chapter covers the detailed aspects of fish freezing including discussions on the relative spoilage rates, freezing process as well as the biochemical changes during freezing; the types of freezing systems and different mathematical models for freezing times; the practice followed for onboard freezing and different onshore processing techniques.

KEYWORDS

- air blast freezing
- automatic plate freezer
- Cleland and Earle equation
- emulsifying ability
- freezer burn

- freezing on board
- immersion freezing
- individual quick freezing
- levy equation
- myofibrillar proteins
- Nagaoka formula
- onshore processing
- oxidation of lipids
- pelagic fish
- Pham method
- Plank's formula
- spray freezing

APPENDIX – A:

Solved Examples for Prediction Models in Section 5.0 of This Chapter

Problem 1: A tunnel freezer of length 1.3 m and width 1.0 m is used to freeze tuna. The cross-sectional area of the produce and the trolley is 0.8 m². If the airflow measured in the open part of the tunnel is 1.8 m³/s, then calculate the average air velocity.

I. Find the cross-sectional area of:
 	
  • Tunnel = 1.3m × 1.0m = 1.3 m²
 	
  • Produce and trolley (shaded areas)
 	
  = 0.8 m²
 	
II. Airflow (obtained from fan rating or measured in open part of the tunnel) = 1.8 m³/s
 	
III. Calculated average air velocity =
 	
  1.8 ÷ (1.3–0.9) = 4.5 m/s

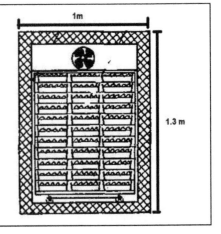

Problem 2: The 100 kg of fish has been frozen, and heat content of every kg of fish (from + 8°C to −30°C) is 80 kcal/kg. If freezing time is 2 h and the fan circulation rate is 2.5 m³/s, then calculate the weight of air circulated during the process and an average rise in air temperature. One may consider the density and C_p of air as 1.45 kg/m³, and 0.24 cal/kg °C, respectively.

The total heat to be extracted 80 × 100 = 8,000 kcal.

Amount of air (kg) circulated during freezing = 2.5 × 3600 × 2 × 1.45 = 26,100 kg

The specific heat of air is 0.24 cal/kg °C, then

Air temperature rise (average) = 8000 ÷ (26100 × 0.24) = 1.28°C

Problem 3: The freezing requirement is 150 kg/h and the freezing time is 18 min. The belt loading density is 4 kg/m², and belt width is 1.3 m. Calculate the belt length.

The freezing requirement is 150 kg/h and the time is 18 min,

The load on the belt = 150 × (18 ÷ 60) = 45 kg/h.

Now, if the belt loading density is 6 kg/m² and belt width is 1.3m; then

Belt loading per unit length = 6 x 1.3 = 7.8 kg/m.

Therefore, the belt length =45 ÷ 7.8= 5.769 m

Problem 4: If the freezing time of a 100 mm thick block of fish is 200 min when the temperature of the refrigerant is −35°C. Calculate the time of freezing if the temperature during operation is −25°C.

Fish freezing temperature is −1°C; therefore effective temperature difference for this system where freezing time is 200 min is 34°C (−1°C−(−35°C)). Based on this, the temperature difference required is 24°C = [−1−(−25)[for freezing a fish of 100 mm thickness. Because the time required for freezing is inversely proportional to the difference in temperature, therefore:

$$t_f = \frac{1}{34} = 200$$

$$T_f = \frac{1}{24} \text{ or } T_f = (200 \times 34) \div 24 = 283 \text{ min.}$$

Now if in this problem, the slab thickness is reduced to 75 mm, what will be the freezing time?

Since freezing time is proportional to the square of product thickness (if the surface heat transfer coefficient is quite high such that the Pa/h factor is sufficiently small so that it can be neglected), then multiplying the square of new thickness (75 mm) and dividing by the square of earlier thickness (100mm) gives the new freezing time (min): 200 x [(75)²/(100)²] = 112 min.

Problem 5: Consider a 5 m long and 0.75 m wide bed used for freezing peas (having diameter of 8 mm) at a rate of 6000 kg/h. Assume that the product freezes at –2°C and has an initial temperature of 12°C. The refrigerating medium is allowed to enter the chamber with a velocity to cause a heat transfer coefficient of 170 W/m²K at –35°C. Given that: density=1050 kg/m³; thermal conductivity=1.0 W/mK and L= 250 KJ/kg; Bulk density within the bed= 525 kg/m³; C_{pu}=3.3 KJ/kgK; C_{pf}=1.8 KJ/kgK. Calculate the freezing time.

The total enthalpy or heat load as calculated from Eq. (5) as follows:

$$\Delta H = 3.3\left(12-(-2)\right)+250+1.8\left(-2-(-20)\right) KJ / kg = 328.6 \, KJ / kg$$

And freezing time from Eq. (4) is calculated as follows:

$$t_f = \frac{1050*328.6*10^3}{(-2-(-35))}\left[1+0.008\left(12-(-2)\right)\right]\left[\frac{8*10^{-3}}{6*170}+\frac{\left(8*10^{-3}\right)^2}{24*1}\right] = 122 \, s$$

Problem 6: Use Plank's equation (modified) to find the freezing time for the fat free beef block of 1 m x 0.6 m x 0.25 m, using the data: h = 30 W/m²K; T_0 = 5°C; T = –10°C; T_m = –30°C; Density =1050 kg/m³; T_f=–1.75°C; ΔH = 248.25KJ/kg. Given that P= 0.3; R= 0.085; C_{pu}= 3.52KJ/kgK; C_{pi}= 2.05KJ/kgK; and k=1.108W/mK.

Using Levy Equation (6), we obtain:

$$t_f = \frac{248.25*1050}{(-1.75-30)}\left(1+0.008\left(5-(-1.75)\right)\right)[\frac{0.3*0.25}{30}+\frac{0.085\left(0.025^2\right)}{1.108} = 22.41 \, h$$

Problem 7: Fish fillet (slab) 0.025 m thick, T_i= 20°C, T= –10°C, T_m = –30°C, density= 1050kg/m³, T_f = –2.75°C, k=1.35W/mK, h_c= 20W/m²K. Calculate the freezing time, if C_{pu}=3 KJ/kgK, C_{pf}= 1.75KJ/kgK, ΔH=240 KJ/kg.

$$\Delta H_{ref} = 240 +1.75\left(-2.75+10\right)= 252.7 \, KJ / kg$$

$$N_{Ste} = \frac{C_{Pf}\left(T_f - T_m\right)}{\Delta H_{ref}} = \frac{1.75\left(-2.75+30\right)}{252.7} = 0.189$$

$$N_{PK} = \frac{C_{Pu}\left(T_f - T_m\right)}{\Delta H_{ref}} = \frac{3\left(20+2.75\right)}{252.7} = 0.270$$

$$P = 0.5\left[1.026 + 0.5808(0.270) + 0.189\left(0.2296(0.270) + 0.105\right)\right] = 0.607$$

$$R = 0.125\left[1.202 + 0.189\left(3.410(0.270) + 0.7336\right)\right] = 0.189$$

Substituting the values in Eq. (8), we get $t_f = 2.290$ h

REFERENCES

1. Butakus, H., (1970). Accelerated dentauration of Myosin in frozen solution. *Journal of Food Science, 35,* 558.
2. Butakus, H., (1974). On the nature of the chemical and physical bonds which contribute to some structural properties of protein foods: A hypothesis. *Journal of Food Science, 39,* 484.
3. Bremner, A., (1984). Quality-an attitude of mind. In: *Australian Fishing Industry Today and Tomorrow* (pp. 244–269). The Australian Maritime College, Launceston, Tasmania, Australia.
4. Bremner, H. A., Olley, J., & Vail, A. M. A., (1987). Estimating time-temperature effects by a rapid systematic sensory method. In: Kramer, D. E., & Liston, E., (eds.), *Seafood Quality Determination* (pp. 413–436). Elsevier, Amsterdam.
5. Cleland, A. C., & Earle, R. L., (1984). Freezing time predictions for different final product temperatures. *Journal of Food Science, 49,* 1230–1232.
6. Cleland, D. J., Cleland, A. C., & Earle, R. L., (1987). Prediction of freezing and thawing times for multi-dimensional shapes by simple formulae, II: Irregular shapes. *International Journal of Refrigeration, 10,* 234–240.
7. Connell, J. J., & Howgate, P. F., (1964). The hydrogen ion titration curves of native, heat coagulated and frozen stored myofibrils of cod and beef. *Journal of Food Science, 29,* 717.
8. Dyer, W. J., & Dingle, J. R., (1961). Fish Proteins with special reference to freezing, In: *Fish as Food* (p. 275). Academic Press, New York.
9. Halliwell, B., (1991). Lipid peroxidation, free radical and human disease (current concept). *The Up John Company* (p. 491). Kalamazoo, Michigan.
10. Heen, E., & Karsti, O., (1965). Fish and shellfish freezing, In: *Fish as Food* (4th edn., pp. 355–418). Academic Press, London.
11. Levy, F. L., (1958). Calculating the freezing time of fish in air blast freezers. *Journal of Refrigeration, 1,* 55–58.
12. McDonald, L., (1982). *Fish Handling and Processing* (2nd edn., pp. 79–87). HMSO, London.
13. Nagaoka, J., Tagagi, S., & Hotani, S., (1955). Experiments of freezing of fish in an air blast freezer. *Proc. Ninth Int. Congr. Refrig. Paris, 2,* 4–10.
14. Olley, J., & Ratkowsky, D. A., (1973). Temperature function integration and its importance in the storage and distribution of flesh foods above the freezing point. *Journal of Food Technology in Australia, 25*(2), 66–73.
15. Peters, J. A., Cohen, E. H., & Aliberte, E. E., (1964). Improving the quality of whiting. *U. S. Fish Wildl. Serv. Circ., 175,* 16–19.

16. Pham, Q. T., (1985). Extension to Plank's equation for predicting freezing times for rectangular blocks of foodstuffs. *International Journal of Refrigeration, 8*, 43–47.

17. Radhakrishnan, A. G., Antony, P. D., & Nair, M. R., (1985). Changes in major protein fractions of oil sardine and mackerel during frozen storage. In: Ravindran, K., (ed.), *Harvest and Postharvest Technology of Fish* (p. 433). Society of Fisheries Technologists, New Delhi – India.

18. Rao, S. B., (1985). Use of polyacrylamide gel electrophoresis technique to confirm the formation of disulfide bonds on frozen storage of *rohuactomyosin*, In: Ravindran, K., (ed.), *Harvest and Postharvest Technology of Fish* (519). Society of Fisheries Technologists, New Delhi – India.

19. Ratkowsky, D. A., Lowry, R. K., McMeekin, T. A., Stokes, A. N., & Chandler, R. E., (1983). Model for bacterial culture growth rate throughout the entire biokinetic temperature range. *Journal of Bacteriology, 154*(3), 1222–1226.

20. Ratkowsky, D. A., Olley, J., McMeekin, T. A., & Ball, A., (1982). Relationship between temperature and growth rate of bacteria cultures. *Journal of Bacteriology, 149*(1), 1–5.

21. Roubal, W. T., & Tappel, A. L., (1966). Damage to proteins, enzymes and amino acids by peroxidizing lipids. *Arch. Biochem. Biophys., 113*, 5.

22. Shenoy, A. V., & James, M. A., (1972). Freezing characteristics of tropical fish, II: Tilapia (*Tilapiamosambica*). *Society of Fisheries Technologists (Cochin, India), 9*(1), 34.

23. Shewan, J. M., (1977). The bacteriology of fresh and spoiling fish and the biochemical changes induced by bacterial action. In: Chatham, N. R. I., (ed.), *Handling, Processing and Marketing of Tropical Fish* (pp. 51–66). Tropical Products Institute, London.

24. Sriker, L. N., & Hiremath, G. G., (1972). Fish preservation I. Studies on changes during frozen storage of oil sardine. *Journal of Food Science and Technology, 9*(4), 191.

25. Srikar, L. N., Seshadari, H. S., & Fazal, A. A., (1989). Changes in lipids and proteins of marine catfish during frozen storage. *International Journal of Food Science and Technology, 24*, 653.

26. Vishwanathan, P. G. N., & Gopakumar, K., (1985). Selective release of fatty acids during lipid hydrolysis in frozen storage of milkfish (*Chanos chanos*). *Fishery Technology, 22*(1), 1.

27. Vishwanathan, P. G. N., Gopakumar, K., & Nair, M. R., (1976). Lipid breakdown in oil sardine during frozen storage. *Fishery Technology, 13*(2), 111.

CHAPTER 7

FERMENTATION METHODS AND FERMENTED FISH PRODUCTS

D. PRISCILLA MERCY ANITHA, S. PERIYAR SELVAM,
IDA IDAYU MUHAMAD, and SHANMUGAM KIRUBANANDAN

ABSTRACT

Fermentation breaks down a substance into a simpler substance. Microorganisms like bacteria and yeasts usually play an important role in the fermentation process and produce antimicrobial substance like bacteriocins to prevent spoilage. The fermentation process is combined with salting (impregnated with salt) and drying to reduce water activity and stop the growth of microorganisms that cause spoilage. Drying is done to reduce the moisture content in the fish. Fermented fish products are considered as a staple food in many countries, which includes fish sauce, garam, patis, bakasang, nam-pla, fish paste, and shrimp paste. Texture, odor, taste, and shelf life depends on the fermented products. Based on their fermentation process, the products are stored in an airtight container or airtight plastic packs to avoid the spoilage and contamination.

7.1 INTRODUCTION

Fermentation is a process in which a substance breaks down into a simpler substance. Microorganisms like bacteria and yeasts usually play a significant role in the fermentation process. Fermented fish is a traditional preparation of fish. It stops the ability of the microorganisms to spoil the fish. In this method, the muscle of the fish becomes more acidic. In other processes, fermentation is controlled by adding salt, where the process is limited to a partial breakdown of the protein. Proteins are broken down into simple compounds with the help of enzymes and microbes. Powerful flavor of fermented fish products is from the proteins and their products that are hydrolytically cleaved such as: peptones, peptides, amino acids, higher fatty acids, and triglycerides.

This chapter focuses on the importance of fermentation and various fermented fishery products, its nutritive values. It also explores how can the waste products like the internal organs, scales of the fishes be utilized.

7.2   PROCESS OF FISH FERMENTATION

Fermentation is one of the fish curing methods that not only preserves fish but also involves the breakdown of fish muscle. Therefore, fermentation is combined with salting (impregnated with salt) and drying to reduce the water activity and stop the growth of microorganisms.

7.2.1   SALTING OF FISH

Salting of fish is a traditional method that is practiced across the globe. It is mostly used in combination with drying and smoking. The presence of an adequate amount of salt (sodium chloride) in fish can prevent, or reduce microbial action. When fishes are placed in the salt solution, which is stronger than the solution of salt in the fish tissue, water will tend to move from the tissue into the salt solution until the strength of the two solutions are equal. Simultaneously, the salt will enter the tissue. This process is called as osmosis. A concentration of between 6 and 10% salt in the fish will prevent the activity of bacteria that cause spoilage [6].

7.2.2   DRYING OF FISH

Salted fish will take up the moisture from the surrounding air depending on the humidity. Therefore, it is necessary to take off the fish from the racks at night and during rain, where humidity is usually high. The fishes are stacked and covered with plastic overnight; consequently, the absorption of water would be minimal. If the fishes are press-piled at night by placing weights on top of the stack, movement of water to the surface of the fish would gradually increase, and subsequently, rate of drying will also be increased. There are also several other methods, such as: mechanical drier, freeze drying and solar driers [6].

7.3   ROLE OF MICROORGANISMS

Fish has its own micro-flora in the slime on the outer part of its body (gut and gills), in its natural environment. Death of these microorganisms and inactivation of enzymes would eventually lead to decaying of the fish. The biochemical reactions of microorganisms in food either result in spoilage or fermentation. Therefore, these reactions are of major concern in food preservation. Bacteriocins are antimicrobial substances produced through microbes, which can withstand salt-rich environment like *Lactobacillus*. They play a major role in the prevention of food spoilage by pathogenic bacteria. Many fermented fish products have bacteriocins, which act against both Gram-negative and positive bacteria. Jeot-gal a Korean fermented fish product has lacticin and Weissellicin 110, an organic compound present in Thai fish sauce named plaasom. It is effective against Gram-positive bacteria [1].

7.4   TYPES OF FERMENTED FISH PRODUCTS

Fermented fish products significantly contribute to the intake of proteins to many world's populations. In many countries, fermented fish items are considered as their essential foods. Depending on the salt content and texture, there are different kinds of fermented fish stuff that are described in this section.

7.4.1   FISH SAUCE

Fish sauce is a golden-yellow colored watery substance obtained through fish fermentation with sea salt. It is generally prepared by adding three parts of fish to one part of salt or in the ratio of 5:1 and natural fermentation is allowed to occur for 2–3 months at 29°C. Bamboo mats are used on the fish to press the mixture, which leads to the breakdown of digestive enzymes and fermentation [9]. Fermented fish sauce is rich in nutrients and minerals that are present in the fish and fish organs which are enhanced by fermentation. Anchovies are one of the fish types most preferably used to make high-quality fish sauces across the globe for its great aroma and color. Bakasang, Nampla, Budu, Patis, Koami are some of the popular fermented fish sauces made in Southeast Asia [12]. It is used as a flavoring agent in various cuisines. It is also a major ingredient in ancient European cuisine [13].

7.4.1.1   GARUM

Garum is a fermented fish sauce that was used in the cuisines of ancient Greece. It is prepared from the blood and intestines of small fishes by microbial fermentation. Fish organs are crushed and mined in the salt and kept under the sunlight for three months. The clear liquid that is formed on the top is taken out and used. When mixed with wine, vinegar or black pepper, garum enhances the flavor of a wide variety of dishes [2].

7.4.1.2   PATIS

Patis is a fishery product made by a fermentation process in the Philippines by mixing three to four parts of small fishes to one part of salt. It is a clear yellowish liquid that is present above the fermented mixture. It has a salty and cheesy flavor. It takes six months to obtain the product [9].

7.4.1.3   BAKASANG

Bakasang is a fermented fish product traditionally prepared from the guts of big, small fishes and fish eggs; and is the staple food of North Sulawesi. It is dark brown fluid with a strong fishy taste. It is usually used as a flavoring agent for many dishes, and it is also blended along with garlic, red chilies, and tomato and blended in coconut oil and consumed along with the hot porridge containing vegetables and rice. Bakasang is also made by using all the body parts of small marine fishes. First, the internal organs are weighed and washed using seawater and completely drained. They are then mixed with 20% salt and transferred into glass bottles. These bottles are kept undisturbed for six weeks for fermentation until a dark and sticky like the product is formed. Conventionally, the bottles are kept near the fire for storage [10].

7.4.1.4   NAM-PLA

Nam-pla is prepared through fish fermentation by adding salt. The full fish is blended in salt in the ratio of 3:1 and left for fermentation for 24 to 38 hours. They are then poured into the tanks, where the temperature is maintained between 35 and 40°C up to twelve months. The liquid that is present on the top is drawn off and strained to obtain the end-product [1].

7.4.2 FISH PASTE

Fermented fish paste is a commonly used product other than fish sauce. They are made from different freshwater fishes and shrimps. It undergoes a chemical reaction, and the muscles are broken down by fermentation process until it becomes soft, creamy purée or a paste. It is mainly used as a flavoring agent. Fish pastes are also made from the fish refuse. They are rich in proteins, calcium and vitamin B. Hentak, Ngari, Tungtap, Bagoong, Belacan, kapi, Miso are some of the fermented fish pastes made in the North-eastern part of India, Malaysia, and Indonesia [2, 3].

7.4.2.1 SHRIMP PASTE

Shrimp paste is a commonly used product in Southeast Asia and Southern China. Fishes and bigger prawns are separated from the original catch. For shrimp paste, smaller prawns are used. The small shrimps are blended with salt in bamboo baskets in the ratio of 4 - 5 kg of salt: 100 kg wet shrimp. Pickled shrimps can dry under the sun by spreading on the mat. The time duration for drying process generally takes places up to 8 hours. Nearly, half the moisture present in the product is evaporated during this process. The mixture is again sorted. The shrimps that were salted are crushed and made into the cream and kept it in the wooden containers. Air bubbles are completely removed at this point. The paste is now allowed to ferment for one week and later plowed from the container to spread on a flat surface to sundry for about 5 hours. Once the drying process is completed, the paste is crushed again and placed back into the wooden tubs, where fermentation takes place for one month, and this process is repeated for the third time, and final product is packed in blocks [9].

7.4.2.2 PLA-RA

Pla-ra is a conventional Thai seasoning prepared through fermenting fish along the rice bran or roasted rice powder. They are fermented by adding salt and stored in airtight jars for180 days. Pla-ra has a very strong odor that is unpleasant to smell. It has a mixed flavor of salt and sour, which depends on the amount of salt added in and also on the production of lactic acid because of the fermentation process by the microorganisms [4].

7.4.2.3   PRAHOK

Prahok is usually made from pounded, salted and fermented mudfish. It was an ancient method of preserving fish that was done for longer periods when fresh fishes were unavailable in large quantities. Due to high salt content and strong flavor, it is taken along with meals and soups as an addition. Prahok has a pungent and apparent smell. It is also called as Cambodian Cheese. It is usually eaten with rice [11].

7.4.3   FISH COD LIVER OIL

Fermented cod liver oil is prepared by fermenting the cod livers. Fermentation allows fat-soluble vitamins and beneficial oils to separate from the liver without causing any damage to the fat-soluble vitamins. Cool temperature fermentation process helps the cod liver oil to maintain the omega-3-fatty acid and vitamins (A and D) content. Based upon the amount of oil and vitamin A present, fish livers are of two types [6]: High vitamin A and low oil product; and Low vitamin A and high oil product.

7.4.4   DRIED FISH CHUNKS

Drying is a traditional method of preserving fish. It is the cheapest method of fish preservation. Dried fish is a very popular and delicious food consumed in many parts of the world. It is prepared through the removal of moisture from edible fishes. The basic principle of underlying in fish drying is the reduction in activity of the muscle enzyme and microorganism to a minimum in which the water content of the fish is traditionally reduced by sun drying. Both bamboo racks and mats are used to spread the fishes for sun drying. The quality of salted and sun-dried fishes is affected by the presence of microorganisms [14]

Several other fermented fish productsthat are used as a flavor enhancer and consumed as protein-rich additions to the diet, which include: crustaceans, mollusks or cartilaginous fish apart from bony finfish. Other fishes that are caught from small streams, lakes, and rivers are used in the preparation of dried fish chunks.

7.5 NUTRITIVE ASPECTS OF FERMENTED FISH PRODUCTS

Fermentation of protein-rich fish food increases the whole protein content and the active ingredients in the food. Many animal studies have shown that fermented food products provide more bioavailable and absorbable form of protein. Lactic acid bacteria (LAB) found in fermented food products produce vitamins, including vitamins B and vitamin K2. Types of vitamins depend on the specific microorganisms that are present in the fermentation process. Bioactive amino acids are given out by the fish proteins at the time fermentation, which confers several health benefits to humanity. Nowadays, bioactive compounds (fermented cod proteins) of fishes are commercially marketed as nutraceuticals which are beneficial for the digestive system. Various other inhibitor amino acids are extracted from fermented fish sauce and consumed a portion of a diet to sustain good blood pressure. The resultant product is highly nutritious with a high amount of protein and amino acids, minerals and B vitamins [6]. Since these products are traditionally fermented (fish sauce and fish paste), they always have positive effects in many ways. In addition, improper handling and processing methods can easily reduce the nutrients, leading to loss of nutrient contents [8].

7.6 QUALITY ASSESSMENT OF FERMENTED FOOD PRODUCTS

Quality assessment is an important step for any fermented food product. The qualities of these products not only reveal the changes in morphological appearance, but also in some of the sensory attributes such as: color, flavor, thickness, stickiness, and smoothness during the storage period.

7.6.1 TEXTURE

For texture analysis, two important categories of textural features are identified. Generally, the product used as food is rigid dried or partially. It is the most employed type of fermented food product in most of the African countries. It is firm in texture and same even after cooking. They are often consumed with starters like soups and sauces. It is widely used as a condiment [5].

7.6.2 COLOR

For any product, the color depends on the type of fish and the treatment process. For full stuff, silvery and metallic appearance is considered as high quality. If the fermentation process is not properly done, products look grayish or dark in color. Fermented products that are spread and sun-dried are light brown in color. Long-term storage of fermented fish products and prolonged drying darkens the product due to the overexposure to the sun [5].

7.6.3 ODOR

Odor of any fermented product plays an important role. The odor of fermented fish stuffs differs from slight to very strong. Similarly, soft and partially dried stuffs have a durable odor. However, dried stuffs have a very slight and subtle odor [5].

7.6.4 SHELF-LIFE

Shelf-life of a fermented fish is the most significant and prime feature that must be evaluated. If the moisture content is increased or decreased, insects would ultimately lay eggs on the product that eventually end up in the development of maggots which destroys the quality of the fish. Under extreme conditions like very dry or lack of high salt content, fermented fish can easily develop dermestes, a species of beetle that grows and feeds on the carrier and other dry animal products. These pests are normally seen in the silk industry in countries like Italy and India. They also thrive on food products like dried fish, dairy products like cheese, animal foods, and poultry. Dried fish stuffs would withstand for nearly 180 days; however, if the products are very tender and partly dried, they remain good for only 90 days [15].

7.7 PRESERVATION

The best way to preserve the food products is by salting, drying, freezing and fermenting. Since the fish products mentioned in this chapter have undergone salting, drying, and fermentation, it is essential to evaluate the products for spoilage periodically. If stored in an airtight container or airtight plastic packs, spoilage and contamination can be avoided [5].

7.8 FERMENTED FISH BYPRODUCTS

Dried shark fins are used in Chinese cookery. The fins, mostly the spinal and caudal, are cut from the animal and any adhering flesh is removed. The fins are dusted with salt in a ratio of one portion of salt to 10 portions of fish, and the cut portions are sprinkled with salt and then kept aside for about 24 hours. The fins are then washed in water and are dried under the sunlight for about one month. The moisture content of the material after drying is generally about 7 to 8%. The dried fins are packed in sacks under pressure so that they are flattened during storage.

In Japan, the cartilage from rays and sharks are prepared for export to China like shark fins. The cartilage of the jaw, fin, and head is cut into medium pieces of 7–9 cm in length and soaked in hot water to remove the adhering meat. The prepared cartilage is then boiled in water and dried in the sun. The product appears in an amber color after drying. The obtained final product is known as meikotsu in Japan [7]. The dried fins and cartilage are generally used as thickening agents in soups.

In many countries, the internal organs of finfish, sea cucumbers, and urchins, etc. are fermented to make sauces and pastes for condiments. In many parts of the world, fish entrails are consumed with the rest of the fish. In some areas, guts are removed and sold separately in the market at a low price. Under other circumstances, the guts are converted into fish meal, silage, etc., for animal feeding [7].

7.9 SUMMARY

Fermentation is the transformation of complex substances into a simpler compound by the action of microorganisms or by the enzymes. In certain processing methods, it is influenced by adding salt to give a desired type of flavor ensuring the preservation of that product simultaneously. Depending on the salt content and texture, there are different types of fermented fish products, such as: fermented fish sauce, fish paste, fish chunks, brined-fermented fish, etc. Fish sauce is a product obtained from the fermentation of a mixture of fish and salt. It is generally prepared by adding three parts of fish to one part of salt and allow for natural fermentation to occur for 2–3 months at 29°C. Bakasang, Nampla, Budu, Patis, and Koami are some of the popular fermented fish sauces made in Southeast Asia. Fermented fish paste is a commonly used product other than fish sauce. Fish pastes are rich in proteins, calcium, vitamin B and other minerals. Hentak, Ngari,

Tungtap, Bagoong, Belacan, and Kapi are some of the fermented fish pastes made in the Northeastern part of India, Malaysia, and Indonesia. Salt dried fishes are also a popular fermented product of India. Fermented byproducts (fishery wastes) are good sources of high-quality proteins and other nutrients. Therefore, these can be used as micronutrients for plant growth and can also be used as animal feed. The dried fins and cartilage are generally used as thickening agents in soups.

KEYWORDS

- amber
- bacteriocin
- cod liver oil
- fermentation
- fish sauce
- garum
- halophilic
- lacticin
- nutraceuticals
- omega-3-fatty acid
- prahok
- preservation
- salting
- shrimp paste
- vitamins

REFERENCES

1. Allison, B., (2016). *Traditional Fermented Fish Products*. The Weston a Price Foundation. https://www.westonaprice.org/health-topics/cod-liver-oil/fermented-fish-foods/. Accessed on January 15.
2. Altair B., & Mudjekee, D., (2013). Species identification of "padas" from fermented fish paste or "bagoong" using DNA barcodes. *Philippine Science Letters*, 6(2), 220–224.
3. Anupam, G., Midori, N., & Toshiaki, O., (2012). Bioactive properties of Japanese fermented fish paste, fish miso, using koji inoculated with *Aspergillus oryzae*. *International Journal of Nutrition and Food Sciences*, 1(1), 13–22.

4. Boon-Long, N., (1985). *Development of Traditional Fermented Fish Product for Small Industries: Processed Development of Pla-ra From Salt Water Fish.* Faculty of agro-Ind., Kasetsart University, Bangkok, http://agris.fao.org/agris-search/search. do?recordID=TH1999000400, Accessed on August 01, 2017.

5. Clucas, I. J., (1985). *Fish Handling, Preservation and Processing in the Tropics: Part 2 (NRI)* (p. 150).Tropical Development and Research Institute, Clerkenwell Road, London, U. K.

6. Curtis, R. I., (1984). Salted fish products in ancient medicine. *Journal of the History of Medicine and Allied Sciences, 4,* 430–445.

7. *Fermented Products,* (2017), Central Institute of Fisheries Technology (CIFT), http:// cift.res.in/uploads/userfiles/file/FERMENTED%20PRODUCTS(1).doc Accessed on July 31.

8. FAO, (2008). *The State of World Fisheries and Aquaculture: Utilization and Trade,* http://www.fao.org/fishery/utilization_trade/en. Accessed on July 31, 2017.

9. Hastings, F., (2015). *Fish Sauce a Versatile Ingredient That Works in Many Cuisines,* http://www.journalnow.com/home_food/food/fish-sauce-a-versatile-ingredient-that-works-in-many-cuisines/article.html. Accessed on July 31, 2017.

10. Helen, J. L., & Langkah, S., (2011). Molecular identification of lactic acid bacteria producing antimicrobial agents from Bakasang, an Indonesian traditional fermented fish product -Lawalata. *Indonesian Journal of Biotechnology, 16*(2), 93–99.

11. Hortle, K. G., (2007). Consumption and the yield of fish and other aquatic animals from the lower Mekong Basin. Mekong River Commission, Vientiane, *MRC Technique Paper No. 16,* p. 87.

12. Nurul, H., (2012). Indonesian fermented fish products. Chapter 42, In: Hui, Y. H., (ed.), *Hand Book of Animal-Based Fermented Foods and Beverage Technology* (pp. 717–737). Taylor and Francis (CRC Press), Boca Raton – FL.

13. Ravipim, C., Smith, P., & Simpson, K., (2013). Acceleration of fish sauce fermentation using proteolytic enzymes. *Journal of Aquatic Food Product Technology, 2*(3), 59–77.

14. Sabiha, S., Sayeed, M., & Barman, F., (2015). Traditional methods of fish drying: An explorative study in Sylhet, Bangladesh. *International Journal of Fishery Science and Aquaculture, 2*(1), 028–035.

15. Samish, M., Argaman, Q., & Perelman, D., (1992). Research note: The hide beetle, *Dermestes maculates* De Geer (Dermestidae), feeds on live turkeys. *Poultry Science, 71*(2), 388–390.

CHAPTER 8

RANGE OF FERMENTED FISH PRODUCTS ACROSS THE GLOBE: SCOPE, USES, AND METHODS OF PREPARATION

SASIKANTH SARANGAM and
KUNDETI SARANYA CHANDANA PRIYA

ABSTRACT

The major classification of fermented fish products is of three types. (i) The fish retains its same shape or preserved in large chunks. (ii) It may be converted into a paste form; or (iii) In liquid form. Although the fermentation procedure of fish differs from region to region, yet the basic steps for process fermentation are: Salting, drying, and various combinations of these two steps. The type of fish used for the process of fermentation also varies across the world, such as: (i) Mostly the fish is taken in the form of small pieces in African countries; (ii) Taken as paste and as a whole in Asian countries. Fermented fish products in the Philippines are limited to fish sauce and paste, which are mainly *Btlloflgisdo* (fermented fish) and *Baloobaloo* (fermented shrimp). In Malaysia, the major sea products are consumed in the form of fresh or frozen; and only a small amount of the marine fish is converted into fermented fish products. Out of these some popular fermented fish products are: Budu, Pekasam, Belasam, Cincaluk. Indonesia is rich in aquatic sources. Processing of fish includes: drying, boiling, and salting. These famous fermented products are consumed locally in huge amounts, and a small amount is being exported. Pedah, Trasi (Shrimp paste), Bekasang are some of the famous fermented fish products in Indonesia. In Thailand, the fermented fish products are mainly classified into three types namely fish with more salt, fish with salt and carbohydrate and fish with salt and fruit. Nom-pta (fish sauce), Kopi (shrimp paste) and Plaa-raa (Fermented fish) are famous fermented fish products in Thailand. In Sri Lanka, 75% of the fish is consumed and processed from "Maldive fish." Jadee fermented fish, Maldive Fish, salted dry fish are major processed fish products.

In Northern European countries, the fermented fish products are classified as low salted, medium salted and highly salted fermented fish products. Maatjes, rakeorret falls under the low salted fermented fish product. Surstromming is under medium salt and sugar salted herring; barrel salted herring without sugar are under high salted products. In Africa, Ndagala, Salanga Gyagawere, adjonfa, Guedj Momone, koobi, kako, ewule, Djegue, jalan are some of the fermented fish products. In Japan, fish sauces are Shottsuru, Ikango, and Konago, Ika Shoyu and Ishiru. Shrimp and oyster products are famous in Taiwan. In Korea, people follow different methods for the process of fish fermentation. Joet-Kal and Joet-Kuk are fish paste and fish sauce, respectively. In South- East Asia, the fermented fresh water fish is classified based on the constituent fish: One is fermented fish made of selected species; the other is made out of miscellaneous small fish which fall in the category of Kampuchea. Narezushi and Shiokara are the famous fermented fish products in Asia.

8.1   INTRODUCTION

The major factors for the spoilage of fish are: bacteria that cause microbiological spoilage; fat oxidation; and autolytic spoilage caused by enzymes. Therefore, various methods (like salting of fish, drying of fish, and smoking of fish, fermentation of fish, canning of fish and freezing of fish are available in order to increase the shelf life of fish.

The process of fermentation involves break down of proteins in the raw material to simpler substances that are themselves stable at room temperature. However, to make it more stable, traditional methods of preservation have been adopted by addition of salt, where the process of breakdown of proteins is controlled to attain the desired flavor and to increase the shelf life of the product. The breakdown of proteins is mainly because of the enzymes and in some cases microorganisms promote the process of fermentation.

When compared with all other preservation methods of fish, fermentation of fish plays a major role to increase the shelf-life. Fermentation of fish is an ancient practice, which has been traditionally followed in many countries. The process of fermentation is mainly applicable for fatty fish, which include species such as: salmon, trout, Arctic char, and herring. These species are not apt for the process of drying, as they contain a large amount of polyunsaturated fatty acids (PUFA). The process of fermentation varies from region to region based on the taste, climatic conditions, arability of salt and fish. Nowadays, the fermentation technology is improving day by day and play a major role in the preservation of fish for a long time. Fermented fish products such as paste and sauce are considered as authentic condiments in many countries.

This chapter explores various fermented fish products all over the globe and the method of preparation of these products.

8.2   GENERIC CLASSIFICATION OF FERMENTED FISH PRODUCTS

8.2.1   FISH SAUCE

Fish sauce is considered as "the mother of condiments," and it is usually prepared by fermenting the fish blood and intestine in a brine solution. The proteolytic and lipolytic enzymes along with lactic acid bacteria (LAB) play a major role in the fermentation of fish by the process of autolysis. In Southeast Asian cuisines, these fermented fish sauce plays a prominent role due to its peculiar odor and umami taste.

8.2.2   FISH PASTE

It is a traditional condiment of south Asia, which is usually prepared by salting and fermentation of fish. The final product attains a soft texture and forms like a paste. This can also be prepared by grinding cooked fish, which is not much preferable as it is less nutritious. This is usually taken with highly flavored foods, soups, rice, etc.

8.2.3   FERMENTED FISH

It is an excellent source of omega-3 fatty acids. In the process, fish is mixed with salt and the enzymatic activity to break down fish proteins, and thereby the transformation of the texture takes place. In the later stages, the excess salt is removed from the fish, and various spices are added according to the local tradition. This fermented fish is also combined with rice in some cultures. Also, the grounded rice powder is added to enhance the taste.

8.3   BENEFICIAL MICROORGANISMS

The most beneficial microorganism that promotes fermentation is a LAB. The flora of the microorganisms promotes the immunity system, digestive system and mental health in humans. It also enhances detoxification in the body.

8.4   NUTRIENT CONTENT

Fish is a rich source of protein and fermentation enhances the overall bioavailability of protein content in food. LAB provide vitamins (such as B, K_2); and the amount depends on the culture media in the process of fermentation.

The factors affecting the microbiological formation and biogenic amines in fermented fish are: Microbial cultures that are present in the process of fermentation, moisture content in the controlled or uncontrolled fermentation process, pH, acidity, salinity, surface area, sanitation of facilities, production time, temperature, additives, and enzymes.

8.5   FERMENTED FISH: CLASSIFICATION BASED ON SALT CONTENT

Fermented fish is fermented with salt until the flesh attains the desired flavor. Table 8.1 represents the fermented fish products in different countries across the globe.

TABLE 8.1   Fermented Fish Products in Different Countries

Country	Fish product
Indonesia	Pedah
	Trasi (shrimp paste)
	Bekasam
	Bekasang
	Fish Silage
	Fermented dry salted marine catch fish
Burma	Nga-pi
Thailand	Nam-pla
	Kapi

TABLE 8.1 *(Continued)*

Country	Fish product
	Khaou-mak (fermented rice)
	Ang-Kak (fermented red rice)
	khem-mak-nad (fish with fruits)
Sri Lanka	Jaadi (high salt)
	Maldive fish (low salt)
	Salted dry fish
Japan	Shottsuru (fish sauce)
	Ikango and Konago (fish sauce)
	IIkaShoyu and Ishiru (fish sauce)
Korea	Yumhae
	Chumaik-o
	Oyuk-chang
	Sikhae
	Eo-ganjang
	jeot- kal (fermented fish paste)
	jeot- kuk (fish sauce)
Taiwan	Shrimp (paste and sauce)
	Oyster (paste and sauce)
	Krill sauce
North European pickled fish products	Maatjes (low salt)
	Schnell- Maatjes (low salt)
	Gravad fish (low salt)
	Rakeorret (low salt)
	Surstromming (Medium salt)
Asia	Narezushi
	Shiokara
Malaysia	Belacan (shrimp paste)
	Cincalok (pickled shrimp)
	Pekasam
	Naniura
	Peda
	Petis
	Picungan
	Terasi

TABLE 8.1 *(Continued)*

Country	Fish product
Africa	Ndagala
	Salanga
	Gyagawere, adjonfa
	Guedj
	Momone, koobi, kako, ewule
	Jalan, djegue
	Yeet, tambadiang, guedj
	Mindeshi, terkeen, fessiekh, kejeick
	Dagaa
Philippines	Bagoong (fish paste)
	BalaoBalao
	BurongIsda

8.6   EQUIPMENTS USED FOR FISH FERMENTATION PROCESS

Fermentation is a low-cost process and does not require costly equipment for processing, and these are easily available and easy to fabricate. The major equipments for the fermentation process are: earthenware pots, barrels, nets, drying mats (which are locally made), jute cans, etc. (which are locally available and affordable); and the equipment usage may differ from process to process that depends on the type of fish and process of fermentation.

8.7   FERMENTED FISH IN EUROPE

Fish is considered as a staple food in European countries. Drying is the ancient preservation technique used for the process of preservation of fish. Mainly species of Salmon, Trout, Charr, and Herring are not used for the process of drying [13].

Preservation of fish using salt is being used before the Christian era in China and ancient Egypt [23]. Later this method of preservation was used by many Europeans. The salt concentration used for the preservation of fish is less; and mainly catch fish is used for this method. The resulting product tastes good and delicious. Although the process of preservation is same all over Europe, yet but fermentation technology varies from region to region based upon the salt concentration used, storage temperatures, storage containers, species of fish used and handling methods [6]. The sauce, which is prepared from the fermented fish, was known as gourmand; this was the main practice of ancient Greeks and Romans. The process of fermentation mainly relies on the naturally occurring enzymatic activity and the bacteria. This results in the development of texture, flavor, and aroma [8].

8.7.1 HAKARL

The preservation of hakarl was an ancient practice for Greenland shark. Method of preparation is shown in Figure 8.1.

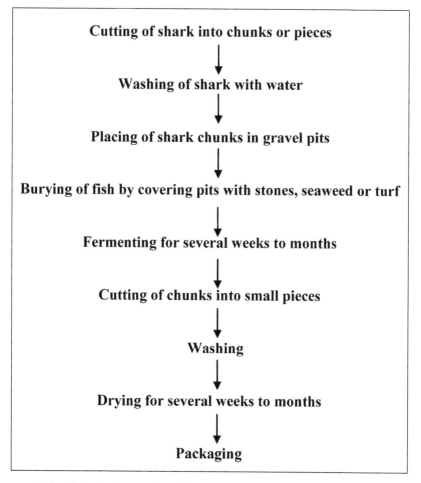

FIGURE 8.1 Method of preparation of *hakarl* in Europe.

In modern methods of fermentation, these shark chunks are placed in closed containers and are fermented for 3–6 weeks. During the process of fermentation, the liquid is drained out through the holes of the containers [22]. The bacterial growth during the process of fermentation is increased greatly, where Moraxella/Acinetobacter groups along with lactobacillus play a major role. The total number of bacterial count is about 10^8/g [25]. The chemical composition of the final product varies from place to place depending on the method of preparation, processing, and part of shrink taken for the process of fermentation. The average chemical composition comprises of 31.6% of water, 44% of fat, 24.8% of protein, 1.4% of salt, 1.7% of ash and 0.7% of ammonia [21].

8.7.2 RAKFISH

Rakfish is available in inlands of Norway, towards the south-west direction from western mid- Sweden [32]. It is produced from freshwater salmonid fish. The production process includes mild salting of the gutted fish, and the salt is placed in the bellies of the fish, which are arranged in layers under pressure. These are stored at low temperatures (3 to 7°C) and packed in tight containers for a span of 3–12 months. During this period, fermentation, and the process of ripening takes place. During the process of fermentation, the fish is totally submerged in the brine solution. The salt concentration in the brine solution is nearly 4–6 % (w/w).

8.7.3 SURSTROMMING

This process is developed in order to store the catch fish (hearing) for a long time with a minimal percentage of salt. The first step is the salting of fish [37]. Pre-salting of the hearing is done for 1 to 2 days. These will be immersed in the saturated salt solution, and stirring is continued. Pre-salting is followed by de-heading and gutting of fish [7]. Then these are filled into barrels having a mild salt solution (17%). Then the barrel is sealed tightly and followed by occasional rotation for first 3 days. The storage temperature of barrels ranges from 15–18°C and the storage period is for about 3–4 weeks. After completion of the process of fermentation, the hearing is transferred into cans along with the brine solution, which was collected during the process of fermentation [4].

8.8 FERMENTED FISH IN THAILAND

Pla-ra is a famous fermented fish product in Thailand. Pla-ra is originated from Mekong Basin. The process includes fermentation of fish along with rice bran or roasted rice powder and salt. This mixture is tightly packed in a container and fermented for nearly 6 months [30]. It is consumed as a side dish. And usually, it is taken with green papaya salad. The odor of the final product is very strong and is unpleasant for some people. And it has a salty and sour flavor depending on the percentage of LAB present in the product [9].

8.8.1 CLASSIFICATION OF PLA-RA IN THAILAND

The classification is based on major ingredients, which were added in Pla-ra. Where the roasted rice powder is considered as the major ingredient, the Pla-ra after fermentation becomes yellow in color with a soft texture and distinctive odor. This is usually consumed in the form of a paste. Catch fish is mainly used in the process of preparation of this product. In the other cases, where the Pla-ra is fermented with rice bran, the final product attains a clear black color with a strong odor. Usually, the fish is soft and small in size, and it is mostly consumed in Northeastern Thailand [9].

8.8.2 METHOD OF PREPARATION OF PLA-RA

In the first phase, salt is added to the fish, and it is fermented until it attains a soft texture (Figure 8.2). In the next phase, based upon the major ingredient either rice bran or the roasted rice powder fermentation is continued by the addition of the major ingredient according to the product type.

In the final product, the chemical composition consists of: protein from 16–18.9%, moisture content from 28–71%, fat from 0.71–3.2%, salt from 5.32–9%, pH from 4.5- 6.2, lactic acid from 0.3–1.9 [15].

8.9 FERMENTED FISH PRODUCTS IN INDONESIA

The traditional methods for processed marine foods include salted, boiled, dried, salted, and fermented fish products [10]. Out of all processed marine foods, fermented fish contributes to 40% for the traditional methods and

50% accounts to total marine catch. These fermented fish products provide unique characteristic properties. With the help of microorganisms (M.O.) and enzymes, the fermentation process progresses leading to the development of flavor and aroma. An enzyme significantly helps in the change of the texture and shape of the fish [14]. The classification of the fermented fish in Indonesia is mainly based on the raw material used in fermentation, types involved in the process of fermentation and form of the final fermented product [15].

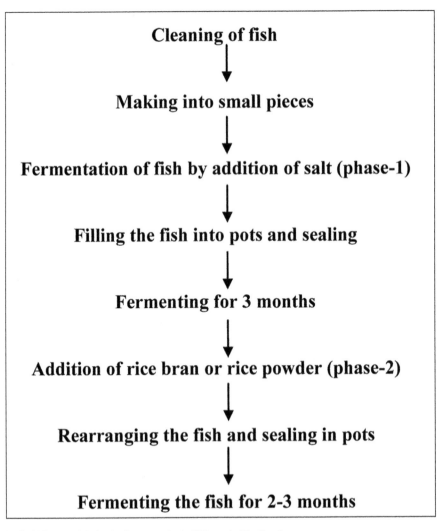

FIGURE 8.2 Method of preparation of Pla-ra in Thailand.

8.9.1 RAW MATERIAL CLASSIFICATION

Raw material classification includes freshwater fish for which bekasam is the fermented fish product. The other is whole gutted fish, which includes fermented fish products such as: terasi, kecapikan, jambal roti, ikantukai, peda, and bekasang. Terasi is available in the form of pounded fish/shrimp and is used to make the fermented fish product.

8.9.2 TYPES OF FERMENTATION

Types of fermentation are mainly based on the ingredient used for the process of formation, which mainly includes salt and carbohydrates. The final product may be in the form of dried fish, moist fish, lumped or pounded fish, liquid or semi-liquid.

8.9.3 PEDHA

Pedha is a fermented fish product in Indonesia, which was imported from Thailand. During the process of transit, the fish develops a distinctive and peculiar aroma during the export as it is not completely dried. This product was popular with the name *Pedah Siam* in Indonesia.

Pedha is prepared from mackerel, which primarily involves salting, drying, and fermentation. Pedha is famous for its characteristic aroma, flavor, and texture, which differ from other salted or fermented fish products.

Pedha originated from salted fish. The shelf life of the fish is less as the product goes bad soon after completion of fermentation. The time for the process of fermentation depends upon the type of processor. There is no set time for the process of fermentation. During the process of fermentation, the fish attains a desirable flavor and texture.

The major steps in the fermentation are as follows: Initially, mackerel fish is cleaned and salted with 1: 3 ratio of salt and fish, respectively, at an ambient temperature of 29°C. Then in the later stage, the excessive salt from the fish is removed and packed. Then the packed fish is meant for fermentation for about 3 months at 29°C. After completion of fermentation, Pedha is ready for export and local consumption. This product is rich in protein. The final product is reddish brown in color, has a slightly pasty texture, and has a particular flavor, which is tasty, cheesy, and salty. It also develops some rancid flavor due to prolonged storage. The microbial study on this product indicates that pedha contains gram-positive coco predominated and small amounts of LAB [14].

8.9.4 JAMBAL ROTI

This is majorly processed from marine catch fish. Pangandaran, Cirebon, Cilacap, and Pekalongan are major production centers in Java. The fermentation process in jambal roti is same at all stages, but they differ in the methods of processing [9]. For Cirebon, the fermentation processing steps of fish are shown in Figure 8.3.

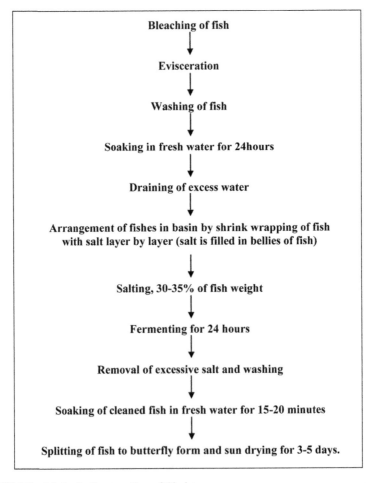

Bleaching of fish

Evisceration

Washing of fish

Soaking in fresh water for 24hours

Draining of excess water

Arrangement of fishes in basin by shrink wrapping of fish with salt layer by layer (salt is filled in bellies of fish)

Salting, 30-35% of fish weight

Fermenting for 24 hours

Removal of excessive salt and washing

Soaking of cleaned fish in fresh water for 15-20 minutes

Splitting of fish to butterfly form and sun drying for 3-5 days.

FIGURE 8.3   Method of preparation of Cirebon.

Fish without icing yields a good quality than fish with icing. New method of processing of jambal roti is done by icing the fish and soaking the iced fish in warm water to get back the normal temperature prior to processing of fish. The soaking of fish is done at 40°C for 60 min to attain normal temperature [29]. The other new method for processing of jambal roti before salting is to soak the fish in a solution of 30% concentration of coconut sugar in which 20% of the fish would be enough for salting [11].

8.9.5 TERASI

This is mainly used as a flavoring agent and consumed in small quantities. The major importers of this product are Netherlands and Suriname. This terasi is usually made from planktonic shrimp [10]. Terasi can be processed in two ways: one is by addition of salt alone, and the other process is by the addition of salt and other ingredients [49].

The terasi processing method involves both salts, other ingredients; and another method involves more or less similar to the process by the addition of salt alone (Figure 8.4). The major difference is in the second pounding stage where tamarind is mixed with the salt solution and coconut sugar before adding rebon. The proportions of tamarind and coconut sugar are 200g and 250 g, respectively for 10 kg mixture of fresh rebon [48].

8.9.6 KECAPIKAN

It is a fish sauce, which is not so famous in Indonesia but it got acceptance in West Kalimantan Province [27]. Kecapikan is usually processed from oil sardine (Figure 8.5). It is traditionally produced by using a high amount of salt; and the fermentation period is prolonged for a long time [31, 40].

8.9.7 IKAN TUKAI

This is one of the famous fermented fish product in West Sumetera. This is mostly processed from barracuda (Figure 8.6) [11].

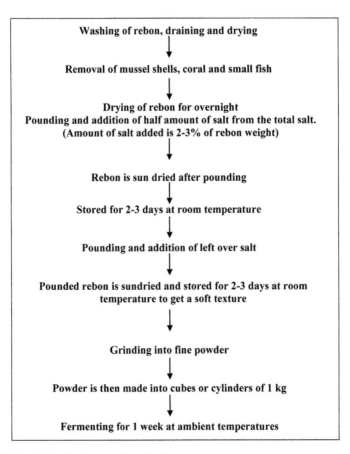

FIGURE 8.4 Method of preparation of rebon.

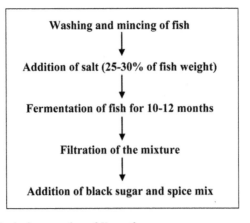

FIGURE 8.5 Method of preparation of *Kecapikan*.

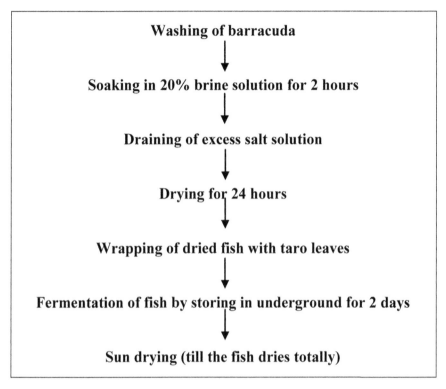

FIGURE 8.6 Method of preparation of *barracuda*.

8.9.8 BEKASANG

This is a prominent fermented fish product in North Sulawesi. In this fermentation process, the Moluccas is processed along with viscera of skipjack. The processing procedure includes (Figure 8.7): the viscera is washed and mixed with salt in 2.5:1 ratio and it is allowed for the process of fermentation for one week. After completion of the fermentation process, the viscera is simmer cooked for 2 hours and then it is filtered. The filtered product is heated and packed [47].

8.9.9 BEKASAM

This fermented fish product is from a freshwater fish, which is found in Sumatera and Central Kalimantan. Bekasam preparation involves the use of carbohydrate sources, which are added to promote the LAB growth by decomposing into simpler components. Some of the major carbohydrates sources are: roasted rice powder, sticky rice, and roasted rice powder. Bekasam product is usually consumed with a combination of chili and sugar [48].

Beheading of fish

↓

Descaling and evisceration

↓

Cut into butterfly form and washing

↓

Soaking of fish in 16% brine solution for 2 days

↓

Drained and addition of cooked rice 50% or sticky rice 25% of fish weight

↓

Sealed in plastic jars

↓

Fermentation for 1-2 weeks

FIGURE 8.7   Method of preparation of viscera of skipjack.

8.9.10   CINCALUK

This is usually processed from rebon and is a traditional fermented fish product in Riau Province. The method of processing depends upon the operator as there is no specified method for the production of cincaluk. In this preparation, fresh shrimp is mixed with boiled rice and salt. For one kg of rice, 200–300 g of salt is added. The product is sealed and left for 4 days. Then this is filled into bottles and sealed. In an alternate method, the shrimp is initially descaled and washed and mixed with tapioca flour, salt, and sugar in the ratio of 20: 1: 1. Tapioca flour is dissolved in water to avoid lumps, and after gelatinization and cooling the shrimps are thoroughly mixed with salt and tapioca flour. This mixture is filled into bottles and sealed. It is allowed for the process of fermentation for a period of 1–2 weeks.

8.10   FERMENTED FISH PRODUCTS IN MALAYSIA

In Malaysia, landed marine fish contributes 63% for fresh consumption; 1% is frozen, 11% is cured, 20% is made into fish meal, and 5% is disposed of by other means, Dried, salted, smoked products come under the category of cured fish. The most prominent fermented fish products in Malaysia are cincalok, pekasam, budu, and belacan [49].

8.10.1   BELACAN

Belacan or shrimp paste is a major fermented fish product in all states of Malaysia. Belacan is almost similar to terasi, which is an Indonesian fermented fish product. Small shrimp of Acetes species or mysid shrimp are two common varieties of fish used for the production of Belacan. This is a thick, salty paste with a strong pungent shrimp odor. It is usually consumed as a flavoring ingredient or a condiment [20]. Belacanhas a typical prominenttypical flavor like salt, shrimp, and meat. It is used as a major ingredient in local dishes namely fried rice with belacan, chili belacan, belacan fried chicken, stir-fried chili paste, spicy noodle soup with belacan, sour, and spicy fish stew with belacan [19]. Malaysian people consume this belacan with other dishes, almost daily [24].

The processing method is shown in Figure 8.8. The finished product is dark in color with strong shrimp odor and taste salty. The 100 kg of wet shrimp yields 40–50 kg of belacan [48]. This product must have less than 24% of protein, not more than 40% water, 15% of salt and 35% of ash [38].

8.10.2   BUDU

Budu is the next fast moving and major consumed product after belacan. It is produced in Kelantan, Terengganu, and east coast states of Malaysia [33, 48]. It is a brown liquid produced (Figure 8.9) by process of fermentation of marine fish anchovy (*Stolephorussp*).

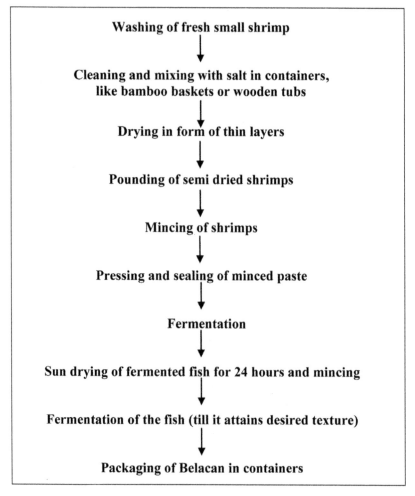

FIGURE 8.8 Method of preparation of Belacan.

The quality of the final product depends on the type of fish used, the percentage of salt added, and duration of fermentation [3]. The percentage of protein must not be <5 and salt must not be <15% [5].

8.10.3 CINCALOK

It is a form of pickled shrimp. This is mainly produced in Melaka; and the raw material for this product is small shrimps of Acetes species (Figure 8.10) or mysid shrimp [21, 49].

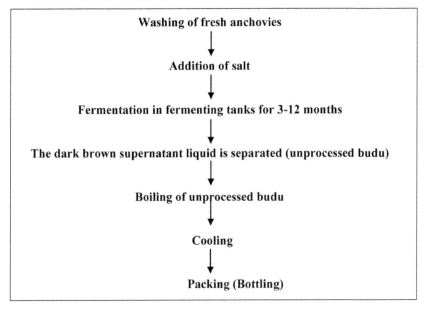

FIGURE 8.9 Method of preparation of Budu.

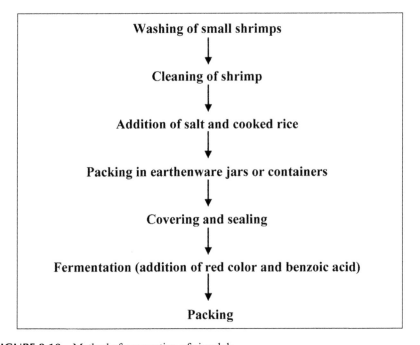

FIGURE 8.10 Method of preparation of cincalok.

8.10.4 PEKASAM

It is made from freshwater fish and usually consumed in deep fried or consumed as a side dish with rice (Figure 8.11). It is mainly produced in Perlis, Kedah, and Perak states of Malaysia [3].

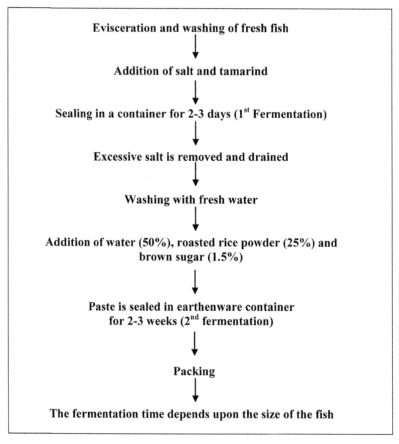

FIGURE 8.11   Method of preparation of pekasam.

8.11   FERMENTED FISH PRODUCTS IN MANIPUR

8.11.1   FERMENTED FISH PRODUCTS

The Brahmaputra River and lakes are major sources of catch freshwater fish in Northeast India. As these lakes are small in size, therefore utilization of freshwater fish is limited. The principal traditional methods of fish processing in this region are namely: salting, drying, and smoking. They can store these fermented fish products for a span of one year. Native dried fish and fermented products are: *Sukakomachha*, *Gnuchi*, *Sidra* and *Sukuti*in Nepal; *Ngari* and *Hentak* in Manipur; *Karoti* and *Bardia* in Assam; and *Tungtap* in Meghalaya.

Sikkim and Darjeeling hills are major regions of fermented fish products consumed by people of these regions [45–49]. Nagari is a popular traditional fermented fish product often consumed with side dish called *Ironba*; and it is a mixture of chilies, potatoes, etc. that are eaten with boiled rice, which is included in the daily diet of all community people in Manipur and in nearby states [41]. In these states, traditional fermentation of fish is followed throughout many generations to identify the culture of these regions.

8.11.2 PREPARATION OF NAGARI

The traditional fermented fish product is from regions of Imphal and surrounding areas where it was imported from Brahmaputra valley of Bangladesh and Assam in which these are usually sun-dried and are non-salted dry fish known as Phoubu (*Puntius sopore*). Fresh fish is usually collected during months October to January and costs between Rs. 60–120/Kg depending on the grade and size. Sun-dried *puntiussophore* is taken and is washed briefly with water. And after one day, the water is drained out, and it is spread and covered with gunny bags. Before the fermentation process, the fish is washed and kept in porous bamboo baskets for overnight in order to drain the entire water. Next day the excess water is removed from the gunny bags by pressing it by legs of laborers and by crushing the bones and head of the fish. The oil obtained by crushing the head initiates the process of fermentation. The holding capacity of an earthen pot is nearly 45- 50 kg, where the inner surface of the pot is coated with a mustard oil. Oil coating must be done for 8- 10 times within an interval of 7- 10 days for new earthen pots (Figure 8.12).

Only one coating is adequate for old pots for the better fermentation process. Anaerobic conditions inside the chamber are maintained through the coating with oil. The fish that was pressed and taken from the gunny bags is dumped into the earthen pots. The pots are sealed with fish scales, oil slurry, polyethylene sheet, cow dung, and mud slurry after the process of packing. For 6–12 months, these packed pots are stored in a dark room at room temperature. The final fermented product *Nagari* comes after the fermentation; and is generally consumed in Myanmar and by all northeastern states.

FIGURE 8.12   (See color insert.) Traditional fermentation method for *Nagari*.

8.11.3   PREPARATION OF HENTAK

Hentak is a fermented fish paste formed during the traditional fermentation process and it is a sun-dried product. The *Esomus danricus* is dried and is crushed to fine powder. Small pieces of *Alocasia macrorhiza* petioles are washed under water, and they are kept under natural drying for one hour. These cut pieces are then crushed with equal weight along with fish powder to make it into a thick paste-like consistency. This paste is made into small balls and stored in earthen pots. This can be used and preserved further after completion of two weeks fermentation in the pots. After a few months, it is hardened and is pounded into a paste with a small quantity of water and preserved as balls for storage.

8.11.4   FERMENTED BAMBOO SHOOT PRODUCTS

Most of the native bamboo shoot fermented products are traditionally consumed by North Eastern Himalayan region such as: *Lung-siej/Syrwa* in Meghalaya, *Mesu* in Sikkim, Soidon, *Soibum* and *Soijin* in Manipur, and *Rep* in Mizoram [18]. Usually, these are prepared during the months of June to September in regions of Manipur [39]. Conventionally, the product is consumed with chutney along with different proportions of green peas, *Colacasia* corns, potato, pumpkins, etc. Also, it is fried with fishes as a supplementary food. According to the abundance of the raw material, the major soibum producing sites in Manipur are localized [12].

8.11.5   PREPARATION OF SOIBUM/SOIJIM

Soibum, obtained from succulent bamboo shoots, is a native fermented food called as: *Bambusatulda* (Utang), *B. vulagaris* and *B. balcooa* (Ching Saneibi), *B. pallia and Dendrocalamus hamiltonii* (Wanap/Unap/Pecha), *D. sikkimensis* and *D. giganteus* (Maribop), and *Melocanabambusoide* (Moubi/Muli) that are mainly consumed in the diet and are more familiar in social customs of Manipur. The prior to the fermentation, traditional method of preparation includes: for the tender, succulent bamboo shoots are defoliated and then chopped, and pressed tightly into the earthen or wooden pots for 6 to 12 months. The preparation of soibum is classified into two types: Noney/Kwatha type and Andro type.

Noney/Kwathais prominent and common type of soibum preparation process. In traditionally designed bamboo chambers, this Noney type process is done in batch fermentation with a more acidic taste.

The lining of the bamboo chamber is done with polyethylene instead of forest leaves, which was done in the old days. Both thin slices of soft and succulent bamboo shoots are packed closely and tightly into this bamboo chamber, and the upper portion of the chamber is sealed with the help of polyethylene after filling it into chamber completely, and weights are placed on the top of the chamber for proper pressing. Production of good quality soibum can be achieved with the help of adequate pressing. As the fermentation progresses, the acidic fermented juice comes out, and it escapes from the chamber with through the perforations at the bottom of the chamber. In solid-state fermentation, the whole setup is left for 6–12 hours. After completion of the fermentation process, the soibum can be stored even for 12 months before sending it into the market.

Andro is another type of preparation of soibum, which is processed by batch fermentation in bulky roasted earthen pots. In this method, a section of the bamboo shoots is filled in the pot and is allowed for fermentation. As the fermentation progresses and the volume of the filled bamboo shoots portion decreases, then new slices of bamboo shoots are filled into the pot by the addition of pressure with the help of hand. This process is continued till the pot is completely filled with bamboo slices, and it is left aside for fermentation for 6- 12 months. In this method, the fermented juice coming out is not drained and however, it is retained in the chamber. In all sorts of soibum preparations, aging plays a major role. The quality of the soibum depends upon the incubation period: higher is the incubation period better is the quality of the product [12]. Partially fermented product is called sobium, and the completely fermented product is known as soibum.

8.12   FERMENTED FISH IN NEPAL

Masular is prepared from dry fish known as Sidra and bottle gourd leaves. It is a famous food product of Tharu community in Nepal. "Masu" means fish and "Lar" means syrup or soup. This product is limited to in house preparation and the final products a dry fishy odor as it is prepared from sidra. The fish is mashed with the leaves in Okhali/dikki, and it is made into round shape, which looks like a pancake. Then the mixture is allowed for sun drying (Figure 8.13). It is consumed as a side dish, or main dish and the shelf life of this product is 2–3 months. This product is rich in proteins, vitamins, minerals, polyunsaturated fatty acids, and essential amino acids [12].

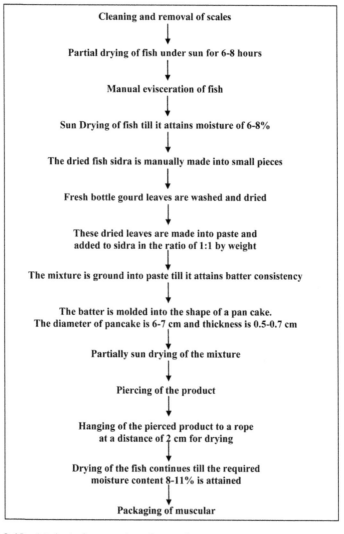

FIGURE 8.13 Method of preparation of muscular.

Bottle gourd has huge medicinal benefits. It is a fruit composed of all essential nutrients required for human health which includes antioxidant activity, anti hyperglycemic, anti diabetics, anti helminthic, and anti-cancer activity. Experimentally it is found that it has a capacity to cure numerous diseases such as sleep disorders, seizure disorders, anxiety, diabetes mellitus, hyperlipidemia, worm infection, and cancer. These leaves are a rich source of fiber, minerals like calcium and iron and rich in vitamin A and C [12].

8.12.1   PREPARATION OF SIDRA *(Figure 8.14)*

Figure 8.14 shows the preparation methods of Sidra. Sidra is an ethnic sun-dried fish product commonly consumed in Nepal, Bhutan, North East India, Darjeeling hills and Sikkim. Fish is collected, washed, dried in the sun for 4–7 days, and stored at room temperature for 3–4 months.

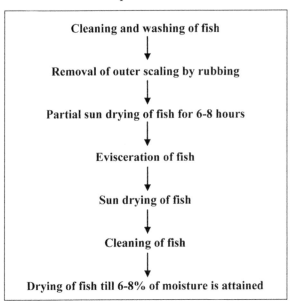

FIGURE 8.14   Method of preparation of Sidra.

8.13   FERMENTED FISH PRODUCTS IN THE PHILIPPINES

Popular fermented fishery products in the Philippines are divided into two categories: (1) the fish is fermented with a high concentration of salt which is about 15–20% of salt in the final product. These include patis (fish sauce) and bagoong (fish paste) which are generally used as condiments; and (2)

Burongisda (fermented rice fish mixture) and buronghipon (also known as balaobalao: fermented shrimp rice mixture). These products after completion of fermentation attain a cheese sort of aroma.

8.13.1   FISH PASTE (BAGOONG)

Bagoong is the undigested deposit of partly hydrolyzed fish or shrimp. It has a salty taste and a cheesy odor. The characteristic properties of the product vary from region to region where it is made and consumed. In the Tagalong provinces, the fish paste is totally fermented and ground by with or without the addition of coloring agent. The products are either partly or fully fermented in the region of Ilocos and Pangasinan provinces. The fish is slightly fermented without liquid in which the fish is hard and contains solid salt, and this process is followed in the region of Visayas and Mindanao [35].

8.13.2   METHOD OF PREPARATION BAGOONG

Anchovies, sardines, herring, silverside, shrimp, slipmouth, freshwater porgy, oysters, clams, and other shellfish are used in the preparation of bagoong. In the first step, the collected fishes are washed completely and drained well. Salt is mixed into the drained samples in different proportions, which vary from 1:3 to 2:7 depending upon the batch size and bulk preparation. Until it develops a characteristic flavor and aroma of bagoong, it is left for fermentation for several months.

In many traditional recipes, Bagoong is consumed in raw or cooked form and is generally used as a flavoring or condiment. It is also consumed as an appetizer by sautéing with onions/ garlic and served with tomatoes or green mangoes. It is usually consumed with vegetables especially in coastal regions. Bagoong is a rich source of protein

Total viable count of bagoong decreases with time. In the beginning of fermentation, aerobic organisms predominate and then it is followed by microaerophilic and anaerobic microorganisms in the later stages [8, 34]. Both desirable and harmful microorganisms are present in the final product.

8.13.3   FISH SAUCE

Both products, bagoong, and fish sauce have equal importance in the Philippines. After fermentation, bagoong is either thoroughly decanted or centrifuged, and the clear supernatant yellow-brown liquid is collected that is known as fish sauce. This is obtained either from fish or shrimp after fermenting it for 1 to 2 years. The quality of the product depends upon the period of the fermentation. If the digestion period is for a long span, then the quality of the product seems better.

The same raw material is used in the preparation of both fish sauce and fish paste; however, the preparation method differs only with respect to the period of fermentation. For fish sauce, the fermentation is prolonged until liquid forms on top of the mixture, after which it is drained and filtered.

There is a rapid decrease in the total bacterial count until 6 months, and after this, there is a slight decrease. Facultative anaerobes are mostly isolated in the process.

The solid material is gradually digested with the protein and solubilized by enzymatic activity, which results in an increase in peptides and amino acids in liquid components. The soluble protein to polypeptide ratio is moderately constant after a span of one month, which suggests the higher occurrence of the photolytic activates in early periods. Lipids break down during fermentation to yield fatty acids, which act as precursors for flavor, aroma, and browning reaction.

8.13.4   FERMENTED SHRIMP AND RICE BALAO BALAO

It is made out with the combination of fermented rice and shrimp (*Penaeus indicus* or *Macrobrachium* species). The mixture is acidic in nature, and the shrimp becomes soft and turns to red in color during the fermentation process. It is consumed either as an appetizer or main dish and is usually prepared in a sautéed form.

In the preparation process, the shrimp are washed and mixed with about 20% of salt and left for 2 hours or overnight. Then the excess water is drained, and the mixture is mixed with cooked rice and fermented for 7–10 days.

8.13.5   FERMENTED FISH AND RICE (BURONG ISDA)

It is a conventional dish in central Luzon and is usually prepared by using freshwater fish. In the process of fermentation, the fish flesh turns very soft, and the bones acquire the characteristic softness of cartilage when cooked. It is placed in oil and garlic/onion before consumption. It is consumed either as an appetizer or as a main dish.

Method of preparation of this product is more or less similar to balao-balao. Fish is scaled, eviscerated, and filleted. It is mixed with salt and kept aside overnight and mixed with cooked rice and allowed for fermentation for a span of 7- 10 days at room temperature [46].

Accumulation of lactic acid from the conversion of carbohydrates, which results in changes in the composition and acidity of the product, has been observed during lactic acid fermentation [1]. Such changes in the attributes are referred to as microaerophiles, but they do not result in the decomposition of the food to its basic components such as CO_2, and H_2O. Lactic acid is a major common end-product of their metabolism.

8.14   EXAMPLES OF FERMENTED FISH PRODUCTS (Figure 8.15)

Selected examples of fish products with their respective common names are indicated below along with their images Figure 8.15):

- *Bagoong*
- *Bottarga*
- Fesikh fish
- *Filetti di Baccalá*
- Fish and seaweed fertilizer
- Fish sauce
- *Garum*
- *Guedj*
- *Hakarl*
- *Hongeo– hoe*

- *Kusaya*
- *PekasamIkan*
- *Plara*
- *Prahok*
- Rakfish
- Seafood sausage
- *Sengmai Nagari*
- Surströmming
- *Tepa Yup'ik*
- *Trassieoedang*

Hakarl

Tepa Yup'ik

Sengmai Nagari

Fesikh fish

FIGURE 8.15   *(Continued)*

FIGURE 8.15   *(Continued)*

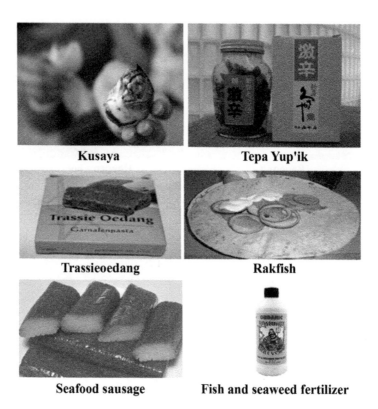

FIGURE 8.15 **(See color insert.)** Selected examples of fermented fish products in the market.

8.15 SUMMARY

Since the fish is a highly perishable food, various methods of processing play a major role in the storage of fish and fish products for a long span for consumption. Fermented Fish is considered a major source of protein. To reduce the water activity in the fish, salting, and drying of fish have been adopted to prevent microbial growth, which is responsible for spoilage of fish. During the process of salting, fermentation occurs that increases the shelf life of fish. In the process of fermentation, umami taste is generated by the digestion of amino acids to impart the peculiar aroma and taste.

Availability of fish for fermentation mainly depends upon the monsoon seasonality, oceanographic, and marine biological factors. In most of the fermented fish products, salt is used as a major ingredient. The present chapter mainly focuses on different types of fermentation processes. This chapter also includes the safety aspects in the handling of the fish from the board to the final stage of processing, nutritive value, microbiological, and chemical parameters of the fermented fish products. The process parameters include salting and drying.

KEYWORDS

- bagoong
- bekasam
- bekasang
- belacan
- budu
- cinculak
- fermented fish
- fish paste
- fish sauce
- gravels
- hakarl
- hentak
- ikantukai
- keapikan
- lactic acid bacteria
- lipolytic enzymes
- nagari
- organoleptic
- pedha
- pekasam
- perishable
- pla-ra
- proteolytic enzymes
- rakfish
- rebon
- salting
- shrimp
- sidra
- smoking
- soibum
- solar dryers
- terasi
- umami

REFERENCES

1. Adams, M. R., (1990). Topical aspects of fermented foods. *Trends in Food Science & Technology, 1*(1), 140–144.
2. Afiza, T. S., Rosma, A., Faradila, B., Wan Nadiah, W. A., & Ibrahim, C. O., (2007). Quality index of Kelantan unprocessed "Budu." In: *Proceedings of the 9th Symposium of the Malaysian Society of Applied Biology: Exploring the Science of Life as a Catalyst for Technological Advancement* (pp. 344–347). Kaula Lumpur, Malaysia.
3. Anonymous, (2009). *Malaysian Food Act and Regulation* (p. 89). MDC Publisher SdnBhd, Kaula Lumpur, Malaysia.

4. Anonymous, (2015). Livsmedel databasen, *The National Food Agency (SE)* (p. 231), Uppsala - Sweden.

5. ALM, F., (2012). Scandinavian anchovies and herring tidbits. *Fish as Food V3: Processing, 2*, 195.

6. Aquerreta, Y., Astiasarán, I., & Bello, J., (2002). Use of exogenous enzymes to elaborate the Roman fish sauce 'garum.' *Journal of the Science of Food and Agriculture, 82*(1), 107–112.

7. Battle, H. I., (1935). Digestion and digestive enzymes in the herring (*Clupeaharengus* L.). *Journal of the Biological Board of Canada, 1*(3), 145–57.

8. Beddows, C. G., (1985). Fermented fish and fish products. In: Brian, J. B., & Wood, E., (eds.), *Microbiology of Fermented Foods* (pp. 416–434). Blackie Academic and Professional, London SET8HN, UK.

9. Boon-Long, N., & Chompreeda, P., (1985). *Development of Traditional Fermented Fish Product for Small Industries: Processed Development of PLA-RA from Salt Water Fish.* Thai National AGRIS Centre, Bankok – Thailand; Progress Report submitted to FAO; http://agris.fao.org/agris-search/search.do?recordID=TH1999000400; pages 118.

10. Budhyatni, S., Murtini, J. T., & Peranginangin, R., (1982). The microflora of terasi powder. *Laporan Penelitian Teknologi Perikanan, 16*, 25–33.

11. Efendi, Y., (1993). Preliminary study on the processing of *ikantukai*. In: *Proceeding of the First Symposium of Indonesia Fisheries* (pp. 152–163). Jakarta, Indonesia.

12. Giri, S. S., & Janmejay, L. S., (2000). Effect of bamboo shoot fermentation and aging on nutritional and sensory qualities of soibum. *Journal of Food Science and Technology, 37*(4), 423–426.

13. Hagen, H., & Vestad, G., (2012). Tre sterke fra innlandet domrake fisk, akevitt og pultost (Three strong inland domrake fish, aquavit and pultost). Brummunddal (Norway), Gutu Forlag, Norway; p. 78.

14. Hanafiah, T. A., (1987). Factors affecting quality of pedahsiam. *PhD Dissertation* (p. 243). University of Washington, Pullman – USA.

15. Hassan, Z., (1980). Pekasam: A fermented fish product (in Peninsular Malaysia). In: *Seminar on Modernization of Malaysian Cottage Food Industries* (p. 5). Serdang, Selangor, Malaysia.

16. Huda, N., (2012). Malaysian fermented fish products. *Handbook of Animal-Based Fermented Food and Beverage Technology, 2*, 709–715.

17. Irianto, H. E., & Irianto, G. I., (1998). Traditional fermented fish products in Indonesia. In: *Report of the 26th Session of the Asia Pacific Fishery Commission-FAO* (p. 112). Beijing, China.

18. Jeyaram, K., Talukdar, N. C., & Rohinikumar, S. M., (2005). Fermented bamboo shoot products of Manipur: Documentation of Indigenous traditional knowledge. In: *Second Int. Conf. Fermented Foods, Health Status and Social Well-Being* (p. 7). Organized by Swedish South Asian Network on Fermented Foods (SASNET), Anand Agricultural University, Anand – India.

19. Jinap, S., Ilya-Nur, A. R., Tang, S. C., Hajeb, P., Shahrim, K., & Khairunnisak, M., (2010). Sensory attributes of dishes containing shrimp paste with different concentrations of glutamate and 5′-nucleotides. *Appetite, 55*(2), 238–244.

20. Karim, M. I. A., (1993). Fermented fish products in Malaysia. In: Lee, C. H., Steinkraus, K. H., & Reilly, P. J. A., (eds.), *Fish Fermentation Technology* (pp. 95–106). United Nations University Press, Tokyo – Japan.

21. Kristinsson, Ö., Efnagreining, Á., & Verkuðum, H., (1993). *Fish Products* (p. 13). Icelandic fisheries laboratories, Reykjavik (Iceland).
22. Kristjansson, L., & Islenskir, S., (1983). *Icelandic Marine and Fish Processing Methods* (pp. 316–395). Bokaútgafa Menningarsjoðs, Reykjavik (Iceland).
23. Kurlansky, M., (2011). *Salt* (p. 13). Random House, New Delhi – India.
24. Leong, Q. L. A. B., Karim, S., Selamat, J., Mohd-Adzahan, N., Karim, R., & Rosita, J., (2009). Perceptions and acceptance of '*Belacan*'in Malaysian dishes. *Int. Food Res. J., 16*, 539–546.
25. Magnússon, H., & Gudbj €ornsdottir, B., (1984). Örveru Og Efnabreytingar Við Verkun Hákarls (Microbial and chemical changes in shark action). Icelandic Fisheries Laboratories, Reykjavik (Iceland); p.15.
26. Martin, A. M., (1998). Fishery's waste biomass: Bioconversion alternatives. In: *Bioconversion of Waste Materials to Industrial Products* (pp. 449–479). Springer, Boston, MA.
27. Mulyokusumo, K., (1974). *Soy Sauce, Peanut Sauce, Fish Sauce* (p. 11). Terate, Bandung -Indonesian.
28. Murtini, J. T., Yuliana, E., & Nasran, S., (1997). Effects of addition of lactic acid bacteria starter in the processing of spotted gouramy (*Trichogaster trichopterus*) bekasam in its quality and shelf-life. *Journal Penelitian Perikanan Indonesia (Indonesia), 5*, 12–18.
29. Nuraniekmah, S. R., (1996). *Effects of Soaking Temperature on Proteolytic Enzyme Activity and Bacteria Growth of Jambal Roti Made of Marine Catfish (Arius thalassinus)* (p. 113). M.Sc. Thesis, Faculty of Fisheries, Bogor Agricultural University, Bogor – Indonesia.
30. Phithakpol, B., (2009). *Food Handling at Village and Household Levels in Thailand, Phase I* (p. 31). Agricultural College, Asian Institute of Technology, Bankok – Thailand.
31. Putro, S., (1993). Fish fermentation technology in Indonesia. In: Lee, C. H., Steinkraus, K. H., & Reilly, P. J. A., (eds.), *Fish Fermentation Technology* (pp. 107–128) United Nations University Press, Tokyo.
32. Riddervold, A., & Heuch, H., (1999). Rakfisk vidunderlig spise (Rakfisk Wonderful Eating). Teknologisk Forlag , Oslo (Norway); p.16.
33. Rosma, A., Afiza, T. S., Wa-Nadiah W. A., Liong, M. T., & Gulam, R. R., (2009). Short communication microbiological, histamine and 3-MCPD contents of Malaysian unprocessed 'budu.' *International Food Research Journal, 16*, 589–594.
34. Saisithi, P., Kasemsarn, R. O., Liston, J., & Dollar, A. M., (1966). Microbiology and chemistry of fermented fish. *Journal of Food Science, 31*(1), 105–110.
35. Sanchez, P., (1983). *Traditional Fermented Foods of the Philippines* (pp. 18–26). UN Workshop on Traditional Food Technologies, FAO, Rome.
36. Sarojnalini, C., & Vishwanath, W., (1994). Composition and nutritive-value of sun-dried *puntiussophore. Journal of Food Science and Technology (Mysore – India), 31*(6), 480–483.
37. 37. Skåra, T., Axelsson, L., Stefánsson, G., Ekstrand, B., & Hagen, H., (2015). Fermented and ripened fish products in the Northern European countries. *Journal of Ethnic Foods, 2*(1), 18–24.
38. Sharif, R., Ghazali, A. R., Rajab, N. F., Haron, H., & Osman, F., (2008). Toxicological evaluation of some Malaysian locally processed raw food products. *Food and Chemical Toxicology, 46*(1), 368–374.
39. Singh, S. K., & Singh, D. K., (2009). *Hepaticae and Anthocer* (p. 15). Great Himalayan National Park (HP), India. Botanical Survey of India, Dehradun, UP - India.

40. Suparno, A., & Silowati, T., (1982). Preparation of fish sauce from mackerel (*Rastrelliger* spp.) by acid hydrolysis. *Laporan Penelitian Teknologi Perikanan, 20,* 29–36.

41. Tamang, B., Tamang, J. P., Schillinger, U., Franz, C. M., Gores, M., & Holzapfel, W. H., (2008). Phenotypic and genotypic identification of lactic acid bacteria isolated from ethnic fermented bamboo tender shoots of North East India. *International Journal of Food Microbiology, 121*(1), 35–40.

42. Thapa, N., & Pal, J., (2007). Proximate composition of traditionally processed fish products of the Eastern Himalayas. *J. Hill Res., 20*(2), 75–77.

43. Thapa, N., Pal, J., & Tamang, J. P., (2004). Microbial diversity in ngari, hentak and tungtap, fermented fish products of North-East India. *World Journal of Microbiology and Biotechnology, 20*(6), 599–601.

44. Thapa, N., Pal, J., & Tamang, J. P., (2007). Microbiological profile of dried fish products of Assam. *Indian J. Fisheries, 54*(1), 121–125.

45. Thapa, N., Pal, J., & Tamang, J. P., (2006). Phenotypic identification and technological properties of lactic acid bacteria isolated from traditionally processed fish products of the Eastern Himalayas. *International Journal of Food Microbiology, 107*(1), 33–38.

46. Vatana, P., & Del-Rosario, R. R., (1983). Biochemical changes in fermented rice-shrimp (*Macrobrachium idella*) mixture: Changes in protein fractions. *Food Chemistry, 12*(1), 33–43.

47. Wudianto, A., Naamin, N., Susanto, K., Irianto, H. E., & Pranowo, S. A., (1996). *A Fishery and Socio-Economic Survey in MCMA of Karakelong-Manado, North Sulawesi Pusat Penelitiandan Pengembangan Perikanan* (p. 30). Indonesia University of Agriculture, Jakarta (Indonesia).

48. Yeoh, Q. L., & Merican, Z., (1978). Processing of non-commercial and low-cost fish in Malaysia. *Proc. IPFC, 18*(3), 572–580.

49. Yunizal, P., (1998). Processing of shrimp terasi. *Warta Penelitiandan Pengembangan Pertanian (In Indonesian), XX*(1), 4–6.

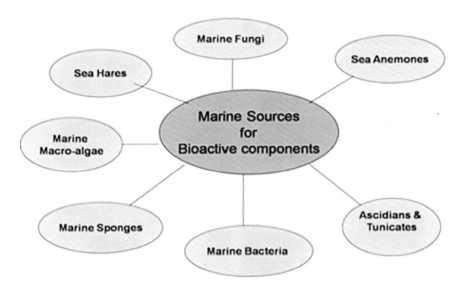

FIGURE 3.1 Marine sources of bioactive components.

FIGURE 8.12 Traditional fermentation method for *Nagari*.

Hakarl

Tepa Yup'ik

Sengmai Nagari

Fesikh fish

Bagoong

Hongeo - hoe

Pekasam Ikan

Fish sauce

Bottarga

Plara

FIGURE 8.15   *(Continued)*

FIGURE 8.15   Examples of fermented fish products in the market.

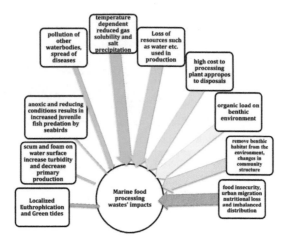

FIGURE 9.2   Negative impacts associated with marine waste.

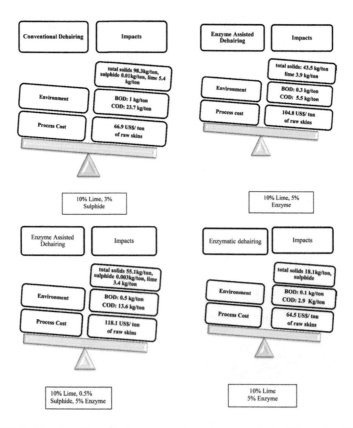

FIGURE 9.3   The dehairing efficiency of protease enzymes derived from the fish wastes compared to a conventional process (lime-sulfide treatment).

FIGURE 9.5   Marine sources of chitin and its structure.

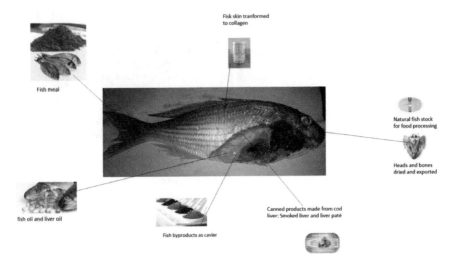

FIGURE 10.2   Different by-products of the fish processing industry.

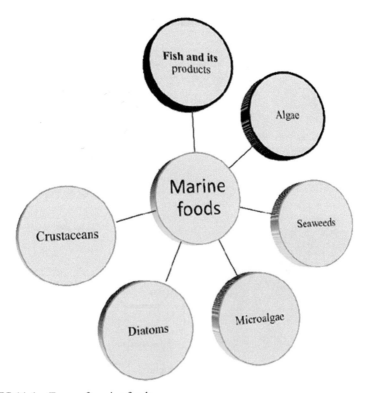

FIGURE 11.1   Types of marine foods.

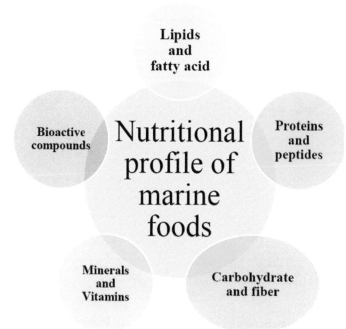

FIGURE 11.2   Nutritional and physiological possessions of marine foods.

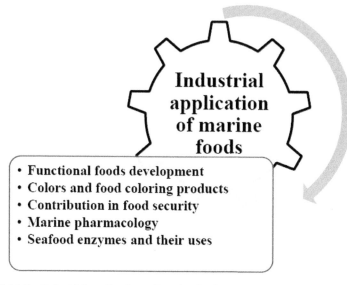

FIGURE 11.3   Industrial applications of marine foods.

Remedy for Biomagnification

On consumption antibiotic gets into humans

Emergence of antibiotic resistant bacteria

← Antibiotics transferred to birds

Probiotic therapy

Organic/Ecological Aquaculture

Soils, algae, water plants gets antibiotics.

Emergence of resistant bacteria (Oppurtunistic pathogens)

Antibiotic treated fish

Probiotic based functional feed fed fish.

FIGURE 12.1 Biomagnification of antibiotics (Left side); Probiotic fish feed application (Right side) in aquaculture.

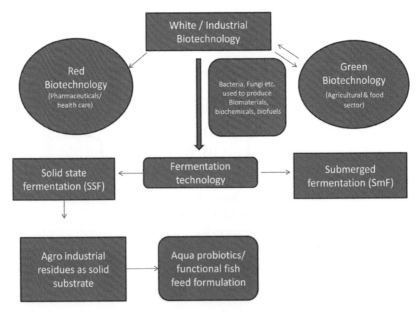

FIGURE 12.3 Flowchart depicting branches of industrial biotechnology.

PART III

Processing and Industrial Applications for Marine Foods

CHAPTER 9

WASTE MANAGEMENT IN PROCESSING OF MARINE FOODS

LOCHAN SINGH, KSHITIZ KUMAR, and VIJAY SINGH SHARANAGAT

ABSTRACT

Marine food production and processing industries based on diversified and enormous ocean richness (fish, mollusks, shellfish, and other seafood) have significantly increased food and diet options for humans. At the same time, the bio-waste generated during these processes has raised serious concerns due to associated problems such as: disposal, economic losses, and environmental challenges. Mitigation of these challenges through alternative utilization of the bio-waste may generate a significant effect on sustainability. Also, the nutritive richness (chitin, omega-3 fatty acid, vitamin-A, calcium, minerals, etc.) and other qualities (bio-active, antioxidant, and enzymatic properties) of the bio-waste have landed a plethora of opportunities for its resource utilization and management. The present chapter focuses on the sources of marine food waste production and the drivers for the management of marine food waste. It also highlights the utilization processes and management strategies to convert the waste to a resource and simultaneously mitigate the associated problems.

9.1 INTRODUCTION

About 70.8% area of Earth is covered with water, which has been a rich source of biodiversity. For many years, the abundant marine lives such as fishes, mollusks, crustaceans, seaweeds, microalgae, etc., have remained as a major food source for several human civilizations due to its affordable price and presence of quality protein. It has been reported that in poor nations, approximately 15% of average animal protein intake to three billion people is provided by fishes [23]. Thus to keep sustainable development and long-term prosperity, drastic growth has occurred in this sector (163 million tons in 2009). The major fish production contribution in 2009 came from marine capture fisheries (49%) followed by freshwater aquaculture (23%), mariculture (21%) and inland capture fisheries (6%). Though the growth rate of this sector has remained static after the 1990s, yet the overfishing and habitat destruction have negatively affected the population of these marine lives, and it has thus become a challenge to meet out the ever-increasing food demand of the human population. Statistics show that worldwide nearly 900 million people remain hungry or underfed [22]. The alternative to emerging crisis lies in preserving the marine foods by extending shelf-life, finding innovative processing techniques, improvising transport and packaging facilities and by reducing the amount of generated waste throughout the supply chain, i.e., beginning from production till the kitchen table.

The first report on "Global Assessment of Fish Discard Generated in the Year 2005" showed wastage of 27 million ton which had dropped to a level of 7.3 million tons [45]. However, the incomparable nature of these figures and the estimates has resulted in a questionable state [34]. The marine capture and aquaculture fisheries annually discard approximately 20 million tons per year worldwide [108]. Another study has shown that out of total fish and seafood supplies throughout the world, 24% kcal of these products are lost [53]. The variations in the assessment of marine waste arise with the region and the type of fishing methods. The estimates of seafood losses in the United States show that nearly 40–47% of edible seafood remains uneaten, of which 51–63% loss occurs at consumption level, 16–32% as discard and 13–16% at distribution and retail levels [54]. Nearly 6.9% harvest and post-harvest losses in India comes from inland fisheries [37]. Globally, 6–8 million tons of crab, shrimp, and lobster shells are wasted, with 1.5 million tons stand-alone contribution comes from Southeast-Asia [114]. Similarly, Norwegian cod fisheries alone generated 232,000 metric tons of byproducts in 2001, of which 107,000 tons were used, and 125,000 tons were dumped.

These wastages are not only loss of nutrients, minerals, and proteins but also lead to a vicious circle of events to hamper our ecosystem as well as the world's social and economic goals of complete sustainability. An understanding of the sources for generation of these wastes, their types, associated impacts and ways of management is thus of immense importance. This will also aid in strengthening the marine food processing industry by highlighting the hidden challenges, gaps in technology and will suggest possible interventions for cutting down these losses.

The present chapter focuses on the sources of marine food waste production and the drivers for the management of marine food waste. It also highlights the utilization processes and management strategies to convert the waste to a resource and simultaneously mitigate the associated problems.

9.2 CLASSIFICATION OF MARINE FOOD WASTE

Marine food waste includes a mixture of diverse plant and animal origin debris generated at different stages of the supply chain. It includes non-target fish species, crustaceans, their skins, bones, head, muscles, shells, blood, processing leftovers or stick-waters, trimmings or cut-offs, and other visceral body parts and organs like spleen, liver, pancreas, gonads, intestines, stomach, frames, fins, tails, offal's, etc. In most cases, no clear distinction between the terms 'marine wastes' and 'marine by-products' occur, opening a broad spectrum of overlapping boundaries and multiple definitions worldwide [79]. In some definitions, non-ordinary saleable products recycled after treatment and with some human utility or value are considered as by-products whereas products of no-utility which are targeted for burning, composting, and destruction are considered as wastes [79]. International Standard Statistical Classification of Aquatic Animals and Plants (ISSCAAP classification) has been used for a long period of time for creating fishery database and statistics (Figure 9.1). Taking this classification as a reference, the marine food waste may be classified into two categories depending on their origin: (a) plant waste; and (b) animal waste.

9.2.1 MARINE PLANT WASTE

The richness in fiber and protein has made marine or aquatic plants and microphytes as an indispensable part of the diet in Japan, Korea, and China. It extensively enjoys a special place in coastal cuisines around the world. The common members of this group are: seaweeds and microphytes. Seaweeds taxonomically include several species and groups of marine, eukaryotic, multi-cellular, and macroscopic algae, which are mainly found at wide ecological niches including a littoral region (sea part near coastal shores) and rocky shores. The massive abundance, non-limitation to water and higher productivity over terrestrial plants further aids in its easy cultivation [41]. The main group members are red, green, and brown algae, which lack a common ancestor. The appearance of seaweed resembles non-arboreal terrestrial plants and possesses structures like thallus (algal body), blade or lamina (flattened leaf-like structure), spore cluster (sorus), stem-like structure (stipe, may be absent or present), floatation assisting organ on blade or between lamina and stipe (fucus and kelp), holdfast, and haptera (specialized basal structure for surface attachment or adherence).

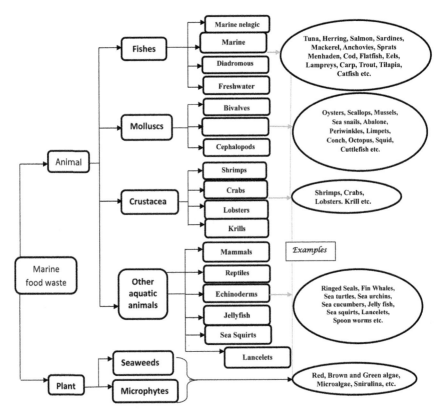

FIGURE 9.1   Classification of aquatic animals and plants.

It has been observed that nearly 8 million tons of wet seaweeds are harvested globally in a year, and 15.5million tons (fresh weight) of macroalgae are produced per year [91]. Also, the annual production of red algae has shown an increment from 5.3 million tons in 2006 to 10.8 million tons in 2011 [48]. *Undaria pinnatifida* is a commonly cultivated brown marine macroalgae in South Korea, with a production of 0.3 million tons in 2014 and possess multiple utilities such as fertilizers, human, and animal nutrition, medicines, and hydrocolloid extraction. However, the difficult commercial handling and non-worthiness of its roots result in its dumping into the sea, which accounts for nearly 40–60% of annual production [41]. Similarly, Kelp macroalgae extensively utilized worldwide for alginate, mannitol, and iodine production and extraction, besides subsidiary food and fertilizers; generates a huge amount of solid waste, which is ultimately dumped into landfill areas. The reason for such large wastage (comprising of crude fiber, residual alginic acid, and protein) is the reduced extraction efficiency (30% only) of alginates and other products from these resources [119]. Per year floating residue generated by alginate industry utilizing *Laminaria japonica* brown algae amounts to tens of thousands of residual materials [30]. Similarly, very limited utilization of seaweed biomass waste has been observed by carrageenan industry amounting to 60–70%, which holds the promising amount of carbohydrates and potential for use as feedstocks and biochemicals [97].

9.2.2 MARINE ANIMAL WASTE

Besides Plants, the vast diversity of marine animals is being consumed worldwide (Figure 9.1). The production of shrimp (ranked second most important among different monetarily traded commodity) in 2007 reached 3,275,726 tons that are equivalent to 16.5% of international fishery revenues. Contrarily, 50% of the whole shrimp is wasted during processing in the form of carapace, tail, and cephalothorax, which could be the source for extraction of bioactive molecules [8]. Depending on the type of species, 48–56% discards are produced from shrimps [80]. Being an important commodity in Indonesia, the total generation capacity of shrimp is more than 500,000 tons/ year while more than 300,000 tons/year solid shrimp waste is produced [35]. Nearly $3.9 billion revenues are generated in Canada through export of 80% caught fishes and seafood, with major contribution coming from processed shrimp. Throughout the process, 50–70% weight of total landings is discarded as waste, which could have been utilized in functional food development and hydrolysates extraction [12]. Fish waste, a by-product of fish market and processing industries ranges, between 30–45% of initial product weight after transformation [107].

Silver mojarras also known as *Diapterus rhombeus* are marine fin-fishes that are abundantly found in coastal estuaries of Atlantic Ocean. In 2006, 2080 tons of these fishes were captured in northeastern Brazil from which 100tons of viscera became part of estimated annual discharge [86]. Viscera, skin, head, bone, and muscle tissues are the important proportions of mackerel fish body, which is generally discarded as wastes even though mackerel are among main 23species fished across the world [18]. Viscera, which is rich in proteins, alone forms a total of 300,000 tons of waste [6]. Similarly, marine invertebrates such as echinoderms which include sea urchins and sea cucumbers have been a part of food and diet for decades. These organisms possess peculiar connective tissues or mutable collagenous tissues which in general forms respective food industry's waste [25]. Not only this, a significant amount of waste liquor is generated after processing and utilization of sea cucumber, which is generally discarded back to the ocean [11]. Cephalopods, i.e. squids, octopuses, and cuttlefishes demand has increased by 15% between 1994 and 1996. Also, the world catches for these organisms was 3.3 million tons higher between 1994 and 2001. Again, the waste generated has been 50% of the fish materials' waste [111].

9.3 SOURCES OF MARINE FOOD WASTE

Marine waste has been observed to generate at various places and stages of the supply chain (Table 9.1), which are discussed in this section.

9.3.1 PRODUCTION OF MARINE FOOD WASTE

Bottom Trawl, Dredge, Gillnetting, Harpooning, Traps, and Pots, etc. are some of the tools widely used by fishermen to land their catch. Each tool has a characteristic working pattern and leaves behind its own peculiar effect on the ocean. The targeted marine fishes often decide the type of catching method to be used. During the process, especially on using nets, the catch obtained comprises not only targeted fishes but also diversified oceanic life-forms, which are usually thrown back dead or dying. On the other hand, the sensitive seafloor habitat is dragged through heavy gear running across the seabed through commonly used Bottom trawl method. It has been observed that out of 27 million ton of the world's marine catches, the proportion of total catch returned to the sea comprising dead, dying or badly damaged remains or discards is significantly high [2]. FAO also in their latest global study of 2005 suggested a drop in these discards to 7.3 million tons, but still, the data are not totally comparable. A weighted global discard ratio of 8% and a significant contribution of catching practices in this discard has also been assessed by the United States' seafood supply chain [54]. It thus becomes essential to understand the working of these

TABLE 9.1 Marine Food Loss and Waste along the Supply and Value Chain

Industry involved	Type of waste generated	Quantity	Country	Reference
Agar–Agar industry	Macroalgae industrial waste	2000–2400 kg/ day	Spain	[26]
Alginate industry	Seaweed waste	tens of thousands of residual materials	China	[30]
Canning	Yellowfin tuna, skipjack tuna, and tongol tuna	Approximately two-third part of the whole fish is utilized. One-third part possesses visceral remains, which becomes waste.	Thailand	[50]
Fish processing	Non-edible or edible parts	2.8 MMT	India	[72]
Processing	Shrimp waste	150,000–175,000 Gg per annum	India	[4]
Processing and production	Shrimp waste	8,250 metric tons	Brazil	[65]
Seafood canning	Captured and cultivated fishes	>35% of the total weight	Spain	[99]
Variable places, different fish markets, etc.	Fish waste	299,128 tons (Persian Gulf); 32,533 t in the Caspian Sea; the total rate in 2003 reached 441,836 tons	Iran	[113]

catching tools and judiciously to select them from gamut [45], to further aid in minimizing the production losses and associated environmental impacts posed by fishing industry.

9.3.2 HANDLING AND STORAGE OF MARINE FOOD WASTE

Loss/waste is not only limited to catch, but losses are also generated on board and due to lack of good handling and storage practices. Marine food comes under perishable food that requires intensive care during handling and storage. Fish and seafood products being rich in protein and spoil rapidly when left unpreserved [33]. Different factors like Rigor mortis, enzyme activity, microbial growth, and storage temperature play a significant role in this spoilage and are majorly active during handling and storage steps of the supply chain.

The muscle tissue of fish on death undergoes various chemical changes and exhibits concomitant physical effects termed as 'Rigor mortis.' Softening of muscles takes place owing to conversion of glycogen into lactic acid and reduction in pH. Oxidative rancidity, which is measured through peroxide value and free fatty acid content, is another indication for spoilage of fatty tissues in fishes. Besides these, a complex series of reactions by enzymes resulting in the breakdown of protein and fat structure creates favorable conditions for the bacterial and other microorganisms' growth. Freshly caught fish, in general, are bacteria free, but may contain bacterial load at surface slime, gills, and intestine. Death of fish creates conditions suitable for these bacteria to attack the flesh owing to decline fish metabolism and resistance, produce undesirable compounds and cause spoilage. The growth and type of bacteria depend largely upon its inhabiting conditions, water from where it is caught and also on the handling methods after its catch. Enzymatic action also affects the components responsible for characteristic fish flavor and odor. Breakdown of ATP generates Hypoxanthine, which on accumulation imparts a bitter taste to fish muscles. Fish gut possesses digestive enzymes, which also contribute to decomposition in the fish through belly bursting process. Similarly, the action of enzyme tyrosinase on tyrosine in shrimps leads to Melanin and black spot formation.

Understanding the mode of action (enzymatic, bacterial, and chemical) responsible for the spoilage of fish and controlling the activity of marine organism through proper handling and optimum storage conditions (immediate lowering of the temperature), the observed losses may be avoided. In fishes, immediate chilling after catch and fish holding at 0°C through proper icing proves beneficial. Similarly, in shrimp's immediate head removal after catch decrements the spoilage rate. For big fishes, enzymatic actions-based spoilage may be reduced through beheading and eviscerating.

9.3.3 PROCESSING AND PACKAGING OF MARINE FOOD WASTE

Processing of fishes and marine organisms is an important requirement due to their perishable nature. This step of supply chain formed 60% of total world fisheries production in 2000 and the share in destined human consumption was 53.7% for fresh fish, 25.7% frozen fish, 11% canned fish and 9.6% cured fish [38]. Before processing, the catch is removed from fishing gears and is transferred to vessels. The different unit operations such as stunning, sorting, grading, bleeding/de-heading/slime removal, gutting or peeling, washing, chilling, and freezing, filleting, curing, unloading, and storing, cooking/ frying, packing, and glazing, breading, brine filling, dipping, seaming, can sterilization, etc. are then performed. As observed, the initial processing steps revolve around the separation of an edible and non-edible part like heads, bones, and viscera. These by-products occurring at both production and processing levels are often rendered into fish meal and oils, which are main ingredients in animal or pet feed and fertilizer [95]. Their further utilization includes the extraction of bioactive compounds. Besides this, a huge number of by-products, i.e. heads, bones, offal, Fish bone, Fish internal organs, Shellfish, and crustacean shells, Fish eggs, Fish skin, Fish frames, and cutoffs are generated throughout processing during different unit operations.

The second step of processing thus focuses on the extension of shelf life of marine organisms and the products as well as on their regular availability for further processing. This goal is thus obtained by proper application of fish freezing technique to facilitate storage and reduce the losses. The amount of waste generated in different marine food processing industries is given in Table 9.2.

9.3.4 TRANSPORT, DISTRIBUTION, AND MARKET CONSUMPTION

Marine food (Primary and secondary processed) distribution and transportation face a lot of challenges. Primarily processed food requires intensive care to maintain desired characteristics. Transportation of marine food requires the desired temperature (below–18°C) to inhabit the growth of

TABLE 9.2 Waste Generated During Processing of Marine Foods

Production	Handling & storage	Processing & packaging	Distribution & storage	User level
		Definition/ Description		
By-catch, On-boards, during or after fishing (also called as farming)	Catch transfer from fishing gear to the vessel, catch holding, Transportation to sea-shores, near-by places or first level-markets, etc. and their storage throughout the places.	Processing of foods at domestic and commercial levels	Transportation to markets (wholesales or retails), food streets, end-users or consumers, etc. and their storage throughout the places.	Losses at end-users level, i.e. households, restaurants, businesses, caterers, and other customers
		Waste is comprised of:		
Discarded fishes (dead, dying or damaged), non-targeted marine organisms	Spoiled or spilled products and products with deteriorated quality, mainly after farming and landing	Separated non-edible parts like heads, bones, and viscera, by-products (Heads, bones, offal, Fish bone, Fish internal organs, Shellfish, and crustacean shells, Fish eggs, Fish skin, Fish frames, and cutoffs) which may or may not be further utilized	Spoiled and deteriorated products	Spoiled and deteriorated products
		Reason for waste generation		
Their non-economic or non-food use, entry of unwanted marine organism in nets, damage or death of the captured organisms on-board or during farming	Microbial spoilage, oxidation, change in odor/ flavor/ texture, the activity of digestive enzymes due to improper ambient conditions	Alginate industry, canning, filleting, freezing, canning or curing activities generate these waste residues in the form of wastewater, trimmings, and by-products which in general are non-saleable	Microbial spoilage, oxidation, change in odor/ flavor/texture, the activity of digestive enzymes due to improper ambient conditions	Depending on consumer preferences, spoilage due to non-favorable storage conditions, bad smell and poor quality, un-organized food sector

the microorganism. Major losses (Spoilage) occur during transportation and storage before distribution. Lack of storage facilities in a retail store is another reason for the spoilage of marine food. Some part of marine food is also wasted at the consumer end in the form of unused marine food. In restaurant and food stalls, customer orders more food and some part of the unconsumed food is also wasted.

9.4   NEGATIVE IMPACTS OF MARINE FOOD WASTE

The adverse impacts of marine food waste on social, environmental, and economic spheres of global sustainability goals have fetched considerable importance in the past few years. The alarming situations have diverted research and development efforts to identify methods to minimize these negative impacts. The adverse impacts of marine food waste (Figure 9.2) are discussed in this section:

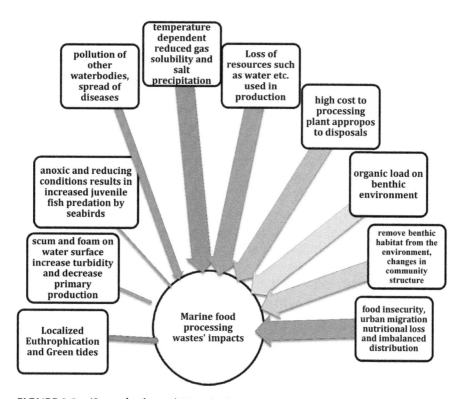

FIGURE 9.2   (See color insert.) Negative impacts associated with marine waste.

9.4.1   ENVIRONMENTAL IMPACT

The marine food waste generated during production and processing has been observed to be rich in total suspended solids (TSS), organic matter and nutrients, fat-oil-grease (FOG) and pathogenic and other microflora along with high biological and chemical oxygen demand (BOD and COD, respectively). The wastewater of variable strength and FOG is generated mainly from trimming and filleting processes while the manufacturing of ice, chilling, and maintenance of cool room throughout the supply chain often results in waste heat generation. Slime, sand, and body fluids are other contributors to wastewater. The uncontrolled dumping, as well as waste flow, may release as high as 27,000 tons of fish daily, which participates in generating approximately 1300–3250 mg/l of COD, 17,000mg/l of TSS and 500–1550 mg/l of BOD. The survey by North Carolina Seafood industry focusing on unloading, washing, sorting, grading, ice separation, and re-icing processes showed an average TSS amount of 2.4 g/kg with a water usage of 1.5 l/kg of fish handled [38].

The nature and the variability of these effluents and waste loads can adversely affect coastal and marine ecosystem, frequently coupled with unwanted fertilization and eutrophication. The aquatic life of such places is deprived of the oxygen due to enhanced BOD levels and the breakdown of organic matter to proteins, nitrogenous; and other compounds release toxic gases like hydrogen sulfide, methane (CH_4), and ammonia. This is accompanied by a short-term increase in nutrient composition bloom plant growth, which will further reduce oxygen levels in water bodies and lead to an initial fish population rise due to advantageous increase in prey items.

Simultaneously, increment and coverage of water surface by phytoplankton makes survival of benthic organisms challenging, and the release of related algal toxins may cause mass mortality of fish species [38]. Also, the critical reduction and anoxic conditions arising adjacent to effluent piles increase predation of juvenile fish species by diving seabirds, marine mammals and higher fishes for meeting their food requirements. The increased turbidity due to seafood waste deposits (scum and foam) on water surface may potentially reduce the primary production [13].

9.4.2   SOCIAL AND ECONOMIC IMPACTS

Fish processing industry is becoming geographically concentrated, vertically integrated, more intensive and extremely linked to distributors, large food processors, and retailers. The industry is a source of livelihood for many people. The waste generated from this industry holds many negative impacts on the environment. As described in this section, species diversification, toxic release, and changes in the population of marine organisms will directly influence the income and livelihood of these associated people. On the other hand, dumping of generated waste at any site may result in foul odor and leaching of compounds, which will degrade soil conditions, sea ambient, and groundwater quality, thereby adversely affecting the health of nearby inhabitants. Another social impact associated with marine waste generation and discard is a loss of nutrients, which if utilized otherwise would have helped in feeding poor and hungry people. For example, loss of seafood may serve as a source of protein, calorie, and eicosapentaenoic (EPA) and docosahexaenoic (DHA) acids (i.e., omega-3 fatty acids)] for 10.1 million men, 12.4 million women, 1.5 million young adults, and 20.1 million adults [54].

9.5 MANAGEMENT OF MARINE FOOD WASTE

Direct disposal, i.e., the conventional waste management method, of marine waste into the oceans and landfill areas, results in huge environmental, social, and economic losses and has been widely discouraged in recent years. Parallel to this, mitigating the disastrous effects of these wastes and simultaneously deriving profit became a challenge for researchers as well as food industries. The abundance and richness of marine food waste (for sundry essential and valuable compounds such as protein, carotenoids, chitin, protein hydrolysate, glycos-amino-glycans, etc.) can be turned into some economic and commercial value through proper extraction techniques. The attempts thus have been made to achieve sustainability by understanding the nutritional and health-promoting properties of these wastes and to reutilize these as resources.

9.5.1 UTILIZATION/ RE-USE: CURRENT STRATEGIES

9.5.1.1 PROTEASE AND ANOTHER ENZYME EXTRACTION

Enzymes are proteinaceous compounds or macromolecules with biocatalytic properties. These catalysts are present in the living cells and are responsible for altering the rate of different metabolic reactions. Under specific conditions, enzymes specifically convert substrates to a product in a reaction by decrementing the amount of activation energy required at the initial phase of the reaction. The fate of different metabolic processes of a cell is largely determined by the set of enzymes involved. Due to the specificity and efficacy of these compounds, they have widely been used at the commercial level for various purposes like: antibiotic's synthesis, biological washing powders, meat tenderization, brewing, cheese, and biscuit making, biofuel production, clarification of fruit juices, etc. Proteases (also known as peptidases and proteinases) refer to those enzymes, which are involved in protein hydrolysis through the breakdown of peptide bonds linking two amino acid groups. Depending on the predominant moiety or catalytic residue, these enzymes are broadly classified into seven groups, namely: serine proteases, cysteine proteases, threonine proteases, aspartic proteases, glutamic proteases, metalloproteases, and asparagines peptide lyases. Currently, microbe-derived enzymes meet 34% demand for animal feed and food industry, 29% of detergent and cleaning sector demand, 11% contribution in leather, textile, and other sectors. Also, the global market for industrial enzymes was around 4.4 billion US$ in 2015 [57]. Microbe-free enzyme preparation and filtration processes are required to generate alkaline proteases from commercially exploited bacterial and fungal sources, but generally, it involves optimized fermentation medium, issues in viable yields and high cost [20].

The marine food wastes, which are generally discarded, were found to be significant sources of sundry enzymes like trypsin, alkaline protease, esterase, etc. The wide temperature stability and activity (50–60°C) of alkaline proteases from marine organisms have fetched special importance in industrial segments like food and laundry [7, 28]. Trypsin (a type of serine protease, which cleaves peptide bond at the carboxyl side of lysine and arginine amino groups) belonging to aquatic organisms are active and stable at pH of 7.5–10, making them apt for commercial purposes like detergent formulations. Also, the specific and controlled proteolysis has enhanced the use of this enzyme in many biochemical and industrial applications. Fish trypsins are of immense interest over mammalian counterparts due to higher catalytic activity [50]. Similarly, esterases are a group of hydrolyzes, which catalyze ester bonds' cleavage and formation.

Esterase and lipase enzymes both participate in inter-esterification, esterification, and trans-esterification reactions but differ in substrate specificity and interfacial activation, which are prime features of lipases only. Lipases favor water-insoluble substrates while triglycerides based fatty acids of a length shorter than six carbons or simple esters are substrates for esterases. Many industries responsible for processing organic chemicals, detergent formulations, oleochemicals, biosurfactants, etc. are dependent on these costly esterase enzymes. The fermentation technology (solid state and submerged) necessary for producing these enzymes require specific media, environment, and microbes. However, in view of sustainable goals, solid-state fermentation (SSF) has been preferred due to their working with residual and waste products such as marine fish processing wastes that are nutrient rich and serve as a potential source of energy and growth for microbes [19].

The common method for enzyme extraction used in various studies includes: (1) Waste homogenization, crude extract preparation and its treatment; and (2) Microbial interaction with waste by-products.

The specificity and activity of the enzymes are further enhanced by purification processes. The different purification techniques are: dialysis, ammonium sulfate precipitation, heat treatment, ultra-filtration, gel filtration chromatography, DEAE-cellulose, and Sephadex G–50 column chromatography, and molecular exclusion chromatography, etc. The homogeneity analysis and confirmation of the type of enzyme form the next step and involve the study of band patterns on SDS-PAGE, inhibitor study, N-terminal sequencing, etc. [50]. Besides these, the activity of the extracted enzyme is highly influenced by several factors like: pH, temperature, salinity, metal ions, surfactants, etc.

Ammonium sulfate precipitation, gel filtration chromatography, dialysis, and ultra-filtration methods were employed to purify Haloalkaliphilic Organic Solvent Tolerant Protease (HAOP) produced by Bacillus sp. APCMST-RS3 from marine shell wastes substrates, and it resulted in 8.49-fold purity and 28.62 U/mg activity of protease with a 22.66% yield. The molecular weight of the protease obtained was 40kDa by SDS-PAGE and zymogram analysis, in contrast to other alkaline proteases, which generally have weight near 15–30 kDa [57]. On the other hand, the alkaline protease generated by *Bacillus alveayuensis* CAS–5 showed a reduction in protease yield recovery with progressive steps of purification attributable to the removal of low activity proteases. The yield of the protease was 46.48, 15.88 and 12.86%with ammonium sulfate treatment, followed by DEAE-cellulose, and Sephadex G–50 column chromatography, respectively. The resultant purified protease (7.7-fold purity) had a specific activity of 518.78 U/mg. It was concluded that the microbial reclamation of marine wastes to produce proteases is a promising approach to the production cost and bio-resource utilization [3].

9.5.1.1.1   Case Study (Feasibility of Fish Wastes Derived Enzymes)

Conventional unhairing operation of the leather industry requires lime (10% w/w) and sodium sulfide (3% w/w) to remove hairs (present on goatskin) and inter-fibrillar material (beneath the skin) and to loosen the epidermis. However, the process adds on pollution load thereby becoming environment constraint process. On the other hand, strong proteolytic enzymes designed for this purpose have been less preferred by tanners due to their high production cost. Thus, adopting a naturally derived enzyme from wastes (such as biomass, marine or fish processing wastes, etc.) may be an economically feasible option.

The dehairing efficiency of protease enzymes derived from the fish wastes, as an environment-friendly alternative to a conventional process (lime sulfide treatment) and their feasibility as a low-cost depilatory agent was thus studied (Figure 9.3). It was observed that partially purified enzyme had a specific activity of 4683 U/mg of protein with pH of 8.0 and 30°C optimum temperature. After dehairing, 28 μg/mL of saccharides and 4300 μg/mL of proteoglycans were released in spent liquor. Also, at par with control leathers, the enzymatically treated leathers shrinked at 110±2°C. Complete hair removal from skin matrix was authenticated by histological and scanning electron microscopic examination. The leathers obtained by the enzymatic process followed the UNIDO norms in physical strength while significantly decrementing pollution loads in the waste stream. Nearly 90%, 81% and 88% reduction in BOD, total solids and COD were observed through this alternative process [81].

9.5.1.2   FUEL DEVELOPMENT

Increasing energy demand and usage of fossil fuels have posed a serious threat to environment owing to release of harmful gases. Also, the generated imbalance in energy management throughout the world has shifted focus to alternative fuel candidates, which can easily be derived from a contemporary biological process involving the conversion of biomass. Treatment of biomass has been performed through thermal, chemical, and biochemical conversions. A consortium of micro-organisms and pre-treatment (such as enzymatic saccharification, pyrolysis, fermentation, anaerobic digestion, etc.) are used for converting the complex macromolecules of waste to low molecule weight compounds (solid, liquid or gas forms) such as CH_4, bio-ethanol, biodiesel, hydrogen, and other volatile fatty acids.

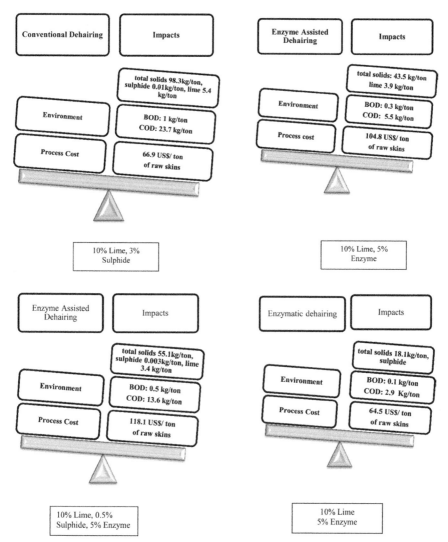

FIGURE 9.3   (See color insert.) The dehairing efficiency of protease enzymes derived from the fish wastes compared to a conventional process (lime-sulfide treatment).

9.5.1.2.1   Biodiesel (Liquid Fuel) Production

Bio-diesel is blended of fatty acid alkyl esters produced through catalyst-driven oil trans-esterification with methanol or ethanol. Similarly, bioethanol is a fuel derived from plant carbohydrates through a fermentation process and may be used as a substitute for gasoline. However, the competition for food sources and discrepancies developed in food-population ratio have raised serious concerns over the use of food and edible crops for biofuel production and has led to the evolvement of sustainable feedstocks for this purpose.

The second-generation bioethanol production includes the delignification process of lingo-cellulosic biomass affecting its commercialization. As a result, better, and viable feedstocks have been targeted as an alternative.

A good feedstock contains more volatiles with larger cellulose content, low lignin, and hemicelluloses, facilitating enzymatic saccharification and other pretreatments, lignin dissolution and transformation of crystalline cellulose to amorphous fractions. It has been observed that seaweeds or macro-alga possess higher generation and carbon resourcing capacity (3–4 times) compared to the terrestrial plant. Also, the capacity of bioethanol production potential for seaweeds is 23,400 L/ha/year, which is 23.16 times more than terrestrial plants (1010 L/ha/year). Production of carrageenan hydrocolloid from seaweeds generates a huge amount of waste biomass, which holds promising value as third generation feedstock for biofuel production. The commonly explored pre-treatment methods for biofuel production include high temperature-long time application, acid dilution, organic acid, lime, and hot water treatment, which incur high energy inputs and cost besides increasing environmental pollution burden.

Uju et al. [97] determined the conversion potential of SWBC (seaweed waste biomass from the carrageenan) industry to biofuel through peracetic acid–ionic liquid pre-treatment. The ionic liquids namely1-ethyl–3-methy-limidazole acetate (Emim OAc), 1-hexylpyridinium chloride ([Hpy Cl] and 1-ethyl–3- methylimidazolium diethylphosphate (Emim DEP) were compared in this study for their conversion efficiency. Different ionic liquids exhibited different reactions with cellulose-rich SWBC (33.6%). Emim OAc darkened the SWBC solution while Hpy Cl swelled the regenerated biomass. Also, yellowish, and whitish products were obtained by Emim OAc pretreatment and Emim DEP or Hpy Cl pretreatments. Cellulose hydrolyzing property was highest and lowest in Hpy Cl and Emim OAc pretreatments. Respectively; while PAA + Emim DEP generated high saccharide content. Nearly 77% cellulose conversion was observed for enzymatic saccharification of untreated SWBC. On the other hand, the enhanced glucose yields, moderate temperature, and pressure requirements and non-residual behavior throughout the conversion process have diverted the focus to the usage of hydrogen peroxide for the biofuel development. The efficacy and catalytic efficiency of this compound were explored by Li et al. [52] on *Ulva prolifera* biomass and solids.

The pronounced effect of pH on the stability and activity of the H_2O_2 was mitigated at 4pH (optimum and like cellulase and cellobiase enzyme reaction conditions), where 5.9 g/L sugar yield (maximum) was observed. Increased pH favored delignification by free-radical development. However, it also facilitated the degradation of H_2O_2 and was a critical parameter for the conversion process. A maximum bioethanol yield of 13.2 g/100 g of residue was obtained when the inoculums size was 10%, 50°C temperature, 12h pretreatment time and 0.2% H_2O_2 concentration at pH of 4. Similar optimization studies have exhibited 99.8% glucose and 55.0% bioethanol yield from seaweed solid wastes (carrageenan industry) at 4.8pH, 50°C temperature, 2% substrate concentration through SHF (separate hydrolysis and fermentation) process; while 90.9% bioethanol was obtained in the SSF (simultaneous saccharification and fermentation) process. SSF process thus saves time, cost, and energy throughout the process [46]. Hence choosing an efficient process is very important to attain sustainable goals.

The other factor determining the efficiency of biodiesel production is the type of catalyst. Slow reaction rate and corrosion phenomenon adhered with acid catalysts while primary catalysts often end up in the formation of soap and decreased biofuel yield. The catalyzed biodiesel often requires water washing step, which contributes further to toxic wastewater generation. The highly reactive metal oxides containing waste materials of steel making industries may serve as low cost, and effective catalyst for the above said purpose [46].

The effect of thermally treated industrial dusts (at varied steel manufacturing stages) such as waste grey dust (slab casting machine, steel converter), waste brown dust (steel converter), waste light pink lining breaking dust (steel converter), waste black lining breaking dust (steel converter), waste limestone and left-over waste off-white paint was analyzed as activated catalysts for biofuel generation from *Ulva fasciata* macro-alga, which develops biodiesel specific precursor triglycerides. The trans-esterification reactions were feasible at 80–100°C and took six hours for completion, with 88% conversion with waste brown dust followed by waste black lining breaking dust. Also, the properties of the fuel obtained exhibited ASTM standards with respect to free fatty acid composition, acid value, cloud point, ash content, kinematics viscosity, etc. [46].

Another crucial factor is harvesting of algal biomass, which despite of its inexpensive nature, limits the downstream applications like biodiesel production, bio-refineries, etc. Microalgae harvesting is performed commonly through filtration, sedimentation, chemical flocculation, centrifugation, co-inoculation with floating agents, etc. because of its small size, but also causes high-cost involvement and pollution load.

Utilization of *Peranema trichophorum* (chlorophytes predator) as a potent flocculating agent for *Chlamydomonas reinhardtii CC–125 cells* showed that 1×10^7 *Chlamydomonas* cells were effectively flocculated by 0.5ml culture filtrate of *Peranema*. The one-liter culture preparation of *Peranema* amounts to less than 1.00 $, making it much economical [82]. Similarly, increasing the concentration of lipids and thereby doubling the biocrude yield may also take place by growing algae hetero-trophically than photosynthetically though the former growing parameter will add-on cost, which could be effectively countered by fuel yield [1].

The deciding factor for the ultimate use of the developed biofuel depends on its chemical composition and its agreement with diesel oil and gasoline fractions of petroleum. For example, bio-oil obtained from fast pyrolysis of waste fish oil contained 4.48% paraffin, 8.31% iso-paraffin, 26.56% olefins, 6.07% naphthenes and 16.86% aromatics. The high amount of olefins and carboxylic acid chromatographic pattern present in the bio-oil matched with that of diesel oil. It has been observed that analytical techniques such as gas chromatography (mass spectroscopy and flame ionization detector) and nuclear magnetic resonance (carbon–13 and proton) were beneficial in characterizing and determining bio-oil properties [109].

9.5.1.2.2 Gaseous Fuel Production

Finite and non-environmental friendly nature of fossil fuels has always been a matter of concern when aiming at sustainable goals. Hydrogen gas which has been considered as a clean source and promising alternative, on other side, depends either on water electrolysis or on steam reformation. But still, it has gained immense importance as hydrogen gas on combustion produces water and heating value of 42kJ/g. Hence, identifying a novel and green approach for the production of this promising fuel has become necessary.

Xia et al. [110] have utilized anaerobic fermentative bacteria to convert mannitol present in seaweeds to hydrogen effectively. In comparison to traditional feedstocks, the marine feedstock offers advantages of easy fermentation (lack of lignin and hemicellulose), minimum substrate pretreatment and high biomass productivity. On the other hand, the marine algal feedstock comprises of mannitol (simple alcoholic sugar, high solubility in water, 20–30% dry weight) as carbohydrate monomers compared to glucose and xylose sugars present in terrestrial plants, because of which its fermentation in anaerobic conditions becomes difficult. Another consideration emphasizes while diverting seaweeds for hydrogen production is the presence of high sodium and proteins, which may further lead to the excess release of ammonia and ammonium salts. It was observed that the conversion efficiency for mannitol-based fermentation was 96.1% ± 7.4%, out of which 17.2% ± 0.4% is hydrogen, 38.3% ± 2.2% is butyric acid and 34.2% ± 3.7% is ethanol. Also, an optimal run gave a specific hydrogen yield of about 224.2 mL H_2/g of mannitol. Hence the seaweed waste may be utilized to produce fuels through this green technology. Similarly, CH_4 is another energy source which has 35793 kJ/m^3 heating values at 101.3 kPa and 0°C (equivalent to 0.76 kg standard coal). And can be produced effectively through anaerobic digestion method.

Anaerobic digestion process provides advantages of working with different temperatures and high organic loads, recovering nutrients essential for plant growth and stabilization after composting. The efficiency of this waste conversion process may further be enhanced by co-digesting different wastes, proper microbial selection (with resistance to inhibitors) and effective removal of toxicants. It was observed that combining strawberry waste with fish waste added organic matter to the mixture and thereby diluted the inhibitors' concentration. Also, an organic load of 22.8–50.6 kg of waste mixture/(m^3 d) generated a CH_4 mean yield of 120 mL/g total volatile solids (VS) at 101.3 kPa and 0°C [84]. In another study, co-digestion of fish waste and sisal pulp (33%: 67%) generated 0.62 m^3 of CH_4/kg VS while a 0.32 and 0.39 of m$^3CH_4$/kg VS were obtained when sisal pulp and fish waste alone were digested [62]. The observed 59–94% increase during co-digestion exhibits high potential of these processes for green fuel production.

9.5.1.2.2.1   Case Study

Iran generated 4.1tcf (trillion cubic feet) natural gas and consumed 4.2tcf in 2008. Also, the increase in population and economic use has further resulted in higher electricity and fuel consumption. To fully meet Iranian needs while maintaining export, projects focusing on renewable energy sources have been undertaken, and the generated fish wastes (299,128 t in the Persian Gulf, 32,533 t in the Caspian Sea, the total rate in 2003 reached 441,836 t) may prove beneficial. In Iran, a machine has been manufactured to separate fish oil from fish wastes. It was observed that 0.8 liters of fish oil are extracted from 7 kg fish wastes through this machine. Also, the biodiesel produced from each liter of fish oil amounts to 0.9 liters. The Cetane number of the generated biodiesel has been determined by the fatty acid composition of oil (i.e., stearic acid, linoleic acid, linolenic acid, palmitic acid, palmitoleic, and oleic acid) [113].

9.5.1.3   ANIMAL FEED AND MEDIA FOR MICROBES

Kelp has been used for various purposes like fertilizer, subsidiary food, the source of alginate etc., leaving behind enormous solid residues rich in crude fiber, residual alginic acid, protein, massive soluble sugars, minerals etc., which when broken down and degraded will release abundant organic nutrients for microbial growth and other cultural purposes. Besides the richness, the seaweeds extracts hold promising bio-stimulation property, making it more suitable for culturing microalgae for biodiesel production.

Study focusing on less cost-intensive method for algal lipid production and on effect of these wastes (8% kelp waste extract, KWE) on four different algal strains (namely, *Chlorella-Arc* (Arctic Chlorella sp.), *S. maxima*, *P. tricornutum* and *C. sorokiniana*) showed an increment of 20.78–25.91% and 1.83–31.86 times in their total lipid content and overall productivity, respectively. It was observed that these wastes mainly possessed nitrogen as a major macronutrient, reducing sugar (84.30% of total sugars), trace elements (15.03mg/L). And higher biostimulant activity of these wastes (8%) was observed for *Chlorella-Arc* (biomass productivity over control, 182.75%) and *C. sorokiniana* (biomass productivity over control, 31.86 times). On another hand, 1% KWE enhanced the *P. tricornutum* cell density over control by 77.55%. These variability results exhibited variable tolerance in algal strains against organic constituents of kelp waste. Also, kelp waste extract had a positive correlation with the amount of neutral lipid content of *Chlorella-Arc*, *S. maxima*, and *C. sorokiniana* while highest value of the same for *P. tricornutum* was obtained at 4% KWE. Additionally, KWE stimulated microalgal biodiesel properties by increasing short-chain fatty acids and monounsaturated fatty acids with simultaneously decreasing degree of polyunsaturated fatty acids (PUFAs) [119].

Marine wastes have also been used as feeds for animals. However according to European Union (EU) guidelines, following precautions must be taken while incorporating these wastes for animal diets:

- First, the year of production and the area of production results in variability in their composition.
- Secondly, the richness of moisture content favors its microbial spoilage and hence, it becomes much essential to avoid spoilage and contamination to prevent animals from catching diseases and sickness.
- Thirdly, it is also important to maintain the digestibility of feed through heat treatment while maintaining nutritive richness and cost of the final product.

Fish waste comprising 1% crude fiber, 19% ether extract, 22% ash and 58% crude protein was used for diet formulations for pigs. The diet was administered at 20%, and the feed impact was compared with control samples (pigs fed on a normal diet). It was observed that fish waste worked as a rich source of minerals mainly calcium, phosphorus, and sodium. Presence of $73.85 \pm 2.45\%$ moisture content exhibited the requirement of giving prior heat treatment to wastes before administering it to pigs. It was also stated that fish wastes provided a large amount of fats to pigs but providing a complete diet with other ingredients will prove much more beneficial than a single feed material. The digestible energy obtained from 1 kg of fish waste was 3540 kcal whereas the cost of drying was estimated to be 0.03 €/kg. It was concluded that an increase of 20% might occur in price, but, converting fish waste to feedstock will benefit the environment and public [21].

In another study, fish flour used as a feed for tilapia feed was substituted with shrimp head waste. Four dietary levels 0, 33.3, 66.6 and 100% was administered to juveniles. It was observed that nutritional composition (except protein) of the pellets exhibited similarity with a control feed (commercial fish feed). No significant variation was observed in weight, length, and growth of juveniles demonstrating the promising future of this waste as fish feed [65].

The waste generated by marine processing industry may further be converted to a fodder through a fermentation process before feeding it to animals (Figure 9.4). This process is called as ensilage, silaging or ensiling and the product generated is called silage. The wastes of fish when subjected to acidic medium and homogenized to maintain acidity results in the formation of a liquid paste called as fish silage. However, the product does not become feasible, firstly due to the cost involved in managing strong acids and secondly due to the non-nutritive nature of the developed product. Hence biological alternative such as lactic acid bacteria (LAB) based fermentation has been explored.

A different LAB like *Lactobacillus plantarum* [71], *Lactobacillus brevis* [96], yogurt-bacteria such as *Lactobacillus bulgaricus* and *Streptococcus thermophilus* [115], native LAB from the digestive tract of freshwater carps [29], as well as *Lactobacillus buchneri* and *Lactobacillus casei* [100] are commonly used for fermentation. Proteolytic enzymes present in the fish hydrolyze protein and fat make it highly digestible. The probiotic characteristics of inoculums, controlled fermentation conditions, environment-friendly and cost effectiveness are some of the favorable parameters responsible for the acceptance of this technique.

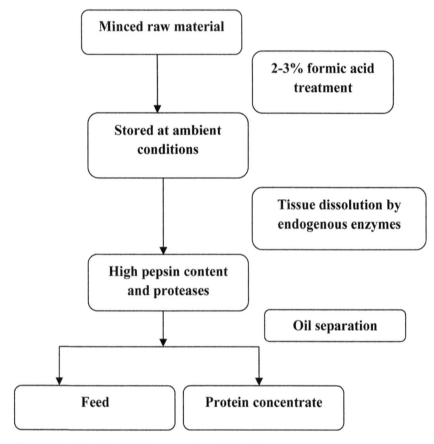

FIGURE 9.4   Procedure for the preparation of fish silage.

The viscera by-products of ray fish, swordfish, and shark along with glucose carbon source were inoculated with different LAB strains; and their growth pattern and metabolites dynamics were evaluated. The shark viscera-based media facilitated higher glucose and protein consumption while rayfish media favored high metabolite production, i.e. lactic and acetic acid. Thiobarbituric acid reactive substances (TBARS, 15 mg/L) and soluble protein (50g/L) were obtained at 216h culture stage. It was also observed that fed-batch culture and re-neutralization resulted in lower yields compared to batch culture [99]. Fermentation of freshwater fish viscera resulted in a recovery of 85% of lipids (19–21%) present. It was observed that factors like type of lactic acid cultures and the degree of hydrolysis (highest 62.3%) did not increase oil recovery. It was also concluded from the study that oil refining forms a necessary requirement after fermentation step and scaling studies must further be carried out tofurther assess the benefits of this method [72] further.

9.5.1.3.1   Case Study

Fish waste in comparison to urban waste is nutritive in nature and may support survival and breeding of many individuals. Kelp gulls (*Larus dominicanus*), which are third-ranked seabird in abundance, includes human activities derived food like fish waste, processing waste at coastal regions, etc. in their diet due to their opportunistic feeding habits. Gulls feeding on fish waste have favored a population expansion. It was also stated that the fish waste generated at three coastal cities of Argentina, i.e. Rawson, Comodoro Rivadavia and Puerto Madryn (6.4, 19.3, 24.1 thousand tons y^{-1}, respectively) is sufficient to support 101,000–209,000 Kelp Gulls [116].

9.5.1.4   FERTILIZER AND SOIL AMELIORATERS (BIOCHAR)

Conversion of biomass to charcoal-like material, i.e. biochar through slow pyrolysis and its application to soil aids in carbon sequestration. Besides this, biochar also facilitates nutrient retention, enhanced water retention and improved soil fertility. The main factors determining the stability and agro-economic utility of biochar are the type of feedstocks and the production conditions, for example, high ash content affects decomposition rate of biomass while high pyrolysis temperature adding up carbon adversely affects the agronomic benefits.

The richness in minerals and other nutrients make marine waste as suitable fertilizers and soil ameliorates. It has been observed that biochar produced from marine seaweed are rich in molybdenum, zinc, potassium, calcium, magnesium, phosphorus, nitrogen, and other exchangeable trace elements with 30–35% carbon content. Studies on *Ulva ohnoi* seaweed have shown that rinsing of biomass increases the carbon content of biochar while increasing pyrolysis temperature adversely affected it. The greatest gravimetric carbon sequestration of about 110–120 g stable C kg^{-1} seaweed was obtained when unrinsed biomass was pyrolyzed at 300°C. Approximately three-fold growth was observed in plants when unpelletized biomass-based biochar was used whereas the germination and growth suppressed with pelletized biomass-based biochar. It was concluded that an effective use of biomass would be obtained by thoroughly understanding the parameters of biochar production [78].

In another study, it was observed that activation conditions for biomass also play a role in final product quality. Activation of biomass through microwave generated ultramicropores while the use of electrical convention furnace resulted in medium-size micropore and mesopore formation. On the other hand, conventional activation of algal meal resulted in the formation of best-activated carbons [26]. The decomposition of fish waste by mixed cultures (5.84×10^5 CFU mL^{-1}) also resulted in amino acid content of 6.91 g 100 g^{-1} and a fertilizing ability like commercial products. Also, the presence of 1% lactate in the same prevented the putrefaction of decomposed broth for 6 months [14]. The potential of marine waste effluents as a fertilizer and thereby in providing essential nutrients to plants mainly *Juncus roemerianus* and *Spartina alterniflora* were estimated [39]. In comparison to unfertilized plants, the response of *Juncus roemerianus* towards finfish and marine shrimp solids was high in terms of biomass and nutrient allocation.

9.5.1.5   *FISH PRODUCTS AND DERIVATIVES: FISH OIL, FISH SAUCE, AND FERMENTED FISH PRODUCTS*

Fermentation is a common food preservation practice to improve nutritional and functional properties of foods. The fermented food products possess enhanced biological properties owing to their rich peptides and amino acids composition [73, 83, 90]. Fish sauce is fermented liquid product prepared by salted fish material at an ambient temperature in closed tanks [17]. One part of salt added to three parts of fish raw material when stored for 6–12 months at ambient tropical temperature undergoes fermentation and produces a fish sauce. Enzymatic and non-enzymatic reaction during fermentation produces a volatile compound that contributes to characteristic flavor and color. The endogenous and microbial enzymes favor protein hydrolysis and then form an amber colored liquid (fish sauce) possessing digested proteins (8–14%) and salt (about 25%). Other than fish, red crab shell, crayfish shell, shrimp by-products, and scampi by-products were used for fermentation.

Another excellent source for generating salt-fermented sauce products are by-products of shrimp processing industry, i.e. head, shell, and tail portions. These portions also comprise of large amounts of nutritive components, extractives, etc. [36, 47, 49]. Other fermented shrimp-based foods developed in the different region are Kapi (i.e., traditionally fermented paste of planktonous shrimp or krill) [68] and Jaloo (indigenous salted krill product), and Koong-Som by small shrimp, etc. Fermentation process through simple but requires sophisticated equipments and large storage space [31] for solubilization and development of desired characteristics in end-product (flavor and color).

Another edible product developed from a marine source and widely used is fish oil. The richness of these products in ω-3 polyunsaturated fatty acids (n–3 PUFAs) such as EPA and DHA known to possess numerous health benefits make these marine lipids unique compared to other lipid sources. This gives fish oils especial importance as commonly used dietary supplements. In addition to that, it is also an option to enrich other food products like egg, bread, margarine, etc. with n–3 long chain fatty acids. However, the application of these PUFAs in food products has been limited due to their susceptibility to lipid oxidation of PUFAs, flavor reduction and decremented nutritional value. Though the oxidation problem is dealt by the addition of widely used synthetic antioxidants but searching natural alternatives for the same will prove beneficial for consumers.

9.5.1.6 CHITIN AND COLLAGEN

Chitin, a glucose derivative, is a long polymeric chain of N-acetylglucosamine, occurring abundantly in nature (Figure 9.5) [92, 112]. It is odorless, tasteless, white or yellowish in color, hydrophobic in nature having excellent biocompatibility and biodegradability [10, 74]. It is a major cell wall component in crustraceans (exoskeleton), insects, cephalopods' shells, fishes' scales, etc. These inelastic, hard, and white polysaccharide share property (like cellulose) such as low solubility, chemical reactivity, and immunogenicity. The chitins present in marine organisms especially crustaceans hold promising economic value due to its versatile biological activities, adherence with proteins and nutrients, and applications in the agrochemical field [16, 40]. Commercially, chitin is obtained from crustacean shells of crabs and shrimps, which are among the main marine food processing wastes [44]. Traditional chemical process using NaOH for deproteinization and HCl for demineralization is expensive and environmentally hazardous. Deproteination process based on enzymatic degradation of chitin, i.e. bioconversion, uses protease enzymes obtained through fermentation and is more eco-friendly than a chemical process. It has been observed that 97.1 and 92.8% deproteinization and 95 and 92% demineralization respectively occurred during bioconversion of shrimp shell wastes [88].

FIGURE 9.5 (**See color insert.**) Marine sources of chitin and its structure.

Collagen is the most prominent protein found in the living body. Skin, bone, and scale found in marine food residues and waste mainly possess structural proteins such as collagen. Being of non-hazardous nature and richness in essential biological nutrients with water retention potential makes them suitable for application in cosmetics, artificial organs, etc. Morimura and his coworkers developed [60] enzymatic hydrolysis process to use collagen from fish and livestock waste. In this method, acidic conditions (pH 3), 0°C temperature and 1 hour duration was used for extracting high molecular weight protein. Isolation of collagen from fish waste (such as fins, material skin, and bone) was performed by Nagai and Suzuk [63], who reported the highest amount of collagen from these resources.

9.5.1.7 CAROTENOIDS AND GLYCOSAMINOGLYCANS

Carotenoids (Figure 9.6) are class of widespread natural pigments complexed with proteins and virtually occurring in all organisms. Besides participating in biological events and imparting color to animals (such as crustaceans), these pigments are suitable as health supplements and additives for animal feed [8]. The null toxicity, natural originality, high versatility, pro-vitamin A activity and recognizable vitamin-like qualities along with variable colorants (lipo- and hydro soluble), make shrimp and other crustaceans' carotenoids highly attractive and desirable. Extraction of these pigments from waste has been performed through different techniques including: traditional solvent extraction, supercritical carbon dioxide extraction (SC-CO$_2$) and enzyme treatment. However, the effective organic solvent extraction method often involves application off extreme heat (for solvent removal) which ultimately results in degradation of heat labile components.

Also, the presence of trace solvent in the product renders it into the non-natural category. On the other hand, the SC-CO$_2$ method overcomes the disadvantage of the solvent extraction method but proves less effective in total carotenoid extraction mainly due to low carotenoid solubility. Babu and his coworkers [4] treated shrimp head waste (wild species of *P. indicus*, *M. monocerous* and *P. monodon*, and cultured *P. monodon*) with different proteolytic enzymes such as pepsin (pH 4.0), trypsin (pH 7.6) and papain (pH 6.2) to maximize the yield of the carotenoids. The recovery percent varied with the type of enzyme and raw material used, highest being produced from trypsin and *Penaeus indicus*. Also, the processed freeze-dried product developed from the frozen protein-carotenoid cake exhibited higher stability of astaxanthin compound (oxidative product of β-carotene), loss of minor carotenoids like β-carotene (due to oxygen and light sensitivity, lipid peroxidation, etc.) and natural antioxidant properties.

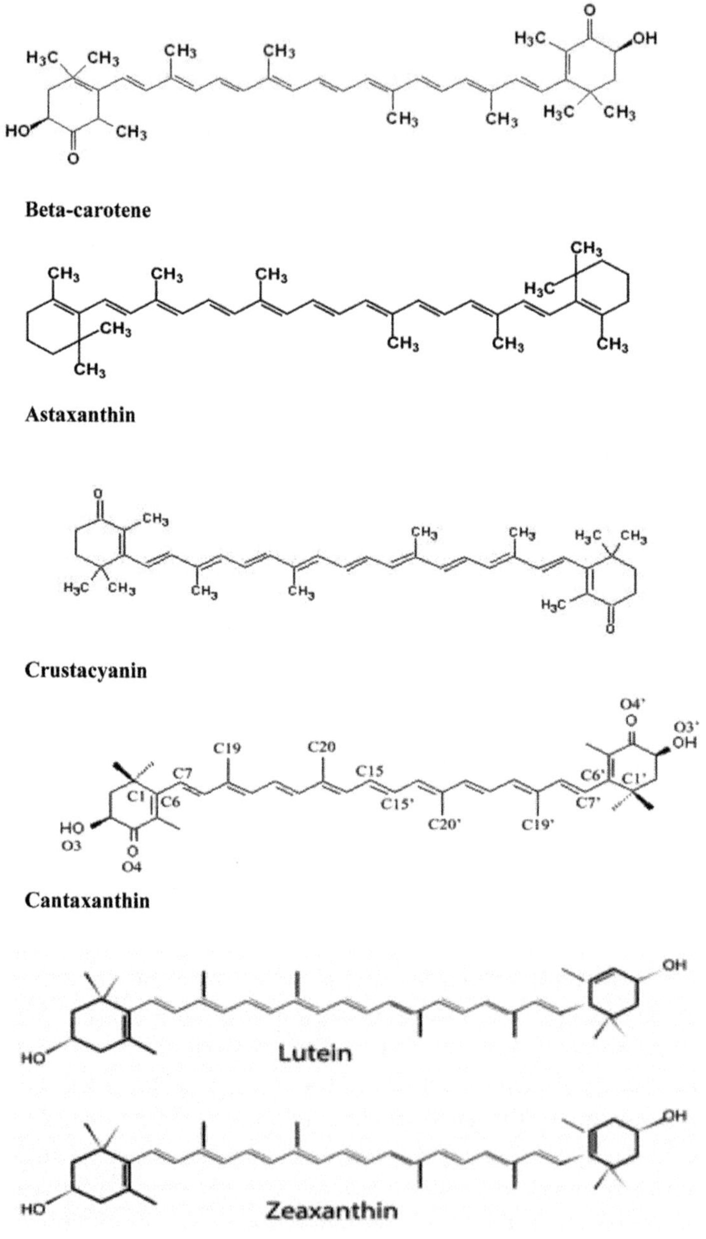

Beta-carotene

Astaxanthin

Crustacyanin

Cantaxanthin

Lutein

Zeaxanthin

FIGURE 9.6 Different types of carotenoids.

Glycosaminoglycan (GAG) also known as mucopolysaccharides are made up of a repeating disaccharide unit of amino acids (N-acetylgalac-tosamine or acetylglucosamine) and uronic (glucuronic acid or iduronnic acid) or galactose sugar, giving it a long-branched structure. The synthesis of these compounds is modulated by processing enzymes involving sulfation, disaccharide construction and changes in molecular mass, which results in high heterogeneity. Except for hyaluronic acid groups, all other three groups of GAGs (namely, heparin/heparan sulfate (HSGAGs), keratan sulfate and chodritin sulfate/dermatan sulfate (CSGAGs)) are sulfated by the help of specific sulfotransferases.

Sulfated GAG is involved in different numerous biological activities, such as: protein-cell surface binding, cell growth, its proliferation and development, anti-coagulation, tumor metastasis, angiogenesis, wound repair, brain development, neurogenic activity, tissue hydration and controlled buffering, regulation of macrophage adhesion and embryo implantation, cell motility, etc.

Recently, shrimp heads have been targeted extensively for extraction of these compounds, whose mammalian analogs hold potential influence on circulatory systems. Integrated recovery and further characterization of this compound along with other materials such as protein, carotenoids, and chitin were performed using enzyme autolysis by endogenous shrimp peptidases, deproteinization, precipitation, and deacetylation, providing it economic feasibility. The process generated 48.1g of protein-calcium concentrate, 44g of lipid concentrate, 85g of precipitate (left after ethanolic extraction), 53g of carapace content and 1.5L of a protein hydrolysate.

Astaxanthin and other carotenoids formed 111.96 and 82.56 mg/kg, respectively, of concentrated ethanolic extract of shrimp heads. Also, 25g of chitin and 17g of chitosan were obtained whereas GAGs amounted to 79 ± 2 mg/kg of the processing waste. The dry matter content of protein hydro-lysate, which may be used for the food flavor and functional ingredients formed 120 ± 0.4g in 9% w/v solution. The use of non-economic enzymes, industrial centrifuge, and integrated approach may generate all the necessary and beneficial compounds from shrimp waste easily in a 1000 L reactor vessel [8]. However, the authors of the study also emphasized the critical role of the shrimp physiology (temperature, salinity, food availability and other seasonal changes)on the enzyme content and composition of shrimp heads and body, which may be mitigated by controlled feed and saline conditions during culturing.

9.5.1.8 BIOACTIVE COMPOUNDS

Extra nutritional constituents with promising health benefits and present in small quantities in foods have been termed as "Bioactive compounds." It has been observed that many constituents in marine food processing residues and waste possess significant biological properties with nutraceutical and pharmaceutical importance. These marine bioactive compounds (such as essential fatty acids, pigments, proteins, polyphenols, carbohydrates, peptides etc.) are natural and have biological functions like anti-inflammation, anti-hypertensive, cardio-protection activities, anti-diabetic, anti-coagulation, anti-thrombotic, and anti-oxidant properties and thus prove safer and lucrative in nutraceuticals and pharmaceutical sector compared to prevalent synthetic drugs. This observed richness and presence of both vitamins, minerals, PUFAs, essential, and non-essential amino acids, polysaccharides, and many other nutrients [43, 101] are found in diverse marine organisms including fish, shellfish, mollusks, cephalopods, crustaceans, echinoderms, seaweeds, and microalgae. Further details on different compounds isolated, their synergetic effect and method of extraction were discussed in a chapter specifically focusing on 'bioactive compounds' in this book.

9.5.1.9 BIOFILMS

Gel-forming ability and film foaming capacity [75] of proteins from marine sources make them suitable for food application. Gelatin from collagen present in the marine source is formed by partial hydrolysis process. It possesses good water holding capacity, gel-forming ability, food product stability, and texture improvement property and hence is used as a food additive [79]. Marine polysaccharides also exhibit similar properties of binding water molecules and dispersing it in food products [5] and hold an important place in the processing industry. For example, confectionary industries utilize agar due to its bland taste, high sugar content and non-flavor nature. It is thus utilized to make custards, jams, jellies, fruit candies and puddings.

Attributes such as foaming, emulsifying, and water binding, which are suitable for modifying textures of diverse food products, are present in carrageenans. The textural modifications in food occur on the interaction of these polysaccharides with other food components [118]. Besides this, marine chitosan, alginate, and fucoidan act as ideal raw materials for developing edible and biodegradable films exhibiting another aspect for utilizing marine food processing residues and wastes. Chitin, chitosan, and their derivatives are also used as nutraceuticals (e.g., increasing dietary fiber, reducing lipid absorption), antimicrobial agents, additives, and edible films [27] and have a variety of food applications.

9.5.1.10 AQUEOUS PURIFICATION

Industrial application of rare element (Lanthanides and heavy metal ions) in metallurgy, ceramic industry, etc. and associated fluid disposal on the river, barrel, and agricultural land leads to bio-accumulation of these elements which cause environment issue (toxicity to the environment and regulating plant growth). Higher cost, the toxicity of element and long-term impact on environment engender the need of apt and eco-friendly treatment method of industrial waste.

Marine food biomass (marine algae, crab carapace, and clam shells, etc.) were identified as potential to absorb various metal ions and used in aqueous purification. Seafood waste, crab carapace, and clam shells have a high affinity toward heavy metals and used for their removal from industrial wastewater [47, 59, 61, 70, 102, 104, 106]. Seaweeds (marine algae) were also identified as potent materials to absorb heavy metal ions and lanthanides due to its cell wall composition (alginate in brown algae is responsible for high metal binding).

Studies have been conducted on seaweeds for the removal of La, Ce, Eu, Pr (III) and Yb from waste fluid [16, 103, 105]. Even with high sorption capacity, the major drawback is swelling of seaweeds during biosorption process. The details of studies focusing on the uses of the marine food waste are given in Table 9.3.

9.5.2 DISPOSAL MANAGEMENT AND MONITORING

Despite of the advancements occurring in the utilization of marine food waste, a large amount of it is still dumped and remains as a cause for the environmental problem. Dumping of marine food waste (organic waste) in an inappropriate manner favors microbiological and parasite activity, disease, and release of toxic compounds into sea and land masses (hydrogen sulfide). Inert materials such as shell aids in habitat recreation owing to increased species diversity and prevention against erosion in intertidal habitats but may prove harmful when care is not taken about the location of the disposal site.

TABLE 9.3　Utilization of Marine Food Waste

Usage category	Targeted marine waste	Output	Challenges	Reference
Aqueous purifiers	Seafood wastes	Removal of priority hazardous substances (Hg and Cd) from water		[59]
Aqueous purifiers	Microalgae-containing microbiota	Fish processing wastewater treatment		[76]
Carotenoids	Shrimp head waste	Carotenoid-protein complex		[4, 8, 80]
Carotenoids	Giant tiger (*Penaeus monodon*) shrimp waste	Astaxanthin		[35]
Chitin	Shrimp heads; White Shrimp (*Penaeus vannamei*) Waste	High-quality protein hydrolysate, chitin, chitosan, carotenoids, sulfated glycosaminoglycans		[8, 87]
Chitin	Marine brown macroalgae	Alginic acid, mannitol, laminarin, fucoidin, cellulose		[41]
Chitin	Blue shark (*Prionace glauca*)	Chondroitin sulfate		[98]
Chitin	Echinoderm connective tissues; the skin of Brown stripe red snapper (*Lutjanus vitta*)	Collagen biomaterials; acid and pepsin-solubilized collagens		[25, 40]
Cryoprotectant	Hydrolysates from different marine sources such as fish, crustaceans, and cephalopods	Prevent protein denaturation in seafoods and seafood products		[64]
Drugs	Atlantic mackerel (*Scomber scombrus*) and its processing by-products	Antibacterial fraction		[18]

TABLE 9.3 (Continued)

Usage category	Targeted marine waste	Output	Challenges	Reference
Enzyme	Viscera of fish pirarucu (Arapaima gigas); yellowfin tuna (Thunnus albacores) spleen; processing waste of the silver mojarra (Diapterus rhombeus)	Trypsin	1) Strategy of protein purification is a limiting factor for such conversions 2) No molecular and biochemical characteristic's information reported for yellowfin tuna spleen proteinases	[28, 50, 86]
Enzyme	Marine crustacean shell waste: treated with Bacillus sp. APCMST-RS3; shrimp shell powder (SSP), crab shell powder (CSP) and squid pen powder (SPP); red scorpionfish (Scorpaena scrofa) viscera; midgut gland of shrimp, viscera of Colossoma macropomum	Proteases	1) Very less literature available on isolation and purification of Haloalkaliphilic Organic Solvent Tolerant Protease (HAOP) from estuarine sediment bacteria 2) Application of these HAOP in bioconversion aspects are very less discussed in the literature	[3, 6, 8, 20, 57, 117]
Enzyme	Different fish processing waste (wastes of ray fish, red snapper, tuna, and red grouper) with Bacillus altitudinis AP-MSU	Esterase		[19, 69, 84]
Feed and media	Kelp waste extracts	Growth and lipid accumulation of microalgae		[119]
Feed and media	Rainbow trout processing waste	Formulated diets for Mozambique tilapia Oreochromis mossambicus		[32]
Feed and media	By-products from Cuttlefish (Sepia officinalis) and wastewaters from marine-products processing factories; fish waste	Media for protease producing bacterial strains; Bacillus proteolyticus CFR3001		[6, 89]

TABLE 9.3 (Continued)

Usage category	Targeted marine waste	Output	Challenges	Reference
Feed and media	Shellfish chitin wastes	Plant growth by Bacillus cereus		[9]
Fertilizer and soil ameliorators	Roots of *Undaria pinnatifida* (Marine macroalgae);	Biochar	1) biochar derived at 600 and 800°C results in increased soil salinity 2) detailed research required on long-term field applications	[41]
Fertilizer and soil ameliorators	Fish waste	Liquid fertilizer	1) pH values are affected by dissolved oxygen and are a critical consideration point for scaling-up	[14]
Fertilizer and soil ameliorators	marine aquaculture solid waste; seaweed and fish waste	nursery production of the salt marsh plants *Spartina alterniflora* and *Juncus roemerianus*; compost for horticulture		[39, 107]
Flavor	Enzymatic hydrolysates of fish by-products	Seafood flavor formulations		[67]
Fuel	marine organic fish waste solids; fish and strawberry residues	Biomethane	1) Understanding microbial community composition 2) Bioconversion affected by high sulfate concentration in marine sludge	
Fuel	Fish industry waste	Biogas		[42]
Fuel	New England fishery waste	Biodiesel		[66]
Fuel	Freshwater and marine algal biomass	Biocrude		[51]

TABLE 9.3 (Continued)

Usage category	Targeted marine waste	Output	Challenges	Reference
Fuel	Spent seaweed Biomass	Bioethanol		[91]
Proteins and hydrolysates	Shrimps cephalothorax, carapace, tail, and remaining tissues	Soluble protein		[8]
Proteins and hydrolysates	Waste liquor of processing sea cucumber	Oils, saponins, proteins, and polysaccharides simultaneously		[11]
Proteins and hydrolysates	Prawn-shell Waste	Single Cell Protein		[77]
Proteins and hydrolysates	Shrimp (*Pandalopsis dispar*) processing byproducts	Enzymatically-produced hydrolysates		[12]
Proteins and hydrolysates	algae protein waste hydrolysate	new anti-oxidative peptide		[85]
Salt	*Ulva ohnoi* and *Ulva tepida*	Seaweed salt		[55]
Sauce	squid processing by-products	low salt fish sauce		[111]
Silages and oils	rainbow trout processing waste	Silage oil		[42]
Silages and oils	Fish industry waste	FW silages were prepared by mixing FW with bread waste (BW) and brewery grain waste (BGW)		

Leaching of harmful chemicals from the disposed waste may affect the groundwater and make it toxic. The dumping site may also release bad or foul odor affecting nearby living organisms adversely including humans. Transmission of diseases may also take place when inter-regional trade occurs, and the waste is disposed-off at a new place. Hence it is advisable to follow appropriate dumping measures and waste management to minimize these impacts. Various legislations have been laid such as the "1992 OSPAR Convention" and the "1972 London Convention", "Food and Environment Protection Act 1985 – Deposits in the Sea (FEPA II)", etc. that focuses on prohibiting waste dumping into the ocean and sea bodies. It is also recommended to use a collaborative approach between the industry and the regulators with input from scientific, technical, and economic expertise to mitigate this problem [58].

9.6   POTENTIAL USE OF MARINE FOOD WASTE

The conversion of marine food wastes to a useful form requires various chemical procedures and critical steps. The feasibility of these processes at a large scale is often governed by various factors like: cost involved throughout the process, conversion efficiency of the method, scaling up parameters, desired end product, etc. Also, care must be taken in assessing whether the alternative pathway chosen to curb the generated waste will be environmentally friendly or will further add up the burden. Hence, different assessment and analysis case studies have been conducted to determine the potentiality of waste management under mentioned circumjacent factors and are discussed here.

9.6.1   CASE STUDY 1: IMPACT OF PRODUCT DISPLACEMENT (PROCESSING AS AN ALTERNATIVE TO WASTE)

The impact of product displacement as well as management of salmon offal and processing waste on eutrophication, acidification, and climate change was assessed through gate to grave (from offal to offal deposition i.e. offal ready for management followed by the mass balance of the biomass leaving the system as ground and discharged offal, stick-water discharges, atmospheric discharges, and co-production) life cycle assessment (LCA) study in Alaska. Carbon dioxide equivalent emissions (CO_2e), sulfur dioxide equivalent emissions (SO_2e) and phosphate equivalent emissions (PO_4e)

were measured for assessing climate change, acidification, and eutrophication, respectively.

The functional units chosen for comparison were the management of 1kg offal and management of approximately 33,000 metric tons of offal intermittently produced in a South-east part of the country in 2010. Grind and discharge, fresh offal processing (conventional and Montlake type) and Mont lake stabilized offal processing management options were included in the scope of the study (Figure 9.7). The displacement was modeled by expanding the system and involving displacement factors such as gross energy and protein content, n–3 fatty acid content and gelatin strength, respectively for meals (animal and plant based), oils (animal and plant based) and pigskin (feedstock to gelatin production).

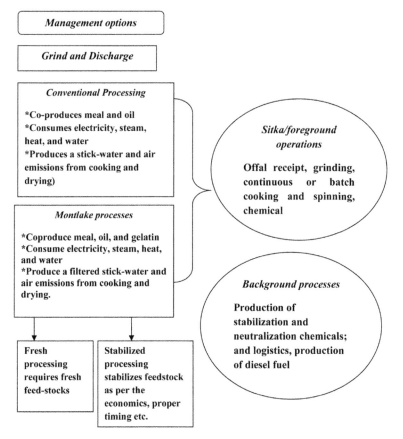

FIGURE 9.7   Management options studied in LCA for salmon processing waste generated in Alaska.

It was observed that in the absence of displacement, the processing practices proved less burdening on the environment compared to the grind and discharge management option. However, the resource addition in Montlake processing contributed in climate change in comparison to conventional processing. The difference between the impact levels of grind and discharge management and processing option (for 1 kg offal) deepened when system boundaries were considered. The stabilization processing, however, exhibited less difference from discharge management option due to increment in resources. The higher contributors to climate change on displacement came from poultry blood meal, poultry fat, and lupin. The contribution of all types of processing for acidification, eutrophication, and climate change ranged as -41 to -0.28 g SO_2e/kg offal managed, $21-10.6$ g PO_4e/kg offal managed, and 2.7 to $+0.12$ kg CO_2e/kg offal managed. On the other hand, $+15$ g PO_4e/kg offal managed contribution to eutrophication was observed for grind and discharge practices. It was also observed that parameters like neutrality, processing capacity, etc. had a more pronounced effect on eutrophication results compared to acidification and climate change.

The study helped in revealing the contributory role of foreground and background systems as well as a choice of displaced products in 1kg offal management. It also highlighted that the grind and discharge process is not completely masked due to processing and may occur in an accompanying form. Despite of favoring of system expansion and process division by ISO standards, a high possibility of mismatch occurs due to use of displaced meal, gelatin, and oil data. The drawback of the study was that it did not seem focused on upstream lifecycle implications and omitted clear majority of LCAs statistical considerations [13]. However, such studies will help in our understanding on how processing methods can reduce the environmental impacts of waste and which path may yield more benefits under limited resources. Hence, similar studies must be undertaken to fulfill a gap in our cognizance for waste management.

9.6.2   CASE STUDY 2: SEASONAL EFFECT

The type of biomaterial used as feedstock for biogas production is a very important parameter determining the yield and quality of the ultimate product. In a study on Irish brown seaweeds (*A. nodosum*), the specific biomethane yield/ weight of seaweed in October harvest was found triple compared with the December harvest. Also, the gross energy yield achieved from October harvest seaweeds amounted to 116 GJ ha^{-1} yr^{-1}. A low biomethane yield in summer months was observed due to low polyphenol accumulation. Ensiling of the harvested product was found to increase biomethane yield by 30% over fresh *A. nodosum*.

It was stated that up to 1.25% of Irish renewable transport energy target could be achieved by *A. nodosum* cultivation from 20,260 ha [93]. Similar seasonal effect was observed with *Laminaria digitata,* brown seaweed species, where 53 m^3 CH_4 t^{-1} yield (highest) was obtained in August compared to December harvest [94]. Substituting the cultivated seaweeds by the seaweed waste may be an alternative and the benefits incurred by producing biofuel needs to be investigated.

9.6.3   CASE STUDY 3: TECHNO-ECONOMIC CHARACTERIZATION FOR PROCESS FEASIBILITY

Economic evaluation plays a critical role in determining the productivity of an enterprise or process. Assessment of technical and economic characteristics of solid waste emerging from marine food processing and other parts of the supply chain thus becomes important.

A study was conducted in Indonesia focusing on utilization of catfish processing solid waste as per the national standards and on estimating the viability and associated environmental impact of this business. Development of this business resulted in a positive impact on local society through the development of related and associated income generating occupations like fish curing, fish filleting, diversification of processed products such as fish-based meatballs, burgers, nuggets, etc., but at the same time, negative impacts arose due to improper waste handling. Utilization of these waste to produce fish oil concentrate, pyloric caeca crude enzyme and bone powder were explored in laboratory experiments and the economic parameters i.e. net present value (NPV), profitability index (PI), benefit-cost ratio (BCR) and internal rate of return (IRR) were found to be 1.60, 2.60, 1.15 and 65.91, respectively, exhibiting feasible and profitable nature of converting catfish processing waste components to valuable product [56].

In another study, economic validity of two processes of bio-ethanol production (i.e., simple pre-treatment and combined pre-treatment from seaweeds) was assessed. The algae-based bio-plant, which are categorized as "waste reduction and resource recovery plants" in asset class 49.5 by U.S. Internal Revenue System (IRS) publication 946, with a capacity of 80,000 ton/ year and 400,000 ton/year undergoing simple and combined process will result in cost-effective bio-ethanol production from brown algae (based on maximum dry seaweed price). The minimum ethanol selling price (MESP) for the simple and combined processes as per the defined capacity was 2.39 and 2.85\$/gal, respectively.

Cellulose composition, laminarin changes and fermentation rate of alginate were the most determining factors for ethanol yield and MESP for both the processes. It was also concluded that the production process is promising economically but is often governed by availability and price [24]. Similarly, the techno-economic characteristics of anaerobic digester of plant size 1.6 MW_{th} and 8.64 tons per day (dry basis) macro-algae feed rate were evaluated. The gas produced from the system was used for the generation of 237 kWe_{net} electricity and 367 kW_{th} heat. The determined breakeven selling price for the electricity was €120/MWh. It was further concluded that systematic studies on cultivation, food supply chain, biogas conversion rates, energy-based applications versus food and cost reductions need to be performed for decrementing the future financial risks associated with this business [15].

9.7   CHALLENGES AND OPPORTUNITIES

The management of marine waste has been in the limelight due to serious sustainability issues. However, several challenges and constraints underlie during the process. It has been observed that very few estimates have been available for marine food wastes. The data available is either very limited or obscure giving only a glimpse of the associated problem. Quantifying waste levels thus becomes mandatory to design the reduction interventions effectively.

Also, very few studies have focused directly on identifying the impacts of generated waste. It has also been observed that lack of managerial and regulatory approaches are driven factors for further application of simple techniques available for reduction of waste loads. The type of utilization processes of the generated waste is further restrained by quality, freshness, and autolytic properties. Studies have shown that waste management possesses immense potential for mitigating the various negative impacts associated with the waste generation and dumping. Proper utilization of waste may aid in a country's economic, energy, and social need. However, determining whether the utilization process is environmentally friendly and promising itself is a major point of consideration and cannot be ignored. LCA is a tool for such studies. However, it itself works under some assumptions and scope boundaries, limiting its usage. An effective tool is thus required to estimate and analyze the situation properly. It has been observed that most of the studies have just begun and given preliminary information about waste utilization methods. Also, very few studies aimed at management and utilization of undefined and mixed type of waste (by-catch or harvested comprising all marine organisms). A huge hiatus still exists for developing our cognizance in this context.

9.8 SUMMARY

Consumption of marine food has increased tremendously owing to its richness in protein, lipids, PUFAs, and bioactive compounds, etc. The marine food production and processing sector generate a lot of waste. The direct dumping of these organic-rich waste has been found associated with many adverse impacts such as loss of nutrients, low groundwater quality, species diversification, euthrophication, bad odor, trans-boundary transmission of diseases, etc. and has become an alarming issue. As a result, sustainability goals, regulations, and legislation have been adopted by governing bodies to mitigate these issues and restriction on the direct dumping of this waste into sea and landmasses have been imposed. This chapter highlights the intensity of the problem (waste generation) and discusses the various ways in which beneficial products may be obtained from waste. The advantages of reutilization of waste and the profitability of the process in economic terms have also been discussed. Various case studies have been added to demonstrate the extent of steps and initiatives taken in the world against the arising problem. Many factors govern the waste to product conversion and need to be evaluated through further research.

KEYWORDS

- **antioxidants**
- **bioactive compounds**
- **biofilms**
- **biofuels**
- **carotenoids**
- **chitin**
- **marine food**

- **monitoring**
- **peptides**
- **protein**
- **seaweed**
- **shrimp waste**
- **sustainability**
- **waste utilization**

REFERENCES

1. Albrecht, K. O., Zhu, Y., Schmidt, A. J., Billing, J. M., Hart, T. R., Jones, S. B., et al., (2016). Impact of heterotrophically stressed algae for biofuel production via hydrothermal liquefaction and catalytic hydrotreating in continuous-flow reactors. *Algal Research, 14*, 17–27.

2. Alverson, D. L., Freeberg, M. H., Murawaski, S. A., & Pope, J. G., (1994). *A Global Assessment of Fisheries Bycatch and Discards* (p. 235). FAO Fisheries Technical Paper No. 339, FAO, Rome.

3. Annamalai, N., Rajeswari, M. V., & Balasubramanian, T., (2014). Extraction, purification and application of thermostable and halostable alkaline protease from *Bacillus alveayuensis* CAS 5using marine wastes. *Food and Bioproducts Processing, 92*(4), 335–342.

4. Babu, C. M., Chakrabarti, R., & Sambasivarao, K. R., (2008). Enzymatic isolation of carotenoid-protein complex from shrimp head waste and its use as a source of carotenoids. *LWT - Food Science and Technology, 41*(2), 227–235.

5. Berna, K., Cirik, S., Turan, G., Tekogul, H., & Edis, K., (2013). Seaweeds for food and industrial applications, food Industry. *InTech.*, pp. 735–748.

6. Bhaskar, N., Sudeepa, E. S., Rashmi, H. N., & Tamil, S. A., (2007). Partial purification and characterization of protease of *Bacillus proteolyticus* CFR3001 isolated from fish processing waste and its antibacterial activities. *Bioresource Technology, 98*(14), 2758–2764.

7. Bougatef, A., (2013). Trypsins from fish processing waste: Characteristics and biotechnological applications - Comprehensive review. *Journal of Cleaner Production, 57*, 257–265.

8. Cahú, T. B., Santos, S. D., Mendes, A., Córdula, C. R., Chavante, S. F., Carvalho, L. B., Nader, H. B., & Bezerra, R. S., (2012). Recovery of protein, chitin, carotenoids and glycosaminoglycans from Pacific white shrimp (*Litopenaeus vannamei*) processing waste. *Process Biochemistry, 47*(4), 570–577.

9. Chang, W. T., Chen, Y. C., & Jao, C. L., (2007). Antifungal activity and enhancement of plant growth by Bacillus cereus grown on shellfish chitin wastes. *Bioresource Technology, 98*(6), 1224–1230.

10. Cheba, B. A., (2011). Chitin and chitosan: Marine biopolymers with unique properties and versatile applications. *Global Journal of Biotechnology & Biochemistry, 6*, 149–153.

11. Chen, D., Yang, X., Cao, W., Guo, Y., Sun, Y., & Xiu, Z., (2015). Three-liquid-phase salting-out extraction of effective components from waste liquor of processing sea cucumber. *Food and Bioproducts Processing, 96*, 99–105.

12. Cheung, I. W. Y., & Li-Chan, E. C. Y., (2010). Angiotensin-I-converting enzyme inhibitory activity and bitterness of enzymatically-produced hydrolysates of shrimp (*Pandalopsis dispar*) processing byproducts investigated by Taguchi design. *Food Chemistry, 122*(4), 1003–1012.

13. Cooper, J., Diesburg, S., Babej, A., Noon, M., Kahn, E., Puettmann, M., & Colt, J., (2014). Life cycle assessment of products from Alaskan salmon processing wastes: Implications of coproduction, intermittent landings, and storage time. *Fisheries Research, 151*, 26–38.

14. Dao, V. T., & Kim, J. K., (2011). Scaled-up bioconversion of fish waste to liquid fertilizer using a 5-L ribbon-type reactor. *Journal of Environmental Management, 92*(10), 2441–2446.

15. Dave, A., Huang, Y., Rezvani, S., McIlveen-Wright, D., Novaes, M., & Hewitt, N., (2013). Techno-economic assessment of biofuel development by anaerobic digestion of European marine cold-water seaweeds. *Bioresource Technology, 135*, 120–127.

16. Diniz, V., & Volesky, B., (2005). Biosorption of La, Eu, and Yb using Sargassum biomass. *Water Res., 35*, 239–241.

17. Dissaraphong, S., Benjakul, S., Visessanguan, W., & Kishimura, H., (2006). The influence of storage conditions of tuna viscera before fermentation on the chemical, physical and microbiological changes in fish sauce during fermentation. *Bioresource Technology, 97*(16), 2032–2040.

18. Ennaas, N., Hammami, R., Beaulieu, L., & Fliss, I., (2015). Production of antibacterial fraction from Atlantic mackerel (*Scomber scombrus*) and its processing by-products using commercial enzymes. *Food and Bioproducts Processing, 96*, 145–153.

19. Esakkiraj, P., Usha, R., Palavesam, A., & Immanuel, G., (2012). Solid-state production of esterase using fish processing wastes by *Bacillus altitudinis* AP-MSU. *Food and Bioproducts Processing, 90*(3), 370–376.

20. Espósito, T. S., Amaral, I. P. G., Buarque, D. S., Oliveira, G. B., Carvalho, L. B., & Bezerra, R. S., (2009). Fish processing waste as a source of alkaline proteases for laundry detergent. *Food Chemistry, 112*(1), 125–130.

21. Esteban, M. B., García, A. J., Ramos, P., & Márquez, M. C., (2007). Evaluation of fruit-vegetable and fish wastes as alternative feedstuffs in pig diets. *Waste Management, 27*(2), 193–200.

22. FAO, (2014). Food wastage footprint: Fool cost-accounting. *Food and Agriculture Organization of the United Nations (FAO)*. http://www.fao.org/3/a-i3991e.pdf (accessed on July 31, 2017.

23. FAO, (2011). Review of the state of world marine fishery resources. *FAO Fisheries and Aquaculture*. http://www.fao.org/docrep/015/i2389e/i2389e.pdf (accessed on July 31, 2017).

24. Fasahati, P., Woo, H. C., & Liu, J. J., (2015). Industrial-scale bioethanol production from brown algae: Effects of pretreatment processes on plant economics. *Applied Energy, 139*, 175–187.

25. Ferrario, C., Leggio, L., Leone, R., Di Benedetto, C., Guidetti, L., Coccè, V., et al., (2015). Marine-derived collagen biomaterials from echinoderm connective tissues. *Marine Environmental Research, In Press, Corrected Proof*, 1–12.

26. Ferrera-Lorenzo, N., Fuente, E., Suárez-Ruiz, I., & Ruiz, B., (2014). KOH activated carbon from conventional and microwave heating system of a macroalgae waste from the Agar-Agar industry. *Fuel Processing Technology, 121*, 25–31.

27. Fiszman, S. M., & Salvador, A., (2003). Recent developments in coating batters. *Trends in Food Science and Technology, 14*, 399.

28. Freitas, A. C. V., Costa, H. M. S., Icimoto, M. Y., Hirata, I. Y., Marcondes, M., Carvalho, L. B., Oliveira, V., & Bezerra, R. S., (2012). Giant Amazonian fish pirarucu (*Arapaima gigas*): Its viscera as a source of thermostable trypsin. *Food Chemistry, 133*(4), 1596–1602.

29. Ganesan, P., Pradeep, M. G., Sakhare, P. Z., Suresh, P. V., & Bhaskar, N., (2009). Optimization of conditions for natural fermentation of freshwater fish processing waste using sugarcane molasses. *J. Food Sci. Technol., 46*(4), 312–315.

30. Ge, L., Wang, P., & Mou, H., (2011). Study on saccharification techniques of seaweed wastes for the transformation of ethanol. *Renewable Energy, 36*(1), 84–89.

31. Gildberg, A., (2002). Enhancing returns from greater utilization. In: Bremner, H. A., (ed.), *Safety and Quality Issues in Fish Processing* (pp. 425–449). Woodhead Publishing Limited and CRC Press LLC, Cambridge.

32. Goosen, N. J., De Wet, L. F., Görgens, J. F., Jacobs, K., & De Bruyn, A., (2014). Fish silage oil from rainbow trout processing waste as alternative to conventional fish oil in formulated diets for Mozambique tilapia *Oreochromis mossambicus*. *Animal Feed Science and Technology, 188*, 74–84.

33. Gram, L., (2010). Microbiological spoilage of fish and seafood products. In: Sperser, W. H., & Doyle, M. P., (eds.), *Compendium of the Microbiological Spoilage of Foods and Beverages* (pp. 87–119). Springer, New York.

34. Gustavsson, J., Cederberg, C., Sonesson, U., Van Otterdijk, R., & Meybeck, A., (2011). Global food losses and food waste: Extent, causes and prevention. In: *International Congress: Save Food* (p. 38). Interpack 2011, Düsseldorf, Germany.

35. Handayani, A. D., Sutrisno, I. N., & Ismadji, S., (2008). Extraction of astaxanthin from giant tiger (*Panaeus monodon*) shrimp waste using palm oil: Studies of extraction kinetics and thermodynamic. *Bioresource Technology, 99*(10), 4414–4419.

36. Heu, M. S., Kim, J. S., & Shahidi, F., (2003). Components and nutritional quality of shrimp processing by-products. *Food Chemistry, 82*, 235–242.

37. *Losses in Agricultural Commodities*, (2017). Press information bureau, Government of India, Ministry of Food Processing Industries. http://pib.nic.in/newsite/PrintRelease.aspx?relid=112181Accessed on 28 November.

38. Islam, S., Khan, S., & Tanaka, M., (2004). Waste loading in shrimp and fish processing effluents: Potential source of hazards to the coastal and near shore environments. *Marine Pollution Bulletin, 49*, 103–110.

39. Joesting, H. M., Blaylock, R., Biber, P., & Ray, A., (2016). The use of marine aquaculture solid waste for nursery production of the salt marsh plants *Spartina alterniflora* and *Juncus roemerianus*. *Aquaculture Reports, 3*, 108–114.

40. Jongjareonrak, A., Benjakul, S., Visessanguan, W., Nagai, T., & Tanaka, M., (2005). Isolation and characterization of acid and pepsin-solubilized collagens from the skin of Brownstripe red snapper (*Lutjanus vitta*). *Food Chemistry, 93*(3), 475–484.

41. Jung, K. W., Kim, K., Jeong, T. U., & Ahn, K. H., (2016). Influence of pyrolysis temperature on characteristics and phosphate adsorption capability of biochar derived from waste-marine macroalgae (*Undaria pinnatifida* roots). *Bioresource Technology, 200*, 1024–1028.

42. Kafle, G. K., Kim, S. H., & Sung, K. I., (2013). Ensiling of fish industry waste for biogas production: A lab scale evaluation of biochemical methane potential (BMP) and kinetics. *Bioresource Technology, 127*, 326–336.

43. Kannan, A., Hettiarachchy, N. S., Marshall, M., Raghavan, S., & Kristinsson, H., (2011). Shrimp shell peptide hydrolysates inhibit human cancer cell proliferation. *Journal of the Science of Food and Agriculture, 91*, 1920–1924.

44. Kaur, S., & Dhillon, G. S., (2013). Recent trends in biological extraction of chitin from marine shell wastes: A review. *Critical Reviews in Biotechnology, 0*, 1–18.

45. Kelleher, K., (2005). *Discards in the World's Marine Fisheries an Update* (p. 131). FAO Fisheries Technical Paper 470, FAO, Rome.

46. Khan, A. M., Fatima, N., Hussain, M. S., & Yasmeen, K., (2016). Biodiesel production from green seaweed Ulva fasciata catalyzed by novel waste catalysts from Pakistan Steel Industry. *Chinese Journal of Chemical Engineering, 24*(8), 1080–1086.

47. Kim, D. S., Koizumi, C., & Hong, B. Y., (2003). Fish sauce. *Bull. Korean Fish. Soc., 27*, 467.

48. Kim, H. M., Wi, S. G., Jung, S., Song, Y., & Jong-Bae, H., (2015). Efficient approach for bioethanol production from red seaweed *Gelidium amansii. Bioresource Technology, 175*, 128–134.

49. Kim, J. S., Shahidi, F., & Heu, M. S., (2005). Tenderization of meat by salt-fermented sauce from shrimp processing by-products. *Food Chemistry, 93*(2), 243–249.

50. Klomklao, S., Benjakul, S., Visessanguan, W., Kishimura, H., Simpson, B. K., & Saeki, H., (2006). Trypsins from yellowfin tuna (*Thunnus albacores*) spleen: Purification and characterization. *Comparative Biochemistry and Physiology - B Biochemistry and Molecular Biology, 144*(1), 47–56.

51. Lavanya, M., Meenakshisundaram, A., Renganathan, S., Chinnasamy, S., Lewis, D. M., Nallasivam, J., & Bhaskar, S., (2016). Hydrothermal liquefaction of freshwater and marine algal biomass: A novel approach to produce distillate fuel fractions through blending and co-processing of biocrude with petrocrude. *Bioresource Technology, 203*, 228–235.

52. Li, Y., Cui, J., Zhang, G., Liu, Z., Guan, H., Hwang, H., Aker, W. G., & Wang, P., (2016). Optimization study on the hydrogen peroxide pretreatment and production of bioethanol from seaweed *Ulva prolifera* biomass. *Bioresource Technology, 214*, 144–149.

53. Lipinski, B., Hanson, C., Lomax, J., Kitinoja, L., Waite, R., & Searchinger, T., (2013). *Reducing Food Loss and Waste* (1–40). World Resource Institute. http://pdf.wri.org/reducing_food_loss_and_waste.pdf Accessed on June 30, 2013.

54. Love, D. C., Fry, J. P., Milli, M. C., & Neff, R. A., (2015). Wasted seafood in the United States: Quantifying loss from production to consumption and moving toward solutions. *Global Environmental Change, 35*, 116–124.

55. Magnusson, M., Carl, C., Mata, L., De Nys, R., & Paul, N. A., (2016). Seaweed salt from Ulva: A novel first step in a cascading bio-refinery model. *Algal Research, 16*, 308–316.

56. Marnis, S., & Fitri, M., (2016). Valuation of economic utilization of fish processing waste Patin (*Pangasius Hypopthalmus*) as an added value for fish processing industry players in the district Kampar, Riau. *International Journal of Economics and Finance, 8*(9), 104–116.

57. Maruthiah, T., Somanath, B., Immanuel, G., & Palavesam, A., (2015). Deproteinization potential and antioxidant property of haloalkaphilic organic solvent tolerant protease from marine Bacillus sp. APCMST-RS3 using marine shell wastes. *Biotechnology Reports, 8*, 124–132.

58. Mazik, K., Burdon, D., & Elliott, A. J., (2005). *Seafood-Waste Disposal at Sea a Scientific Review* (pp. 1–76). Report to the Sea Fish Industry Authority, Institute of Estuarine and Coastal Studies, University of Hull, Yorkshire, UK.

59. Monteiro, R. J. R., Lopes, C. B., Rocha, L. S., Coelho, J. P., Duarte, A. C., & Pereira, E., (2016). Sustainable approach for recycling seafood wastes for the removal of priority hazardous substances (Hg and Cd) from water. *Journal of Environmental Chemical Engineering*, *4*, 1199–1208.

60. Morimura, S., Nagata, H., Uemura, Y., Fahmi, A., Shigematsu, T., & Kida, K., (2002). Development of an effective process for utilization of collagen from livestock and fish waste. *Process Biochem.*, *37*(12), 1403–1412.

61. Morris, A., & Sneddon, J., (2011). Use of crustacean shells for uptake and removal of metal ions in solution. *Appl. Spectrosc. Rev.*, *46*(3), 242–250.

62. Mshandete, A., Kivaisi, A., Rubindamayugi, M., & Mattiasson, B., (2004). Anaerobic batch co-digestion of sisal pulp and fish wastes. *Bioresource Technology*, *95*(1), 19–24.

63. Nagai, T., & Suzuki, N., (2000). Isolation of collagen from fish waste material: Skin, bone and fins. *Food Chemistry*, *3*, 277–281.

64. Nikoo, M., Benjakul, S., & Rahmanifarah, K., (2016). Hydrolysates from marine sources as cryoprotective substances in seafoods and seafood products. *Trends in Food Science & Technology*, *57*, 40–51.

65. Oliveira, C. J. M., Oliveira de Souza, E., & Bora, P. S., (2007). Utilization of shrimp industry waste in the formulation of tilapia (*Oreochromis niloticus* Linnaeus) feed. *Bioresource Technology*, *98*(3), 602–606.

66. Palmer, J. D., & Brigham, C. J., (2016). Feasibility of triacylglycerol production for biodiesel, utilizing *Rhodococcus opacus* as a biocatalyst and fishery waste as feedstock. *Renewable and Sustainable Energy Reviews*, *56*, 922–928.

67. Peinado, I., Koutsidis, G., & Ames, J., (2016). Production of seafood flavor formulations from enzymatic hydrolysates of fish by-products. *LWT - Food Science and Technology*, *66*, 444–452.

68. Phithakpol, B., (1993). Fish fermentation technology in Thailand. In: Lee, C. H., Steinkraus, K. H., & Reilly, P. J. A., (eds.), *Fish Fermentation Technology* (pp. 155–166). Seoul: United Nations University Press, Seoul.

69. Quinn, B. M., Apolinario, E. A., Gross, A., & Sowers, K. R., (2016). Characterization of a microbial consortium that converts mariculture fish waste to biomethane. *Aquaculture*, *453*, 154–162.

70. Rae I. B., Gibb, S. W., & Lu, S. G., (2009). Biosorption of Hg from aqueous solutions by crab carapace. *Journal of Hazardous Materials*, *164* (2/3), 1601–1604.

71. Raghunath, M. R., & Gopakumar, K., (2002). Trends in production and utilization of fish silage. *Journal of Food Science and Technology*, *39*, 103–110.

72. Rai, A. K., Swapna, H. C., Bhaskar, N., Halami, P. M., & Sachindra, N. M., (2010). Effect of fermentation ensilaging on recovery of oil from fresh water fish viscera. *Enzyme and Microbial Technology*, *46*(1), 9–13.

73. Rajapakse, N., Mendis, E., Jung, W., Je, J., & Kim, S., (2005). Purification of a radical scavenging peptide from fermented mussel sauce and its antioxidant properties. *Food Res. Int.*, *38*, 175–182.

74. Ramírez, M. Á., Rodriguez, A. T., Alfonso, L., & Peniche, C., (2010). Chitin and its derivatives as biopolymers with potential agricultural applications. *Biotecnologia Aplicada*, *27*, 270–276.

75. Rasmussen, R. S., & Morrissey, M. T., (2007). Marine biotechnology for production of food ingredients. *Adv. Food Nutr. Res.*, *52*, 237–292.

76. Riaño, B., Molinuevo, B., & García-González, M. C., (2011). Treatment of fish processing wastewater with microalgae-containing microbiota. *Bioresource Technology*, *102*(23), 10829–10833.

77. Rishipal, R., & Philip, R., (1998). Selection of marine yeasts for the generation of single cell protein from prawn-shell waste. *Bioresour. Technol.*, *65*, 255–256.

78. Roberts, D. A., & De Nys, R., (2016). The effects of feedstock pre-treatment and pyrolysis temperature on the production of biochar from the green seaweed Ulva. *Journal of Environmental Management*, *169*, 253–260.

79. Rustad, T., (2003). Utilization of marine by-products. *Electronic Journal of Environmental, Agricultural and Food Chemistry*, *2*(4), 458–463.

80. Sachindra, N. M., Bhaskar, N., Siddegowda, G. S., Sathisha, A. D., & Suresh, P. V., (2007). Recovery of carotenoids from ensilaged shrimp waste. *Bioresource Technology*, *98*(8), 1642–1646.

81. Saranya, R., Prasanna, R., Jayapriya, J., Aravindhan, R., & Tamil, S. A., (2016). Value addition of fish waste in the leather industry for dehairing. *Journal of Cleaner Production*, *118*, 179–186.

82. Sathe, S., & Durand, P. M., (2015). Low cost, non-toxic biological method for harvesting algal biomass. *Algal Research*, *11*, 169–172.

83. Sathivel, S., Bechtel, P., Babbitt, J., Smiley, S., Crapo, C., & Reppon, K., (2003). Biochemical and functional properties of herring (*Clupea harengus*) byproducts hydrolysates. *Journal of Food Science*, *68*, 2196–2200.

84. Serrano, A., Siles, J. A., Chica, A. F., & Martín, M. Á., (2013). Agri-food waste valorization through anaerobic co-digestion: Fish and strawberry residues. *Journal of Cleaner Production*, *54*, 125–132.

85. Sheih, I. C., Wu, T. K., & Fang, T. J., (2009). Antioxidant properties of a new antioxidative peptide from algae protein waste hydrolysate in different oxidation systems. *Bioresource Technology*, *100*(13), 3419–3425.

86. Silva, J. F., Espósito, T. S., Marcuschi, M., Ribeiro, K., Cavalli, R. O., Oliveira, V., & Bezerra, R. S., (2011). Purification and partial characterization of a trypsin from the processing waste of the silver mojarra (*Diapterus rhombeus*). *Food Chemistry*, *129*(3), 777–782.

87. Sjaifullah, A., & Santoso, A. B., (2016). Autolytic Isolation of chitin from white shrimp (*Penaues Vannamei*) waste. *Procedia Chemistry*, *18*, 49–52.

88. Sorokulova, I., Krumnow, A., Globa, L., & Vodyanoy, V., (2009). Efficient decomposition of shrimp shell waste using *Bacillus cereus* and *Exiguobacterium acetylicum*. *Journal of Industrial Microbiology & Biotechnology*, *36*, 1123–1126.

89. Souissi, N., Ellouz-Triki, Y., Bougatef, A., Blibech, M., & Nasri, M., (2008). Preparation and use of media for protease-producing bacterial strains based on by-products from Cuttlefish (*Sepia officinalis*) and wastewaters from marine-products processing factories. *Microbiological Research*, *163*(4), 473–480.

90. Steinkraus, K. H., (2002). Fermentations in world food processing. *Compr. Rev. Food Sci.*, 23–32.

91. Sudhakar, M. P., Merlyn, R., Arunkumar, K., & Perumal, K., (2016). Characterization, pretreatment and saccharification of spent seaweed biomass for bioethanol production using baker's yeast. *Biomass and Bioenergy*, *90*, 148–154.

92. Suleria, H. A. R., Masci, P., Gobe, G., & Osborne, S., (2016). Current and potential uses of bioactive molecules from marine processing waste. *J. Sci. Food Agric., 96*, 1064–1067.

93. Suleria, H. A. R., Masci, P. P., Gobe, G. C., & Osborne, S. A., (2017). Therapeutic potential of abalone and status of bioactive molecules: A comprehensive review. *Critical Reviews in Food Science and Nutrition, 57*(8), 1742–1748.

94. Suleria, H. A. R., Gobe, G., Masci, P., & Osborne, S. A., (2016). Marine bioactive compounds and health promoting perspectives, innovation pathways for drug discovery. *Trends in Food Science & Technology, 50*, 44–55

95. Tacon, A. G. J., & Metian, M., (2008). Global overview on the use of fish meal and fish oil in industrially compounded aquafeeds: Trends and future prospects. *Aquaculture, 285*, 146–158.

96. Uchida, M., Amakasu, H., Satoh, Y., & Murata, M., (2004). Combinations of lactic acid bacteria and yeast suitable for preparation of marine silage. *Fisheries Science, 40*, 507–517.

97. Uju, W. A. T., Goto, M., & Kamiya, N., (2015). Great potency of seaweed waste biomass from the carrageenan industry for bioethanol production by peracetic acid-ionic liquid pretreatment. *Biomass and Bioenergy, 81*, 63–69.

98. Vazquez, J. A., Blanco, M., Fraguas, J., Pastrana, L., & Perez-Martin, R., (2016). Optimization of the extraction and purification of chondroitin sulphate from head by-products of *Prionace glauca* by environmental friendly processes. *Food Chemistry, 198*, 28–35.

99. Vázquez, J. A., Nogueira, M., Durán, A., Prieto, M. A., Rodríguez-Amado, I., Rial, D., Gonzalez, M. P., & Murado, M. A., (2011). Preparation of marine silage of swordfish, ray and shark visceral waste by lactic acid bacteria. *Journal of Food Engineering, 103*(4), 442–448.

100. Vázquez, J. A., Docasal, S. F., Prieto, M. A., González, M. P., & Murado, M. A., (2008). Growth and metabolic features of lactic acid bacteria in media with hydrolyzed fish viscera. An approach to bio-silage of fishing by-products. *Bioresource Technology, 99*, 6246–6257.

101. Venugopal, V., (2005). Availability, consumption pattern, trade and need for value addition. In: *Seafood Processing: Adding Value through Quick Freezing, Retortable Packaging and Cook-Chilling* (pp. 1–23). CRC Press, Boca Raton, FL.

102. Vijayaraghavan, K., Arun, M., Joshi, U. M., & Balasubramanian, R., (2009). Biosorption of As(V) onto the shells of the crab (*Portunus sanguinolentus*): Equilibrium and kinetic studies. *Ind. Eng. Chem. Res., 48*(7), 3589–3594.

103. Vijayaraghavan, K., & Jegan, J., (2015). Entrapment of brown seaweeds (*Turbinaria conoides* and *Sargassum wightii*) in polysulfone matrices for the removal of praseodymium ions from aqueous solutions. *Journal of Rare Earths, 33*(11), 1196.

104. Vijayaraghavan, K., Palanivelu, K., & Velan, M., (2005). Crab shell-based biosorption technology for the treatment of nickel-bearing electroplating industrial effluents. *Journal of Hazardous Materials, 119*(1–3), 251–254.

105. Vijayaraghavan, K., Sathishkumar, M., & Balasubramanian, R., (2011). Interaction of rare earth elements with a brown marine alga in multi-component solutions. *Desalination, 265*, 54.

106. Vijayaraghavan, K., Winnie, H. Y. N., & Balasubramanian, R., (2011). Biosorption characteristics of crab shell particles for the removal of manganese(II) and zinc(II) from aqueous solutions. *Desalination, 266*(1–3), 195–200.

107. Vives, M., Labandeira, S., & Mosquera, L., (2013). Production of compost from marine waste: Evaluation of the product for use in ecological agriculture. *Journal of Applied Phycology*, *25*(5), 1395–1403.

108. Wilson, J., Hayes, M., & Carney, B., (2011). Angiotensin-I-converting enzyme and prolyl endopeptidase inhibitory peptides from natural sources with a focus on marine processing by-products. *Food Chemistry*, *129*(2), 235–244.

109. Wisniewski, A., Wiggers, V. R., Simionatto, E. L., Meier, H. F., Barros, A. A. C., & Madureira, L. A. S., (2010). Biofuels from waste fish oil pyrolysis: Chemical composition. *Fuel*, *89*(3), 563–568.

110. Xia, A., Jacob, A., Herrmann, C., Tabassum, M. R., & Murphy, J. D., (2015). Production of hydrogen, ethanol and volatile fatty acids from the seaweed carbohydrate mannitol. *Bioresource Technology*, *193*, 488–497.

111. Xu, W., Yu, G., Xue, C., Xue, Y., & Ren, Y., (2008). Biochemical changes associated with fast fermentation of squid processing by-products for low salt fish sauce. *Food Chemistry*, *107*(4), 1597–1604.

112. Xu, Y., Bajaj, M., Schneider, R., Grage, S. L., Ulrich, A. S., Winter, J., & Gallert, C., (2013). Transformation of the matrix structure of shrimp shells during bacterial deproteination and demineralization. *Microbial Cell Factories*, *12*(90), 1–12.

113. Yahyaee, R., Ghobadian, B., & Najafi, G., (2013). Waste fish oil biodiesel as a source of renewable fuel in Iran. *Renewable and Sustainable Energy Reviews*, *17*(6), 312–319.

114. Yan, N., & Chen, X., (2015). Sustainability: Don't waste seafood waste. *Nature*, *524*, 155–157.

115. Yoon, H. D., Lee, D. S., Ji, C. I., & Suh, S. B., (1997). Studies on the utilization of wastes from fish processing, 1: Characteristics of lactic acid bacteria for preparing skipjack tuna viscera silage. *Journal of the Korean Fish Society*, *30*, 1–7.

116. Yorio, P., & Caille, G., (2004). Fish waste as an alternative resource for gulls along the Patagonian coast: Availability, use, and potential consequences. *Marine Pollution Bulletin*, *48*(7/8), 778–783.

117. Younes, I., Nasri, R., Bkhairia, I., Jellouli, K., & Nasri, M., (2015). New proteases extracted from red scorpion fish (*Scorpaena scrofa*) viscera: Characterization and application as a detergent additive and for shrimp waste deproteinization. *Food and Bioproducts Processing*, *94*, 453–462.

118. Yu, S., Blennow, A., Bojko, M., Madsen, F., Olsen, C. E., & Engelsen, S. B., (2002). Physico-chemical characterization of Floridian starch of red algae. *Starch/ Starke*, *54*, 66–74.

119. Zheng, S., He, M., Jiang, J., Zou, S., Yang, W., Zhang, Y., Deng, J., & Wang, C., (2016). Effect of kelp waste extracts on the growth and lipid accumulation of microalgae. *Bioresource Technology*, *201*, 80–88.

CHAPTER 10

UTILIZATION OF FISH AND SHELLFISH BYPRODUCTS FROM MARINE FOOD INDUSTRIES: BENEFITS AND CHALLENGES

RAMESH SHRUTHY and RADHAKRISHNAN PREETHA

ABSTRACT

Food industry residue, refuse, and wastes comprise about 30% of worldwide agricultural productivity. Fruits and vegetable wastes, lingo-cellulosic materials, and animal and fisheries operation refuses are the main food industry based wastes. The food wastes are in general of good quality and have potential benefits, and can be used to produce other renewable and reprocessed value-added products. Mainly in the field of fish and fisheries, the processing of shrimp and crustaceans will produce a huge quantity of wastes. These rejects and discards from the fisheries can be reused as by-products. Up to 30% of the total landing in the fisheries industries considered as underutilized, by catch and unconventional or unexploited, give rise to over 22 million tons of fish waste annually. Although a proportion of these gets reprocessed, yet several tons end up as waste and requiring disposal. The by-products from the fish industry are generally defined based on harvesting and processing methods as well as fish species. This chapter summarizes important fish by-products that include: fish guts and viscera, trimmings, minced fish from fish frames, fish liver and oil, fish skin, napes, and fish roes, fish heads, and shrimp shells and fish protein hydrolysates. Moreover, this chapter focuses on the technologies used to process them in a manner, which will not pollute the environment.

10.1 INTRODUCTION

Fish production has grown globally at a constant rate during the past five decades, with a study increase in fish food supply with median rate (3.2%) annually, along with global population growth of 1.6%. Per capita, fish consumption globally has increased from 9.9 kg (in the 1960s) to 19.2 kg [39] on an average. The fifty million tons of fish have been harvested from the Pacific Ocean compared to20 million tons from Atlantic Ocean [30]. In the Atlantic Ocean, cod harvesting of around 1 million ton is being harvested. China is the world leading fish-producing nation in the world, having more than 16million tons, followed by Peru, United States, Russia, Norway, and Iceland. Due to overfishing, the wild fish stocks are diminishing, and it may create many problems in the ecosystem. A report on state world fisheries and aquaculture (SOFIA) was published by FAO in 2014 [36].

Some of the major fishing countries with fish outputs are given in Figure 10.1. In these countries, fish production is mainly dependent on large water bodies, marine coastlines, rivers, and lakes [37]. Annual fish production (Billion kg) is as follows:

| Chile | 3.82 | Japan | 4.77 | Myanmar | 5.05 |
| Philippines | 4.69 | Russia | 4.40 | US | 5.41 |

- The FAO [36] indicates that 52% of world fish stock is fully utilized and more than 25% is either overexploited. The overexploitation (feasible level of maximum production level) can disturb sustainability and can lead to the collapse of the system. Moreover, 90% of large predatory fish stocks have already expired worldwide [36]. These issues can lead to serious problems, such as:
- The imbalance of the entire ecosystem due to the extinction of the species results in the risk of collapse and stress to the ecological unity of the oceans.
- Depletion of seafood has a negative impact on population globally since many people are dependent on seafood due to the social, economic or dietary reasons.

This is mainly because of modernization and increase in capacity of the fishing vessel. Sustainability of the stock is for a constant supply of resources, and it is important for the environment. The structure and function of the ecosystem can be represented graphically using classical pyramid [13] as suggested by Georg Borgstromin 1972 and harvesting a slice of the pyramid should be proportional to the abundance of resources at different levels of the pyramid for sustainability [19]. The pyramid is valid, and it would be more stable if the species of steps of the pyramid are used for human consumption.

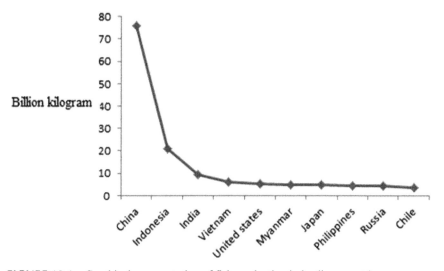

FIGURE 10.1   Graphical representation of fish production in leading countries.

In the small-scale fishing industry, fish is often not processed all before being offered to the consumers. In Spain, Portugal, and Norway, the fishes are processed at sea, and then the waste may commonly be discarded or may be brought ashore and are subsequently processed in plants for making by-products for a useful purpose.

The possibilities of production of more food from limited resources have been focused more in the last couple of decades. The improved usage of by-products is suggested due to reasons including environmental and economical limitations [39]. The nature/type of by-products may vary according to the harvesting and processing method used. The main by-products which are obtained from the fish are: head, backbone, trimmings or cutoffs, skin, gut, roe, liver, and oil (Figure 10.2).

The land based-operation and offshore processing are the main methods of fish harvesting. In land-based operation, small boats are used for fishing trips, and the process will take a few hours to few days or week. Therefore, the processing of the fish and by-products will take place at the land. However in case of offshore processing, the dragger (which is a commercial fishing vessel with onboard freezing design to store/harvest fish up to a month without any deterioration) is used; and the recovery of the product and by-product will take place thereafter [4, 30].

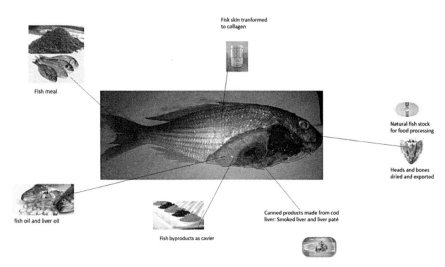

FIGURE 10.2   (See color insert.) Different by-products of the fish processing industry.

The main reasons behind the limitation of byproducts utilization from the marine resources are the unavailability of raw materials from a species for a prolonged period and higher cost of production [2, 15]. The use of remarkable properties of enzymes and proteins from the fish species was replaced by the genetically engineered microbes to reduce the cost of these specialized products. Moreover, chemical synthesis of these proteins also contributed to the restriction of isolation/purification of the proteins from fish process industry waste [42].

This chapter evaluates the benefits and challenges for the utilization of fish and shellfish by-products from marine food processing industries.

10.2   BY-PRODUCTS RECOVERY FROM FISH AND SHELL FISH

There are various by-products that are obtained from the fish and shellfish (Figure 10.3), and are mainly classified into three categories according to the source: Firstly, we will look at utilizing the viscera, then frames cut-off (backbone, skin, belly flaps, etc.), and finally utilization of head and shrimp shell. For example, globally tuna processing industry handles four million ton tuna species annually. By-products such as viscera, head, backbone, skin, belly flaps and muscle have traditionally been transformed into fertilizers and fish meal called value-added products [30].

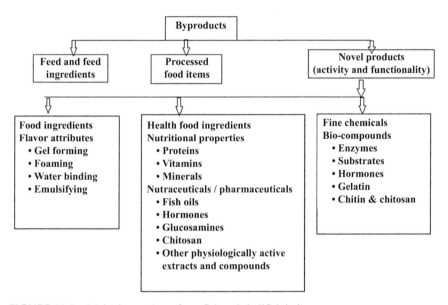

FIGURE 10.3   Major by-products from fish and shellfish industry.

The dark meat of the fish has higher nutritional quality and has a good amount of long chain omega-3 fatty acids, vitamins, iron, and copper. One of the main by-products is the cod liver oil, which is an important source of vitamins A and D [8]. Table 10.1 indicates the nutrient level for fish by-products and allowable daily intake (ADI). Fish by-products consist of protein, lipid fractions, vitamins, and minerals. However, they are an important source of environmental contamination. The waste management of the marine industry will resolve the problems related to the environmental issues. The production of fish oil and fish meal is one of the by-products obtained after waste management [13].

10.2.1 FISH VISCERA

Normally fish viscera (including liver, roe, and milt) constitute 10–20% of the total weight of fish. The digestive enzyme of large quantities has been seen in the fish intestine, stomach, and the pyloric caecum. The enzymes from the fish intestine must be isolated, purified, and undergo characterization, and this will highlight the potential application in the field of medicine and research [30, 39, 42]. The enzymes obtained from fish are mainly used for food processing and in detergent production, leather processing, and chemical modification.

TABLE 10.1 Fish By-Products Nutrient Level, Daily Need for Children

Fish and byproducts	Nutrient level per 100 g of fish	Allowable Daily Intake (ADI) for children and deficiency
Bones from herring	Zinc: 19 mg	5.6mg (of moderate bioavailability); 80,000 child deaths per year
Cod fillets	Iodine: 250 µg	120 µg; 2 billion people suffer from deficiency.
Cod liver oil	Vitamin A: 5000 µg	500 µg; 250 million infants suffer from deficiency.
Dried tuna frames	Iron: 35 mg	8.9 mg (at 10% bioavailability); 1.6 billion people suffer from deficiency.

The handling procedure for by-product production is important. The improper handling and conservation of fish viscera can be disintegrated by oxidation, enzymatic changes, and microbial spoilage oxidation. The overall logistical operation for the huge quantity of the product is therefore mandatory for proper utilization [12]. In the field of food processing, the main application of enzyme includes: chill-proofing of beer, biscuit industry, softening of meat, processing of minimally processed fruits/vegetable product, and hydrolysis of numerous related food proteins such as: vegetable proteins, collagen, and gelatin.

The main hydrolytic enzymes [10] are produced from cod serine proteases including trypsin [11], elastin [5, 8], serine collagenases [28], chymotrypsin [5, 8], and chymotrypsin [5, 8].

10.2.2   LIVER AND LIVER OIL

The liver is one of the by-products of the fish industry, which is mainly used for making liver oil. Cod liver oil is extracted from the liver of Atlantic cod (*Gadus morhua*). Although it is not acceptable to most people, yet the traditional population believes that the cod liver oil provides diverse health-promoting factors. Cod liver oil has a higher percentage of omega-3 fatty acids, which can reduce heart failures [25].

The liver oil has Omega-3s and Vitamin D that play an essential role in reproductive, hormonal, cardiovascular, neurological health and immune, neurological health. Hence both adults and children can gain these health benefits, if they consume cod liver oil [33]. According to the USDA1972, one teaspoon of cod liver oil has about 41 calories and 4.5 g of fat that includes monounsaturated, saturated, and polyunsaturated fatty acids [7, 35]. Cod liver oil processing includes the extraction of the oil from the cod liver by cold pressing; and after extraction, the oil is subjected to high heat during different steps such as the degumming (212°F), deodorization (374°F or more), and molecular distillation (392°F). During this high heat processing, denaturation of unsaturated fatty acids and trans-fat formation can happen and also chances of heavy metal contamination from the source is also high [3, 7, 33]. Hence after processing manufacturer should ensure that the product is free from both heavy metals and trans-fat. Similarly, due to the high content of saturated fatty acids, it easily gets spoiled by rancidification reactions (lipid oxidation) [7].

The omega-3 fatty acids are also obtained from oily fish or seafood such as salmon, mackerel, and sardines. Fish liver oil is rich in vitamin D. The Omega-3s and vitamin D are favorable for preserving brain health and lowering inflammation and preventing depression and blood clotting. The arthritis inflammation can be reduced by the consumption of cod fish liver oil [33].

Nowadays, canned cod liver and capsules have a high demand for the protection of human health.

10.2.3 FISH OIL

In recent years, herring salting or freezing of the fish has been mostly used for human consumption and 95% of harvested capelin species has been used for oil production and fish meal [3]. The fish species called Pelagic is used for the preparation of fish body oil, where the whole fish is grounded during the processing of Pelagic fish body oil. The animal feed is the major byproduct of fish oil processing [3].

The processing of the fish for the production of the fish oil is performed at the fish meal plant [40, 41], where the fish is cooked and is pressed to separate the oil and solid meal; and the remaining solid particles are removed by three-phase centrifugations. The solid meal, which contains bones, are utilized as a bone meal for animal feed and as fertilizer. The dry bone meal[35] is a high-quality by-product with more desirable functionality and bioactive for the aquaculture growth.

The general alkaline refining method for fish oil usually make-up the consecutive steps [22]. After recovering the oil, the steps for the refining of the vegetable oil are indicated in Figure 10.4, and as follows:

Step 1: The removal of hydratable phospholipids and metals by the process water degumming.

Step 2: The removal of excess amount of non-hydratable phospho-lipids, salts of Magnesium and Calcium by adding phosphoric acid or citric acid.

Step 3: Addition of sodium hydroxide solution in order to neutralize the free fatty acids, followed by washing with hydrated phos-pholipids and soap.

Step 4: The decomposition of hydro-peroxides and color absorption by bleaching using natural or acid-activated clay minerals.

Step 5: Deodorization process by steam distillation at low pressure (2–6m bar) and the temperature of 180–220°C to remove vola-tile compounds such as aldehydes and ketones.

10.2.4 FISH ROE

Fish roe is also one of the important and costly by-products with high nutrient content. During spawning season of the fish, the fish roe is collected. In the United States and Russia farming a white sturgeon (*Acipensertrans-montanus*) is carried out for the production of highly priced so-called Black caviar (salt-cured fish-eggs from Atlantic salmon, *Catarci*) [2]. Generally, the harvesting of the fish roe is from Norway and Iceland, which will undertake by the sensory evaluation and electronic nose analysis to find the quality of fish and also may reflect the quality of the products made from roe [17, 30, 39]. Once caught, the sturgeon (family *Acipenseridae*) will be transferred to a large boat, where workers slit her open and collect the roe in tubes, sometimes stored with salt in insulated plastic tubes and they undergo curing during storage.

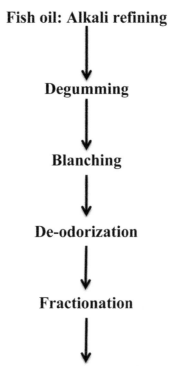

FIGURE 10.4 Refining of the fish oil.

In Japan, caviar variety products are used for traditional sushi cuisine dish called masago, where the curing technique and maturing process are used as per Japan tradition [39].

10.2.5 FISH FRAMES, COLLARS AND TRIMMINGS

The fillet like products with textural characterization has been prepared by using trimmings obtained from the fish.

The low-price mince of dark color from the collar and the frame are separated after the fillet collection. Gutted cod of one ton after processing produce 100–180kg minced fish. The dark color frames and the mince from the nape are from the cutoffs or trimmings, which are of poor quality. Frame mince obtained from the kidney of the fish will be dark reddish in color and which is seen beneath spinal cord or the backbone, and by the abrasive handling of the frame, the reddish color gets mixed with the flesh [42]. The efficient waterjet technology is used to clean the flesh from the backbone of the round fish to make flesh white in color. The above water-cleaned fish flesh with the good physicochemical property is used to prepare food such as surimi (minced fish paste, [41]. Other than that, this has been used for the preparation of value-added products such as fish sauce, fish meal, and fish and animal feed. The fish bone is rich in nutrient contents [40, 41], as shown is in Table 10.2.

TABLE 10.2 Nutrient Contents of Fish Bone

Nutrient Contents in fish bone	Type of fish			
	Cod	Salmon	Herring	Mackerel
Ash (g/100g)	58	50	51	44
Calcium (g/100g)	19	14	16	14
Iodine (mg/100g)	0.4	0.3	0.1	0.2
Iron (mg/100g)	5	3	6	7
Protein (g/100g)	39	47	44	59
Zinc (mg/100g)	10	23	19	13

10.2.6 FISH SKIN, GELATIN, AND PEARL ESSENCE

The fish skin is mainly made of gelatin and scales [16, 18]. The skin of fish can be stored under the refrigerated condition and frozen for a specific time without any adverse effect [16, 30]. The pearl essence is obtained from the scales of the fish, which is a shimmery substance generally used in cosmetics such as lipstick and blush. Fish gelatin offers the application as a food ingredient, and it can be used in food where mammalian gelatin is not acceptable [39]. The extraction process of gelatin from catfish is explained in Figure 10.5 [16, 28]. The fish gelatin has a molecular weight from 95 to 138 kDa with good rheological properties. The gelatin contains 14 amino acids and has high Proline content. Fish gelatin is dissolvable in cold water hence it plays an important role in the quality of frozen products and is also used in the cosmetic industry for manufacturing of shampoo along with protein [43].

A brown-banded bamboo shark, blacktip shark, yellow tuna are also used for gelatin production by different pre-treatment and extraction procedures [21].

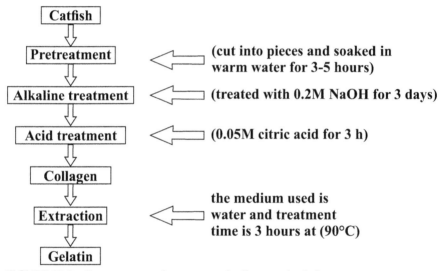

FIGURE 10.5   Common extraction process of collagen and gelatin.

In a recent research report, the fish skin of Tilapia was used to treat burn injuries in 32 patients in Brazil. The fish skin is collected and sterilized in chlorhexidine, which is an antiseptic, and then transferred into glycerol at various concentrations, and finally, radio-sterilization is done to sterilize against viruses. The healing process is completed in 9–11 days. This is one of the cost-effective and eco-friendly treatment method [6] for burn injuries.

10.2.7 ISINGLASS

The dried swim bladders of the fish are by-products, which are used to produce isinglass. It is the type of collagen, which used for the clarification or fining of beer and wine [21]. Another major use of the isinglass is for the preparation of specialized glue [30, 34].

The production of isinglass originated from the Dutch huizenblaas – huizen, which is a type of sturgeon. Isinglass was originally made exclusively from sturgeon, which is no longer used as a source of isinglass, especially after introducing cod as a cheap substitute in 1795 by William Murdoch. This was enormously used in Britain instead of Russian isinglass. The bladders were extracted from the fish then processed and dried, and are transformed into various shapes for further use such as a coating of goldbeater's skin. Isinglass is similar to gelatin, but it is unique because the dried film can be reactivated with moisture and hence can be used as adhesive. Compared to other adhesives, it has high adhesive strength, and therefore it can be used for the repairing of parchment [35].

10.2.8   FISH HEAD

Fish head, cheeks, tongue are considered as delicacies, which have been salted and sold in the fish markets as dried fish snacks. One of the important by-product from the head and shell from the codfish industry after processing eis astaxanthin, and bone meal for animal feed. However, in Norway, codfish head was dried and was replaced for whole fish as a raw material for preparing dry fish for export [30]. There are different drying methods [4, 40], such as:

- The process of computer-aided control air blast cabinet drying.
- Geothermal heating.

These two methods are commonly used to produce high quality dried cod heads. The commercial pigment Astaxanthin (a keto-carotenoid, $C_{40}H_{52}O_4$) obtained from the crustacean waste processing is used in salmon aquaculture feed and food for crustacean species [13]. However, globally 95% of astaxanthin supply is made synthetically at a low tariff.

10.2.9   SHRIMP OFFAL/ SHELLS

The shrimp processing will generate large quantities of waste (Figure 10.6). The slow degradation kinetics of the shell causes environmental issues. Shrimp offal consists of protein, chitin, and the coloring agent astaxanthin

[35, 23] and also an unavoidable ingredient for salmon feeds. The chitin is the most abundant biopolymer next to cellulose, in nature. Crab and shrimp shell contains high chitin content, 32–17% on the dry basis [9, 20, 29].

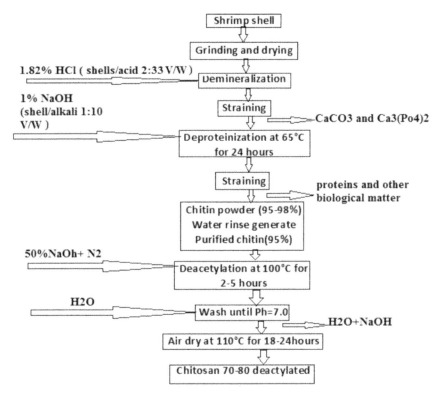

FIGURE 10.6   Processing steps for the manufacture of chitin and chitosan [23].

10.3   BIOCHEMICAL PROPERTIES

For the treatment of wastewater, the chitosan is of utmost importance because chitosan is a biodegradable polysaccharide. Chitosan is also used as an additive in food, cosmetics, agriculture, and medical application (wound healing property). Chitosan and chitin greatly rely on chemical properties of the polymers: many interesting reactions have been reported due to its cationic nature, and at the same time other biopolymers are anionic under similar conditions [22, 38].

These psychrophilic enzymes from fish waste, which show high activity at low temperature compared to their bacterial counterparts and mammalian cells, which are highly sensitive to heat. These characteristics are advantageous for their industrial applications.

10.3.1 BIOACTIVE PEPTIDES

The antimicrobial agent such as protamine is used in food processing industries and preservation purpose, and it is prepared from more than 50 fish species. The growth of Bacillus spores can be prevented using protamine. Antioxidant activity has been reported for isolated peptides [17, 24].

10.3.2 FISH PROTEIN HYDROLYSATES

In the case of the food processing industry, wastes such as viscera, head, and tail of the fish are generally considered as low-value products. However, the recovery of protein is done by enzymatic hydrolysis of these low-value products, which is the efficient method and increases the commercial value of biomass [11, 12].

Fish protein hydrolysis (FPH) attracts more attention because it can be used for the preparation of bioactive peptides for disease treatment [1]. Even though this method is more time consuming and costly, yet FPH obtained from this method is more nutritional and offers a vast array of applications including animal nutrition, food additive, pharmaceutical, and cosmetics. Several commercially available protease enzymes have been used to hydrolyze fish protein, such as: Alcalase, Flavourzyme, Papain, Neutrase, and Bromilane. Papain had been chosen and reported as a protease enzyme that is widely used to produce protein hydrolysates. Scientists [14, 26, 31] have reported the use of 1% papain to digest King fish protein at 37°C for 6 hours; and the study compared the by-product with a large spectrum of protease that was obtained after the enzymatic hydrolysis. FPH obtained by this method is an ideal choice of protein used in biomedicine and commercial applications. However, the high cost of the enzyme is still one of the issues that make its production in industrial scale economically difficult [30].

10.4 SUMMARY

The demand for fish, fish products, and fish byproducts are increasing worldwide for boththe human and animal consumption. The production of byproducts from the fish waste follows the simple and low-cost techniques. The usage of fishmeal and fish oil for the aquaculture industry is growing since it has high nutrient value and economically feasible. The major challenges and constraints concerned with byproducts are: The isolation of the specific compounds by using the present techniques and inadequate supply of the superior quality of by-products will not be practical because of high cost apart from omega-3 fatty acids. Complete utilization of the fishery waste should be encouraged since it can lead to environmental protection.

KEYWORDS

- chitin
- chitosan
- fish meal
- fish oil
- food processing industries
- liver oil
- marine food
- protein hydrolysates
- shellfish

REFERENCES

1. Alasalvar, C., & Taylor, T., (2002). *Seafoods: Technology, Quality and Nutraceutical Applications* (pp. 1–5). Springer, Verlag, Heidelberg, Germany.
2. Adler-Nissen, J., (1986). *Enzymic Hydrolysis of Food Proteins* (pp. 365–404). Elsevier Science Publication, London.
3. Aparicio, R., McIntyre, P., Aursand, M., Eveleigh, E., Marighetto, N., Rossell, B., et al., (1998). Fish oil. In: Aparicio, R., (ed.), *Food Authenticity-Issues and Methodologies* (pp. 218–213). Eurofins Scientific, Nantes.
4. Arason, S., Thoroddsson, G., & Valdimarsson, G., (1992). The drying of small pelagic fish-the Icelandic experience. In: Burt, J. R., (ed.), *Pelagic Fish: The Resource and its Exploitation* (pp. 291–298). Fishing News Books, https://openlibrary.org/publishers/Fishing_News_Books_Ltd.
5. Asgeirsson, A., & Bjarnason, J. B., (1989). Purification and characterization of trypsin from the poikilothermic Gadusmorhua. *European Journal Biochemical, 180*(1), 85–94.

6. BBC, (2017). India, *Business Insider India*: Fish skin treatment clinical trials by Brazilian doctors, http://www.bbc.com/news/av/world-latin-america./can-fish-skin-help-treat-burns, Accessed July 31, 2017.

7. Bourre, J. M., Bonneil, M., Clement, M., Dumont, O., Durand, G., Lafont, H., Nalbone, G., & Piciotti, M., (1993). Function of dietary polyunsaturated fatty acids in the nervous system. *Prostaglandins, Leukotrienes and Essential Fatty Acids, 48*(1), 5–15.

8. Bragado, M., & Kristbergsson, K., (2002). Seasonal changes in chemical composition and quality parameters of capelin, *Journal of Aquatic Food Product Technology, 11*(3), 87–103.

9. Dautzenberg, H. G., (1998). Physicochemical characterization of chitosan in dilute solution. *Advanced Chitin Science, 2*(1), 429–436.

10. De la Parra., Rosas, J. P., & Lazo, M. T., (2007). Vian. Partial characterization of the digestive enzymes of Pacific bluefin tuna *Thunnus orientalis* under culture conditions. *Fish Physiol. Biochemistry, 33*(3), 223–231.

11. De Vecchi, S., & Coppes, Z., (1996). Marine fish digestive proteases- relevance to the food industry and the South-West Atlantic region- A review. *J. Food Biochemistry, 20*(1), 193–214.

12. Gbogouri, G. A., Linder, M., Fanni, J., & Parmentier, M., (2006). Analysis of lipids extracted from salmon (*Salmo Salar*) heads by commercial proteolytic enzymes. *European Journal of Lipid Science and Technology, 108*(9), 766–775.

13. Gessner, J. S., & Spratte, K., (2011). *Historic Overview on the Status of the European Sturgeon (Acipensersturio) and its Fishery in the North Sea and its Tributaries With a Focus on German Waters*, 195–219

14. Gildberg, A., (1994). Enzymic processing of marine raw materials. *Journal of Bioprocessing & Biotechniques, 28*(1), 1–15.

15. Gildberg, A., (2002). Enhancing returns from greater utilization. In: Bremner, H. A., (ed.), *Safety and Quality Issues in Fish Processing* (pp. 425–449). Woodhead Publishing Limited and CRC Press LLC, Cambridge.

16. Gudmundsson, M., & Hafsteinsson, H., (1997). Gelatin from cod skins as affected by chemical treatments. *Journal of Food Science, 62*(1), 37–39.

17. Haard, N. F., Simpson, B. K., & Sikorski, Z. E., (1994). *Biotechnological Applications of Seafood Proteins and Other Nitrogenous Compounds* (pp. 195–216). Chapman and Hall, New York.

18. Hafsteinsson, K., (1997). Gelatin from cod skins as affected by chemical treatments. *Journal of Food Science, 62*(1), 37–39.

19. Holding, J., Bundy, A., Van Zwieter, P. A. M., & Plank, M. J., (2016). Fisheries: The inverted food pyramid. *ICES Journal of Marine Science, 4*, 16–97.

20. Ibrahim, H. M., Salama, M. F., & El-Banna, H. A., (1999). *Shrimp's Waste: Chemical Composition, Nutritional Value, and Utilization*, Nahrung, Springer, *43*(1), 418–423.

21. Kaewdang, O., Benjakul, S., & Kaewmanee, T., (2014). Characteristics of collagens from the swim bladders of yellowfin tuna (*Thunnus albacores*). *Food Chemistry, 155*, 264–270.

22. Kim, S. E., & Mendis, E., (2006). Bioactive compounds from marine processing byproducts – a review. *Food Research International, 29*(4), 383–393.

23. Kjartansson, G., Zivanovic, S., Kristbergsson, K., & Weiss J., (2006). Sonication-assisted extraction of chitin from shells of freshwater prawns (*Macrobrachium rosenbergii*). *Journal of Agriculture Food Chemistry, 40*(1), 3317–3323.

24. Klomklao, S., Benjakul, S., & Visessanguan, W., (2004). Comparative studies on proteolytic activity of splenic extract from three tuna species commonly used in Thailand, *Journal of Food Biochemistry*, *28*(5), 355–372.

25. Kołakowska, A., Stypko, K., Domiszewski, Z., Bienkiewicz, G., Perkowska, A., & Witczak, A., (2002). Canned cod liver as a source of n-3 polyunsaturated fatty acids, with a reference to contamination. *Journal of Agricultural Food Chemistry*, *46*(1), 40–45.

26. Liaset, B., Nortvedt, R., Lied, E., & Espe, M., (2002). Studies on the nitrogen recovery in enzymic hydrolysis of Atlantic salmon (*Salmo Salar,* L.) frames by Protamex MT protease. *Process Biochemistry*, *37*(2), 1263–1269.

27. Leather, R. V., Sisk, M., & Dale, C. J., (1994). Analysis of the collagen and total soluble nitrogen content of isinglass finings by polarimetry. *Journal of Inst. Brew*, *104*(1), 100–334.

28. Norland, P. E., (1989). Fish gelatine. In: Voigt, M. N., & Botta, J. R., (eds.), *Advances in Fisheries Technology and Biotechnology for Increased Profitability* (pp. 325–333). Technomic Publishing Co. Inc., Lancaster, Pennsylvania, USA.

29. Rao, M. S., Nyein, K. A., Trung, T. S., & Stevens, W. F., (2007). Optimum parameters for production of chitin and chitosan from squilla (*S. empusa*). *Journal of Applied Polymer Science*, *103*, 3694–3700.

30. Rustad, T., (2003). Utilization of marine by-products. *Journal of Environmental Agriculture and Food Chemistry*, *43*(4), 458–463.

31. Sachindra, N. M., Bhaskar, N., & Mahendrakar, N. S., (2005). Carotenoids in different body components of Indian shrimps. *Journal of Food Science and Agriculture*, *85*, 167–172.

32. Santos, M. H., Silva, R. M., Dumont, V. C., Neves, J. S., Mansur, H. S., Heneine, L. G. D., (2013). Extraction and characterization of highly purified collagen from bovine pericardium for potential bioengineering applications. *Materials Science and Engineering, 33*(2), 790–800.

33. Skeie, G., Braaten, T., Hjartåker, A., Brustad, M., & Lund, E., (2009). Cod liver oil, other dietary supplements and survival among cancer patients with solid tumors. *International Journal of Cancer*, *125*, 1155–1160.

34. Sherman, V. R., Wen, Y., & Meyers, M., (2015). The materials science of collagen. *Journal of Mechanical Behavior Biomedical Material*, *05*, 23, E-article, doi: 10.1016/j.jmbbm.

35. Sikorski, Z. E., Gilberg, A., & Ruiter, A., (1995). Fish products. In: Ruiter, A., (ed.), *Fish and Fishery Products* (pp. 315–346). CAB International, Wallingford, U.K.

36. *The State of World Fisheries and Aquaculture* (2009), FAO, Fisheries and Aquaculture Department, http://www.fao.org, Rome, p. 23.

37. STATICE, (2017). *Statistical Series Fisheries* 90:61 (2005:6) 1–16ISSN 0019–1078, www.hagstofa.is/ and www.statice.is/series. Accessed July 31, 2018.

38. Tabarestani, H. S., Maghsoudlou, Y., Motamedzadegan, A., & Mahoonak, A. R. S., (2010). Optimization of physicochemical properties of gelatin extracted from fish skin of rainbow trout (*Onchorhynchus mykiss*). *Bioresource Technology*, *48*(2), 6207–6214.

39. Valdimarsson, G., (2003). Utilization of selected fish byproducts in Iceland: Past, present and future. *Advances in Seafood Byproducts*, *23*(1), 71–78.

40. Valdimarsson, G., (1998). Developments in fish food technology – implications for capture fisheries. *Journal of Northwest Atlantic Fisheries Science*, *23*, 233–249.

41. *Weeratunge*, W. K. O. V., & Perera, B. G. K., (*2016*). Formulation of a fish feed for goldfish with natural astaxanthin. *Chemistry Central Journal, 56(3),* 10–44.

42. Whittle, K. J., & Howgate, P., (2000). Recovered meat from Pacific whiting frame. *Journal of Aquatic Food Product Technology, 11*(1), 5–18.

43. Wibowo, S., Velazquez, G., Savant, V., & Torres, J. A., (2005). Surimi wash water treatment for protein recovery: Effect of chitosan-alginate complex concentration and treatment time on protein adsorption. *Bioresource Technology, 96,* 665–671.

CHAPTER 11

MARINE FOODS: NUTRITIONAL SIGNIFICANCE AND THEIR INDUSTRIAL APPLICATIONS

AAMIR SHEHZAD, ASNA ZAHID, ANAM LATIF,
RAI MUHAMMAD AMIR, and HAFIZ ANSAR RASUL SULERIA

ABSTRACT

Marine environment is a foremost reservoir of vitamins, minerals, and bioactive compounds that possess health benefits and have possible applications in medicine, technology, and food. Marine environment has organisms, such as: seaweeds, macro or microalgae, crustaceans, fish, and their products, some fungi and bacteria. These organisms are a source of healthy and functional food ingredients such as: polysaccharides, lipids, proteins, and peptides, chitin, pigments, minerals, vitamins, and phenolic compounds. Nonetheless, marine foods are indeed a source of components possessing high nutritional and physiological activities. They are also known for their positive effects on cardiovascular diseases owing to the existence of omega-3 fatty acids, polyunsaturated fatty acids (PUFA) and low content of saturated fatty acids. During the past 50 years, key compounds substantiating their potential for industrial development are: functional foods, nutritional supplements, pharmaceuticals, enzymes, therapeutic agents from marine sources. In addition to this, a significant contribution to food security has been made predominantly in developing countries. European Medicines Agency (EMEA) and Food and Drug Administration (FDA) have approved eight marine-based drugs and other several compounds. Only three drugs Yondelis®, Prialt®, and Carragelose® have been introduced in the market without any modification of original natural molecules. Marine enzymes are known for their applications in the food industry; and subsequently, they are distinctive protein molecules not present in terrestrial sources, or they may also be recognized as enzyme from terrestrial organisms but with unique activities.

11.1 INTRODUCTION

The ocean is a medicine cabinet of Mother Nature. Marine environment is a foremost reservoir of vitamins, minerals, and bioactive compounds that have potential applications in medicine, technology, and food. Marine environment possesses a diverse range of salinity, temperature, light, nutrients, and pressure. Due to these characteristics, marine organisms produce protective metabolites and mechanisms to survive in harsh environments [12].

Currently, the public is aware of the positive relationship between health and consumption of seafood. Consumers distinguish that fish and by-products of fish are wholesome and nutritious. Seafood is an excellent source of lipids with unsaturated fatty acid content and high-quality proteins. Likewise, seafood is easily digestible and bioavailable within the body, in addition being a source of a vast range of vitamins and minerals. Moreover, marine foods possess functional and nutraceutical properties. Nutraceuticals include: omega 3 fatty acids, fish protein hydrolysates, chitosan, chitin, algae, and microalgae, carotenoids, taurine, antioxidants, etc. [24].

This chapter focuses on the nutritional significance and industrial applications of marine foods and their by-products.

11.2 TYPES OF MARINE FOODS

The functional foods tend to improve the state of well-being or health and reduce the risk of disease, besides providing nutritional effects. Functionality can be an intrinsic feature of a food matrix or may be incorporated to enhance health and decrease adverse health effects accomplished by [11]:

1. Enhancing the bioavailability of health-promoting food ingredients.
2. Excluding a chemical alteration of a harmful constituent or promoting useful changes.
3. Increasing concentration of existing food constituents.
4. Supplying effective concentrations of food components and probiotic microorganisms.

Functional food ingredients exist both in marine and terrestrial environments. Terrestrial functional food ingredients have been investigated more than the marine functional constituents. Marine food sources are distinguished for their remarkable biodiversity, offering robust evidence for their exploitation as healthy foods and functional food ingredients possessing biological properties [12]. Organisms of potential interest in the marine environment include: seaweeds, macro or microalgae, crustaceans, fish, and fish products, some bacteria and fungi (Figure 11.1). The marine organisms are a source of healthy food and functional food ingredients such as: polysaccharides, lipids, proteins, peptides, chitin, pigments, minerals, vitamins, and polyphenolic compounds [38].

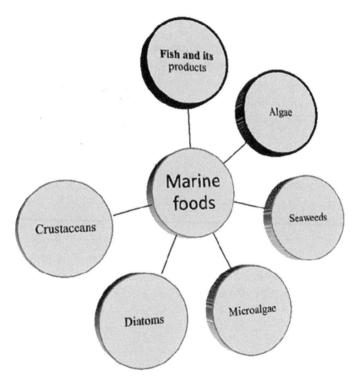

FIGURE 11.1 **(See color insert.)** Types of marine foods.

11.2.1 FISH AND BY-PRODUCTS OF FISH

The fish and their by-products are a major source of bioactive ingredients. Fish has a high quantity of omega-3 fatty acids and polyunsaturated fatty acids (PUFA). Other active ingredients include: Calcium from bones of fish, hydrolysates of proteins, biologically active peptides, and amino acids like taurine that possess antioxidant properties having positive effects on the cardiovascular system [15, 42, 51]. Different parts and organs of fish like heads of fish, skin, tail, viscera, blood, and shells hold an abundance of compounds, which can be used as functional foods [46]. Isolation of bioactive compounds from marine fish and byproducts resulting from their processing can be attained through different techniques of extraction and purification of these peptides. These extracted and purified bioactive ingredients are employed in various biotechnological applications [12].

11.2.2 ALGAE

Algae, particularly edible marine algae like macroalgae and innumerable seaweeds possess inordinate taxonomic diversity. About 10,000 species of algae have been identified out of which 5% is used for the edible purpose [6]. Also, more than hundreds of species of macro-algae have their utilization as sea vegetables, especially in Asian countries [11]. Algae animate in composite environments and are therefore exposed to harsh environmental impacts; therefore they adapt to new environmental conditions quickly. For their survival, they produce biologically active secondary metabolites [20], which include phenols and polyphenols, acetogenins, derivatives of amino acids, terpenes, and other bioactive compounds.

Algae are classified into three divisions based on the type of pigments: Brown algae, green algae, and red algae. Brown algae possess polysaccharides including: Fructans, cellulose, and laminarins. It also possesses bioactive and pharmacological properties like antitumor, anti-inflammatory, anti-oxidant, antifungal, cytotoxic, and nematocidal properties [8, 23].

11.2.3 SEAWEEDS

Seaweeds are a source of potential medicinal properties. Seaweeds, when incorporated in foods, possess cholesterol reducing, anti-obesity, and anti-hypertensive properties. Moreover, these are also known to scavenge free radicals and stimulate healthy digestion [12].

Nutritionally, seaweeds are low in calories and are a major source of polysaccharides, antioxidants, vitamins, minerals, and dietary fiber [31]. Additionally, biological properties of seaweeds include: anticoagulant, anti-inflammatory, antibacterial, antioxidant, apoptotic, antioxidant, and antiviral activities [39]. Consumption of such compounds has shown to possess high bioavailability. Furthermore, extraction, and isolation of such compounds are easy and cost-effective [31]. The seaweeds as a functional food provide alternative therapies for human health [36].

11.2.4 MICROALGAE

Microalgae are microscopic entities, which exist in marine milieu; and are usually present in littoral and benthic habitats. Microalgae play an imperative role in ocean productivity and are a source of bioactive compounds [8]. Microalgae can improve nutritional profile as they have an abundant quantity of PUFA and pigments like chlorophylls and carotenoids [8]. Microalgae possess large diversity; and can be used as a nutritional supplement and natural food coloring agent [34].

11.2.5 DIATOMS

These are specific organisms dominating phytoplankton of nutrient-rich and cold water. Diatoms easily adapt to different environments possessing a high degree of tractability. Due to this trait, diatoms are of interest in the field of biotechnology and are capable of producing biocompounds for functional foods [4]. Some prominent functional food ingredients include: docosahexaenoic acid (DHA), eicosapentaenoic acid (EPA), omega-3 fatty acid, fucoxanthin and chlorophyll [12].

11.2.6 CRUSTACEANS

Crustaceans in the marine environment are a source of extractable chitin and chitosan. Chitin is recognized as the second most copious natural polymer whilst chitosan is a biodegradable polymer derived from chitin [11]. These polymers are derived from shrimp, crab, cuttlefish, and processing of bones and shells. It is used as a food preservative due to its antibacterial properties [16]. Polymers derived from crustaceans show biological activities such as: anti-carcinogenic, antioxidant, hypocholesterolemic, and immune boosting properties [51].

11.2.7 MARINE BACTERIA AND FUNGI

Fungi and bacteria in the marine environment are imperative natural compounds that have drawn the attention of researchers worldwide [21]. Bacteria imitative from the intestinal tract of marine organisms are considered to be strains of probiotics [22]. Similarly, marine fungi are being investigated for their metabolites [3]. Marine fungi are believed to have greater concentrations of amino-butyric acid (GABA). Additionally, these fungi being rich in bioactive compounds possess anti-inflammatory, anti-carcinogenic, antiviral, anti-bacterial, and anti-spasmodic activities [7].

11.3 NUTRITIONAL AND PHYSIOLOGICAL CHARACTERISTICS OF MARINE FOODS

Massive biodiversity existing in the marine environment makes it difficult to draw inferences on nutritional and physiological properties of marine foods. Nonetheless, marine foods are indeed the source of components possessing high nutritional and physiological activities [28]. Marine environment contains several types of organisms, as discussed in this chapter, thus being a source of high profile food ingredients providing functional and bioactive properties such as: fatty acids, peptides, and proteins, polysaccharides, minerals, and vitamins (Figure 11.2).

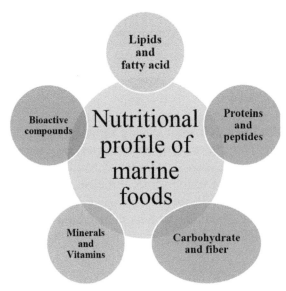

FIGURE 11.2 **(See color insert.)** Nutritional and physiological possessions of marine foods.

11.3.1 MARINE FATTY ACIDS AND LIPIDS

Fatty acids derived from marine foods possess decreased content of saturated fatty acids and high content of omega-3 and PUFA. Marine foods are known for their constructive effects on cardiovascular diseases due to the presence of omega-3 and PUFA and lower quantity of saturated fatty acids. Moreover, these foods also contain eicosapentaenoic acid (EPA) and docosahexaenoic acid (DHA) that is beneficial for inflammatory response and functionality of membranes [44]. Seafoods, in addition, also comprise of species such as: cod, salmon, herring, and halibut, etc. which possess mono-unsaturated fatty acids (MUFA) [28].

11.3.2   PROTEINS AND PEPTIDES

Marine food's health benefits have been attributed to the lipid content present in them. *In vitro* and *in vivo* studies indicate that peptides and proteins from marine sources possess bioactive properties [50]. Peptides from the marine source are encrypted in their native proteins and absorbed during digestion or processing of foods. Moreover, bioactive from proteins and peptides are also produced *in vivo* by enzymatic hydrolysis or chemical extraction procedures like solvent extraction [25]. Nonetheless, bioactive properties of marine peptides and proteins depend on the composition of their amino acids, their structure, length, and sequence [25].

11.3.3   CARBOHYDRATES AND DIETARY FIBER

Only a small amount of carbohydrates and fibers are present in muscle based marine foods; however, greater quantities of complex carbohydrates and dietary fibers are available in edible microalgae. Additionally, seaweeds contain sulfated polysaccharides that possess medicinal and antioxidant properties. Chitosan and chondroitin sulfate that is derived from the exoskeleton of crustaceans have pharmaceutical, medicinal, and biotechnological applications. Moreover, chondroitin sulfate is also beneficial for patients of osteoarthritis. Similarly, chitosan is beneficial for obesity and dyslipidemia patients [28].

11.3.4   MARINE MINERALS AND VITAMINS

The content of vitamins and minerals in marine organisms is far greater than that in meats and plants. Among minerals, selenium, and iodine are the most prominent. All fat-soluble vitamins are present in marine foods, whilst, oily fish is a tremendous source of vitamin D that is usually deficient in mammalian meat and plants [28].

11.3.5 OTHER BIOACTIVE COMPOUNDS

Other bioactive compounds present in marine foods include: phytosterols, polyphenols, carotenoids, and taurine. These compounds possess a diverse range of medicinal and pharmaceutical properties [28].

11.4 INDUSTRIAL APPLICATIONS OF MARINE FOODS

Several thousand years ago it was discovered that marine species have potent activities. In addition to providing fish products, the oceans can act as dwelling places for several types of reef corals, seaweeds and many other microorganisms as bacteria. Fish, mollusks, and crustaceans from oceans or other marine areas are traditionally used as food, although the other elements of the sea, like coral reefs, seaweed, corals, and micro-organisms can also produce a great diversity of nutraceuticals, new enzymes, drugs, industrial, and bioactive substances for food as well as for health care. During the past 50 years, key compounds substantiating their potential for industrial development as functional foods, nutritional supplements, pharmaceuticals, enzymes, therapeutic agents, have been provided by marine sources [48]. Figure 11.3 elucidates some major industrial applications of marine foods.

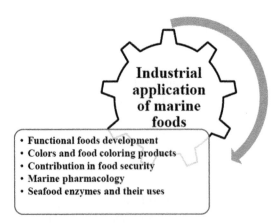

Industrial application of marine foods

- Functional foods development
- Colors and food coloring products
- Contribution in food security
- Marine pharmacology
- Seafood enzymes and their uses

FIGURE 11.3 (See color insert.) Industrial applications of marine foods.

11.5   DEVELOPMENT OF FUNCTIONAL FOODS

All marine resources are extremely rich in value-added complexes having high biological activities and can serve as functional food constituents. Numerous forms of polysaccharides as chitosan, chitin, sulfated polysaccharides, omega 3 oils, protein, and protein hydrolysates, short chain peptides, amino acids (taurine), carotenoids, and several other bioactive ingredients are few instances of compounds added at various stages of food production process, from processing to storage [18, 24]. Biologically active marine-derived functional constituents establish a wide diversity for research. Side by side, many foods are deemed as a potential carrier for the incorporation of marine-derived functional constituents to augment their value at both economic and nutritional level [13].

Natural health products and functional foods are an evolving field in Food Science because of increasing acceptance among sensible health consumers and provide new avenues for research [40, 49]. Products constituting marine-derived chitosan, chitin, or omega-3 fatty acids are certain food products being commercialized and sold in the market all around the world including USA, Japan, and other European countries [24].

Various nutritional and health benefits can be attained by the consumption of functional foods. The most studied and deliberated health benefits of functional foods include the immune stimulation and augmentation of antioxidant activity [11].

Now, marine-derived functional compounds such as fish proteins, fish oils, and seaweeds have been used in bakery, pasta, confectionery, and dairy products. These are being added as nutritional enrichment and as a fortificant in different foods, leading to the development of functional foods. Kadam et al. [24] described that the more and continuous efforts to explore and enterprise of innovative, functional foods centered on marine functional constituents is required to reduce the health glitches with diet. Food manufacturing and delivery companies in Germany are on track to market functional foods with cyanobacteria and microalgae for example: bread, pasta, soft drinks, and yogurt. The same type of developments can also be seen in Japan, France, USA, Thailand, and China [13].

11.5.1   COLORS AND FOOD-COLORING PRODUCTS

Microalgae and marine species enclose a profusion of pigments related to the incidence of light beside chlorophyll, the principal photosynthetic pigment.

These different pigments, i.e., phycobiliproteins, enhance the effectiveness of light energy utilization, whilst carotenoids protect plants from solar radiation and correlated effects. The carotenoid pigments from microalgae and other marine sources have an important position in market, e.g., β-carotene obtained from Dunaliella acts as a vitamin A precursor; lutein, canthaxanthin, and zeaxanthin are used for chicken skin coloring; and astaxanthin from Haematococcus for fish muscles coloration in aquaculture and for other pharmaceutical purposes [27]. The phycobiliproteins, phycoerythrin, and phycocyanin are distinctive to algae, and some of their preparations are being exploited for food and cosmetics. This progress in marine biodiversity will go outside the applications in photodynamic therapy and diagnostics and outspread to nutrition, pharmacy, and cosmetics [41].

11.6 CONTRIBUTION TO FOOD SECURITY

Globally, seafood, particularly fishes serve as major sponsor of national food security through direct consumption or export and trade. In the coastal areas of developing countries, fish is the major contributor to the daily diet of people. The countries with low GDP per capita are more likely to eat usually low valued fish, satisfying their animal protein requirements [37]. Throughout the world, almost one million persons are dependent on fish as a major provenance of animal protein; particularly in coastal belts, the reliance on fish is greater. Only ¼ of the world's population meets around 20% of their animal protein requirement from fish. However small island States depend utterly on fish [48].

Generally, the marine products are thought to be inexpensive in comparison to those obtained from terrestrial animals, and their products globally are as an imperative protein source. Numerous customary fish products as cured or dried fish can possibly be appropriately transported and stored, hence makes their readily available to the consumer. Owing to these facts, the contribution of fish in the human diet has exceeded above 25% in several developing nations and even greater in inaccessible parts of inland and coastal areas of certain countries [32]. The comparative share of fish to total animal protein differs significantly in different regions: the highest rate is in Oman, Egypt, Yemen, and Morocco ranging from about 15–25% [9]. Many evidences suggest that fish can perform an imperative role in neonatal, fetal, and maternal nutrition. Consuming fish 2 or 3 times per week are being stimulated as a part of a healthy and balanced diet both for childbearing females as well as the whole family [1, 37].

TABLE 11.1 Some Functional Foods Based on Integration of Marine-Derived Ingredients

Product	Marine-derived ingredient	Description	Major accomplishments	Reference
Beef Patties	Seaweed (*Undariapinnatifida*)	Low fat and low salt beef patties with 3.3% of *Undariapinnatifida*	Patties were softer, had less cooking losses and lower thawing. Higher mineral contents. High sensorial, technological, and nutritional properties.	[30]
Beverage (Milk, Fruit juice)	Chitosans; chitooligosaccharides (COS)	Cow milk and apple juice with 0:5 w:v % and COS less than 3 and 5 kDa	Antimicrobial effectiveness against E. Coli and Staphylococcus aureus. In low pH foods effective as food preservatives.	[35]
Bread	Marine-derived polysaccharide (Sodium alginate k-carrageenan)	Bread made with varied hydrocolloids of 0:10:5% (flour based).	Reduced moisture loss throughout bread storage. Exhibited antistaling properties are hindering crumb hardening.	[5]
Salad dressing	Cod liver oil	Enriched with 10 w=w % net fish oil or 20%.	Higher peroxide value than in oil emulsion en-riched milk after a storage period of 25 days.	[29]
Spaghetti	Micro-encapsulated EPA and DHA.	Spaghetti incorporated with 0:6, 1:2 and 1:8% (flour basis) of oil.	Higher fat, EPA, and DHA in spaghetti. Spaghetti with 1:2% microencapsulated oil had lesser loss of DHA and EPA (<10%).	[19]

11.7 MARINE PHARMACOLOGY

Evidence from archeological research shows that, in ancient times, drugs used were primarily extracted from marine organisms along with plants and animals. Marine algae are mostly used for intestinal disorders while sponges from the marine environment are used as hemostat in vaginal hemorrhages [43]. Rickets, usually known as deformity of bones during infancy is thought to be cured by cod liver oil. McCollum in 1992 extracted two fat-soluble vitamins A and D from cod liver oil. After tremendous research on cod liver oil, pharmaceutical companies commercialized vitamin D preparation made from cod liver oil against rickets. Similarly, fucol, an oil containing iodine, was commercialized [17].

The pharmacology of marine natural products (MNPs) is more often concomitant with MNPs chemistry instead of mainstay pharmacology. In early era, the MNP pharmacology has developed from wide-ranging investigations of marine life for unique MNPs providing numerous innovative chemical units irrespective of their pharmacological action into what now is reflected as the target line in drug discovery, aiming the specific diseases (e.g., cancer, inflammation, etc.), and other molecular targets as specific receptors or enzymes [14, 47]. Nonetheless, the universal pharmaceutical line remains very dynamic and embraces eight European Medicines Agency (EMEA). Food and Drug Administration (FDA) approved drugs and other several compounds. Only three drugs Yondelis®, Prialt®, and Carragelose® have been introduced in the market without any modification of original natural molecules [33].

11.8 POTENTIAL INDUSTRIAL IMPORTANCE OF SEAFOOD ENZYMES

Industrial enzymes are currently contributing to the world economy around 2 billion US$, out of which about 50% is contributed by food enzymes. In recent years a sound upsurge in number, applications, and annual turnover from enzymes has been observed. Marine source food enzymes assisting food processing have apprehended the interests of both food processors and regulators in most of the developed and industrialized nations. The accessibility of diversified enzymes in marine milieu has made the seafood and their by-products a prospective source for enzyme recovery [26].

According to one approximation, global fish and fish products processing ends in a 63.6 million metric tons annual discard, accounting for 45% of worldwide fish production. Enzymes being recuperated from seafood and byproducts support the ethical and environmental concerns regarding discards and enhance the economic position of the seafood companies [2]. Different enzymes from marine sources have found their applications in the food industry; subsequently, they are distinctive protein molecules not present in terrestrial sources. The characteristics of marine enzymes are different from their homologous protease obtained from animals that are warm-blooded, due to high tolerance of temperature, increased concentration of salts, low nutrient availability, and high pressure. These aspects of marine enzymes are linked to the predominant circumstances in their habitat, such as oceanic waves and hydrothermal apertures [26].

11.9   GLOBAL SEAFOOD SUPPLY CHAIN PROCEDURE

Seafood supply chain involves a lot of mediators between fisherman and the final consumer (Figure 11.4). Seafood supply chain starts from the fisherman or producer and ends with a consumer. End buyer, which comes before consumer, supplies fish to national and international markets, restaurants, and other foodservice organizations.

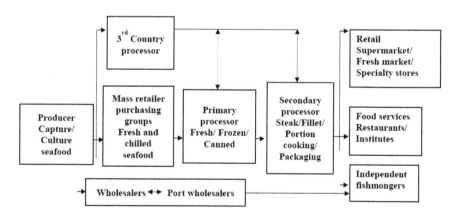

FIGURE 11.4   Global seafood supply chain procedure.

A third country processor is recently emerging that is a country to which products are exported to be processed. In some localities, fishermen bypass whole supply chain and entirely sell their products to the consumer on the beach or door to door. Hence, the seafood that is sold in the market possesses a supply chain consisting of many mid-chain players, who are involved in transformation, packaging, and transportation of product to final sale point [45].

11.10 SUMMARY

Marine food sources possess biodiversity, offering strong footings for their possible application as healthy foods and functional food ingredients that possess biological properties. Organisms present in marine environment include: seaweeds, macro or microalgae, crustaceans, fish, and fish by-products, some fungi and bacteria. These organisms are a source of healthy food and functional food ingredients including polysaccharides, lipids, proteins, and peptides, chitin, pigments, minerals, vitamins, and phenolic compounds. Moreover, industrial application of marine foods includes the development of functional foods, nutritional supplements, pharmaceuticals, enzymes, therapeutic agents providing benefit to humans. Marine food sources can also contribute significantly to uplifting food security issues, particularly in developing countries. The utilization of these foods can be increased through the provision of state of the art supply chain and management facilities including cold stores so that the products can reach the end users in benefitting manner.

KEYWORDS

- chitin
- crustaceans
- enzymes
- fish
- functional foods
- lipids
- macro or microalgae
- nutraceuticals
- omega-3 fatty acids

- peptides
- pharmaceuticals
- phenolic compounds
- polysaccharides
- polyunsaturated fatty acids
- proteins
- seafood
- seaweeds
- therapeutic agents

REFERENCES

1. Béné, C., (2003). When fishery rhymes with poverty, a first step beyond the old paradigm on poverty in small-scale fisheries. *World Development, 36*(1), 945–975.
2. Bhaskar, N., Sachindra, N. M., Suresh, P. V., & Mahendrakar, N. S., (2010). Microbial reclamation of fish industry by-products. In: Montet, D., & Ray, R. C., (eds.), *Aquaculture Microbiology and Biotechnology* (pp. 249–275). Science Publisher Inc., Enfield NH, USA.
3. Bhatnagar, I., & Kim, S. K., (2010). Immense essence of excellence: Marine microbial bioactive compounds. *Marine Drugs, 8*(10), 2673–2701.
4. Bozarth, A., Maier, U. G., & Zauner, S., (2009). Diatoms in biotechnology: Modern tools and applications. *Applied Microbiology and Biotechnology, 82*(2), 195–201.
5. Chang, H. C., & Wu, L., (2008). Texture and quality properties of Chinese Fesh egg noodles formulated with green seaweed (*Monostromanitidum*) powder. *Journal of Food Sciences, 73*(8), 398–404.
6. Chojnacka, K., Saeid, A., Witkowska, Z., & Tuhy, L., (2012). Biologically active compounds in seaweed extracts: Prospects for the application. *The Open Conference Proceedings Journal, 3*(1), 20–28.
7. Ebel, R., (2010). Terpenes from marine-derived fungi. *Marine Drugs, 8*(8), 2340–2368.
8. El-Gamal, A. A., (2010). Biological importance of marine algae. *Saudi Pharmaceutical Journal, 18*(1), 1–25.
9. FAO, (2006). *The State of World Fisheries and Aquaculture,* Food and Agriculture Organization of the United Nations, Rome, http://www.fao.org/3/a-i5555e.pdf Accessed on January 28, 2017.
10. Fleurence, J., Morançais, J., Dumay, J., Decottignies, P., Turpin, V., Munier, M., Garcia-Bueno, N., & Jaouen, P., (2012). What are the prospects for using seaweed in human nutrition and for marine animals raised through aquaculture? *Trends in Food Science & Technology, 27*(1), 57–61.
11. Freitas, A. C., Rodrigues, D., Rocha-Santos, T. A., Gomes, A. M., & Duarte, A. C., (2012). Marine biotechnology advances towards applications in new functional foods. *Biotechnology Advances, 30*(6), 1506–1515.

12. Freitas, A. C., Pereira, L., Rodrigues, D., Carvalho, A. P., Panteleitchouk, T., Gomez, A. M., & Duarte, A. C., (2015). Marine functional foods. In: Kim, Se-Kwon, (ed.), *Springer Handbook of Marine Biotechnology* (pp. 969–994). Springer, New York.

13. Freitas, A. C., Rodrigues, D., Carvalho, A. C., Pereira, L., Pant, T., Gomez, A. M., & Duarte, A. C., (2015). Marine functional foods: Omega 3 fatty acids produced from microalgae. In: Kim, Se-Kwon, (ed.), *Springer Handbook of Marine Biotechnology* (pp. 969–994). Springer, New York.

14. Glaser, K. B., & Mayer, A. M., (2009). Renaissance in marine pharmacology: From preclinical curiosity to clinical reality. *Biochemical Pharmacology, 78*, 440–448.

15. Guérard, F., & Decourcelle, N. C., (2010). Recent developments of marine ingredients for food and nutraceutical applications: A review. *Journal des Sciences Halieutiqueet Aquatique, 2*, 21–27.

16. Hayes, M., Carney, B., Slater, J., & Brück, W., (2008). Mining marine shellfish wastes for bioactive molecules: Chitin and chitosan, Part B: applications. *Biotechnology Journal, 3*(7), 878–889.

17. Holick, M. F., (2011). Photobiology of vitamin D. *Vitamin D, 1*, 13–22.

18. Hurst, D., (2006). *Marine Functional Foods and Functional Ingredients*, Marine Institute, https://oar.marine.ie/bitstream/handle/10793/92/FINAL_MFF_Briefing   Doc15Jan07 ForesightSeries5.pdf?sequence=1&isAllowed=y Accessed on 21 Feb 2018.

19. Iafelice, G., Caboni, M. F., Cubadda, R., Di Criscio, T., Trivisonno, M. C., & Marconi, E., (2008). Development of functional spaghetti enriched with long-chain omega3 fatty acids. *Cereal Chemistry, 85*(2), 146–151.

20. Ibañez, E., & Cifuentes, A., (2013). Benefits of using algae as natural sources of functional ingredients. *Journal of Science of Food and Agriculture, 93*(4), 703–709.

21. Imhoff, J. F., Labes, A., & Wiese, J., (2011). Bio-mining the microbial treasures of the ocean: New natural products. *Biotechnology Advances, 29*(5), 468–482.

22. Itoi, S., Yuasa, K., Washio, S., Abe, T., Ikuno, E., & Sugita, H., (2009). Phenotypic variation in *Lactococcuslactis* subsp. lactis isolates derived from intestinal tracts of marine and freshwater fish. *Journal of Applied Microbiology, 107*(3), 867–874.

23. Je, J. Y., Park, P. J., Kim, E. K., Park, J. S., Yoon, H. D., Kim, K. R., & Ahn, C. B., (2009). Antioxidant activity of enzymatic extracts from the brown seaweed *Undariapinnatifida*by electron spin resonance spectroscopy. *LWT-Food Science and Technology, 42*(4), 874–878.

24. Kadam, S., & Prabhasankar, P., (2010). Marine foods as functional ingredients in bakery and pasta products. *Food Research International, 43*(8), 1975–1980.

25. Kim, S. K., & Wijesekara, I., (2010). Development and biological activities of marine-derived bioactive peptides: A review. *Journal of Functional Foods, 2*(1), 1–9.

26. Kim, S. K., & Ratih, P., (2012). Potential role of marine algae on female health, beauty, and longevity. Chapter 4, In: Se-Kwon Kim, (ed.), *Marine Medicinal Foods: Implications and Applications-Animals and Microbes* (pp. 41–55). Elsevier B. V., New York.

27. Klotz, U., (2006). Ziconotide: A novel neuron-specific calcium channel blocker for the intrathecal treatment of severe chronic pain: Review. *International Journal of Clinical Pharmacology and Therapeutics, 44*(10), 478–483.

28. Larsen, R., Eilertsen, K. E., & Elvevoll, E. O., (2011). Health benefits of marine foods and ingredients. *Biotechnology Advances, 29*(5), 508–518.

29. Let, M. B., Jacobsen, C., & Meyer, A. S., (2007). Lipid oxidation of milk, yogurt, and salad dressing enriched with neat fish oil or pre-emulsified fish oil. *Journal of Agriculture and Food Chemistry, 55*(19), 7802–7809.

30. López-López, I., Cofrades, S., Yakan, A., Solas, M. T., & Jimenez-Colmenero, F., (2010). Frozen storage characteristics of low-salt and low-fat beef patties as affected by Wakame addition and replacing pork backfat with olive oil-in-water emulsion. *Food Research International, 43*(5), 1244–1254.

31. Lordan, S., Ross, R. P., & Stanton, C., (2011). Marine bioactives as functional food ingredients: Potential to reduce the incidence of chronic diseases. *Marine Drugs, 9*(6), 1056–1100.

32. Maclean, N., & Laight, R. J., (2000). Transgenic fish: An evaluation of benefits and risks. *Fish and Fisheries, 1*(2), 146–172.

33. Martins, A., Vieira, H., Gaspar, H., S., & Santos, S., (2014). Marketed marine natural products in the pharmaceutical and cosmeceutical industries: Tips for success. *Marine Drugs, 12*(2), 1066–1101.

34. Mata, T. M., Martins, A. A., & Caetano, N. S., (2010). Microalgae for biodiesel production and other applications: Review. *Renewable and Sustainable Energy Reviews, 14*(1), 217–232.

35. Mikuš, L., Kováčová, M., Dodok, L., Medved'ová, A., Mikušová, L., & Šturdík, E., (2013). Effects of enzymes and hydrocolloids on physical, sensory, and shelf-life properties of wheat bread. *Chemical Papers, 67*(3), 292–299.

36. Mohamed, S., Hashim, S. N., & Rahman, H. A., (2012). Seaweeds: Sustainable functional food for complementary and alternative therapy. *Trends in Food Science & Technology, 23*(2), 83–96.

37. Mozaffarian, D., & Rimm, E. B., (2006). Fish intake, contaminants, and human health: Evaluating the risks and the benefits. *Journal of American Medical Association, 296*(15), 1885–1899.

38. Ngo, D. H., Wijesekara, I., Vo, T. S., Van Ta, Q., & Kim, S. K., (2011). Marine food-derived functional ingredients as potential antioxidants in the food industry: An overview. *Food Research International, 44*(2), 523–529.

39. O'Sullivan, L., Murphy, B., McLoughlin, P., Duggan, P., Lawlor, P. G., Hughes, H., & Gardiner, G. E., (2010). Prebiotics from marine macroalgae for human and animal health applications. *Marine Drugs, 8*(7), 2038–2064.

40. Özer, B. H., & Kirmaci, H. A., (2010). Functional milks and dairy beverages. *International Journal Dairy Technology, 63*(1), 1–15.

41. Pulz, O., & Gross, W., (2004). Valuable products from biotechnology of microalgae. *Applied Microbiology and Biotechnology, 65*(6), 635–648.

42. Rateb, M. E., & Ebel, R., (2011). Secondary metabolites of fungi from marine habitats. *Natural Product Reports, 28*(2), 290–344.

43. Rauter, A., Palma, F. B., Justino, J., Araújo, M. E., & Dos Santos, S. P., (2013). Natural products in the new millennium: Prospects and industrial application. In: Rauter, et al., (eds.), *Chemical Variability of the Volatile Metabolites From the Caribbean Corals of the Genus Gorgonia* (pp. 463–474). Springer Science & Business Media. New York.

44. Riediger, N. D., Othman, R. A., Suh, M., & Moghadasian, M. H., (2009). Systemic review of the roles of n–3 fatty acids in health and disease. *Journal of the American Dietetic Association, 109*(4), 668–679.

45. Roheim, C. A., (2008). Seafood supply chain management: Methods to prevent illegally-caught product entry into the marketplace, *IUCN World Conservation Union-US for the Project PROFISH Law Enforcement, Corruption, and Fisheries Work*, https://cmsdata. iucn.org/downloads/supply_chain_management_roheim.pdf Accessed on Jan 7, 2017.

46. Samarakoon, K. W., Elvitigala, D. A. S., Lakmal, H. C., Kim, Y. M., & Jeon, Y. J., (2014). Future prospects and health benefits of functional ingredients from marine bio-resources: A review. *Fish and Aquatic Science, 17*(3), 275–290.

47. Schoffski, P., Dumez, H., Wolter, P., Stefan, C., Wozniak, A., & Jimeno, J., (2008). Clinical impact of trabectedin (ecteinascidin–743) in advanced/metastatic soft tissue sarcoma. *Expert Opinion Pharmacotherapy, 9*(9), 1609–1618.

48. Suleria, H. A. R., (2016). Bioactive compounds from Australian Blacklip Abalone (*Haliotisrubra*) processing waste. *PhD Thesis, School of Medicine* (p. 152). The University of Queensland, https://espace.library.uq.edu.au/view/UQ:406669, Accessed on 27 Sep 2016.

49. Suleria, H. A. R., Osborne, S., Masci, P., & Gobe, G., (2015). Marine-based nutraceuticals: An innovative trend in the food and supplement industries. *Marine Drugs, 13*(10), 6336.

50. Wijesekara, I., & Kim, S. K., (2010). Angiotensin-I-converting enzyme (ACE) inhibitors from marine resources: Prospects in the pharmaceutical industry. *Marine Drugs, 8*(4), 1080–1093.

51. Xia, W., Liu, P., Zhang, J., & Chen, J., (2011). Biological activities of chitosan and chitooligosaccharides. *Food Hydrocolloids, 25*(2), 170–179.

CHAPTER 12

FORMULATION OF PROBIOTIC-BASED FUNCTIONAL FISH FEEDS: ECO-FRIENDLY APPROACH

KEERTHI THALAKATTIL RAGHAVAN and
GLINDYA BHAGYA LAKSHMI

ABSTRACT

Fish farming sector of aquaculture is the fastest growing food producing sector. Intensification to maximize production results in disease outbreak; and common disease management uses veterinary medicines like antibiotics and another chemotherapeutic. The misuse creates antimicrobial resistance, causing a serious threat to the entire humanity. This threat can be overcome using probiotics. Probiotics are healthy microorganisms and substances, which contribute to intestinal microbial balance. Solid-state fermentation using locally available agricultural residue as major feed ingredient supports the growth of probiotics and therefore it can be employed for the formulation of functional feeds. The formulation, steady supply and daily administration of probiotic-based functional feed are a safe method for disease prevention, management of aquaculture. Various other functions due to use of probiotics in aquaculture are: growth promotion, inhibition to pathogens, and improvement in nutrient digestion, immunity, and water quality. In addition to probiotics, the administration of prebiotics, synbiotics, and cobiotics in aquaculture as dietary supplement boosts the functional property of the feed and promotes environment-friendly aquaculture.

12.1 INTRODUCTION

According to Food and Agriculture Organization of United States [14], "*aquaculture is considered as a lucrative industry for high-quality animal protein producing sector and have the potential to meet the increasing market for aquatic food in the world.*" Globally, aquaculture is expanding both in intensified and diversified directions with ago alto optimize the efficiency of production to maximize profit. Aquaculture also minimizes the gap between supply and demand of fish [15]; and aquaculture aims to produce a blue revolution.

Aquaculture gained relevance in the economic activities of many countries with extremely diversified species of fish, technologies, and farming systems employed [7]. Pollution and impairment, depletion of resources and disease outbreak are major constraints that arise with the rapid progress of aquaculture [37]. The report of World Aquaculture 2012 forecasts that greater than 50% of world's edible fish come from aquaculture and the growth rate is due to several factors such as [16]: (a) Maximum sustainable exploitation of many fisheries; (b) More demand for super quality, safe, healthy aquatic products; and (c) Minimum greenhouse gas emission from aquaculture [16]. By 2050, it is estimated that the aquaculture sector will have the potential to feed 9 billion persons on our planet [28].

Main technological advances that contributed for increasing aquaculture production in a sustainable manner are: improving the disease prevention and control; and improving feed formulation, manufacture, and feeding practices. Among these constraints, fish, and shellfish disease is a serious threat to the commercial success of aquaculture. Veterinary medicines are commonly used as preventive measures to control these diseases. The use and abuse of antimicrobial drugs in human medicine, agriculture, and in aquaculture exert pressure on the microbiota and thus create anti-microbial resistance [17]. Another major issue is related to the consumption of antibiotic-treated fish, the antibiotic residue accumulation within the fish, and related health problems.

Generally, in the aquaculture system, pathogens can go in and out and interact with wild populations. As a result, alien species are introduced and may cause a major problem for the local species. Animal health management depends on veterinary medicines and therapeutic products, and their optimum doses are safe, but the overuse and misuse create problems. A wide range of antibiotics are used in aquaculture to treat bacterial diseases. Research has confirmed that the regular preventive application of antibiotics in aquaculture has led to the development of bacterial resistance. Biomagnification is another problem caused by antibiotic use, and the release of antibiotic into nature/ecosystem may enter the food chain (Figure 12.1). The bacteria pathogenic to humans could be resistant to the antibiotic; therefore bacterial resistance may create risk not only to the aquaculture sector but also to human health. Because of the legal prohibition on antibiotics, an alternative strategy has been introduced to replace antibiotics for the control of bacterial infection in fish aquaculture [7]. The use of biocontrol agents (like viruses, algae, and bacteria) for disease management is still in the infant phase. Environmentally acceptable, economically feasible, socially equitable approach to animal health management in aquaculture can contribute to its sustainable development.

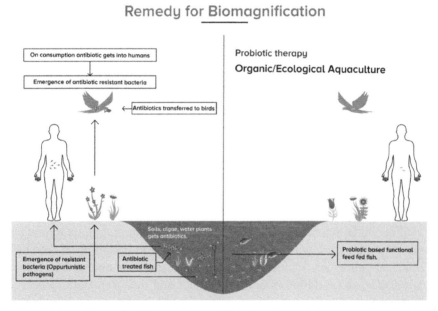

FIGURE 12.1 (**See color insert.**) Biomagnification of antibiotics (Left side); Probiotic fish feed application (Right side) in aquaculture.

This chapter focuses on the alternative strategy for antibiotic treatment in aquaculture, introducing the concept of the use of probiotics, prebiotics, synbiotics, and cobiotics. The chapter also discusses the formulation of probiotic-based fish feed adapting solid-state fermentation using locally available agro-industrial residue as the solid substrate. Promoting organic aquaculture will be a blessing for our environment.

12.2 IMPORTANCE OF NORMAL FLORA

Normal flora is continuously interacting with the environment, and its disturbance leads to disorders and can later develop as a disease. An interface exists between the internal environments especially the alimentary canal and the external body.

Gastrointestinal (GI) microbiota has a significant effect on the health of both humans and animals. The multi-functions performed by intestine are digestion, absorption, water, and electrolyte equilibrium, metabolism, and immunity. New scientific developments have opened to know more about the complex and delicately balanced microbiota of the GI tract. Physiological functions contributed by the normal flora to the overall growth are: degradation of certain food components, production of certain B vitamins, stimulation of the immune system and production of digestive and protective enzymes [33]. Microorganisms present in the water and in the sediments (sewage/feces) influence the microbiota of fishes and make changes in the gills and digestive tract. At the egg/larval stage, the colonization of microorganisms starts and continues to the developmental stages of fish [43]. Only sketchy information is available on the importance and role of bacteria associated with the healthy fishes. Opportunistic pathogens are responsible for much of the fish diseases. The interrelationships between aquatic organisms and its normal flora are complex and finely tuned [43]; Compared to terrestrial animal's aquatic organisms, normal flora shows a closer relationship between the surrounding environment and intestinal microbiota. Major functions played by the GI microbiota of fishes are: digestion and immunity to prevent diseases, competition with opportunistic bacteria and inhibition of the colonization of pathogens. Variability among GI microbes is one of the specialties in fishes, and it differs in both freshwater and seawater. With respect to salinity and species, the microbiota is different. It constitutes aerobic, facultative aerobic and obligate anaerobic bacteria. These bacteria can be divided into indigenous or autochthonous and transient or allochthones. Indigenous bacteria can colonize within the fish gut where transient cannot. Factors responsible for the variability of microorganisms in different fish species are: intestinal microenvironment, nutrition, age, and environmental factors, etc. Intestinal microbiota produces inhibitory compounds, which has the capacity to prevent or control the colonization of potential pathogens in fish. Studies have revealed that it is possible to manipulate the microbiota of developing fish by probiotics and prebiotics, to a certain extent [6].

12.3   WHAT IS PROBIOTICS?

"Probiotika" the original word for 'Probiotic' used as an antonym for antibiotic etymologically comes from the Latin word "pro" means favor and Greek word "bios" means life. Parker in 1974 first used to describe probiotics as microbial feed/food supplement, and he defined probiotic as "*organisms and substances, which contribute to intestinal microbial balance.*" Commonly used terms for probiotic are 'beneficial,' 'friendly,' 'healthy' bacteria or 'probiont' [62]. Fuller in 1989 gave a widely accepted definition for probiotics as [22] "*live microbial feed supplement which beneficially affects the host animal by improving its intestinal microbial balance.*"

Knowledge about probiotic has increased, and its application in aquaculture is recent. The aquaculture probiotics include both Gram-positive and Gram-negative bacteria. The first application of probiotics in aquaculture was by Kozasa, and he used *Bacillus toyoi* spores as a feed additive for *Seriolaquinqueradiata* (Yellow tail) [36]. He extended the definition of probiotics to 'water additives' in aquaculture. Probiotic organisms come under *Generally Regarded as Safe* (GRAS) category [41]. Many researchers have dealt with probiotics from empirical use to scientific approach.

Initially, aquaculture uses probiotics prepared for terrestrial animals. The incorporation of probiotics in farm animal feed supplements started in 1970, but its use in aquaculture is relatively a new concept [18]. Different products with varying success rate are being made available in commercial aquaculture to maximize productivity. Majority of the microbes present in the aquatic animals are transient in nature and may change suddenly with the microbial load from food and water. Application of probiotics in aquaculture must resolve some legal issues, such as: treating probiotic with the animal (fish), introducing probiotics to the water, and the method of application should be cost-effective.

12.3.1 HOW DOES PROBIOTIC THERAPY DIFFER FROM USE OF ANTIBIOTICS?

Antibiotics are effective against treating diseases but cannot solve the basic problem. On the other hand, use of antibiotic and chemical destroys much of the useful bacteria residing within the aquatic organism and the bacteria present in the water column in the pond. During the primary level of infection, the pathogen gets into the host and needs to adhere to the tissue surfaces. This can be prevented by creating competition for adhesion sites on GI tract or other tissue surfaces. Probiotic acts by producing an antimicrobial effect by secreting antibacterial substances such as bacteriocins and organic acids, prevent pathogen adhesion to the intestine, competing for nutrients required for the pathogen to exist, modulating the immune mechanism, regulating allergic response and reducing multiplication of cancer in mammals. Based on this, it is postulated that probiotics provided at sufficient concentration and viability, will beneficially affect the health of the host in a friendly and effective manner. Aquaculture probiotics also degrade the organic matter, and as a result, the water quality would improve, reduce unpleasant odors, increase zooplanktons and ultimately increase aquaculture production. Only those bacteria, which can adhere to enteric mucus and wall surfaces establish in fish intestine. In aquaculture, probiotic application reduces the disease outbreak.

12.3.2 WHY LAND PROBIOTICS NOT USED IN AQUACULTURE?

Microorganisms play a crucial role in aquaculture such as: maintaining oxygen levels by the activities of algae and bacteria, affecting water quality factors (pH, ammonia), contributing to the food web, disease control, and environmental impact of the effluent [42]. In the beginning, commercial probiotic for terrestrial animals was used in aquaculture, but the existence becomes uncertain in the GI tract of aquatic animal [1]. Therefore, the probiotics used for aquaculture must be highly specific and should be different from land animals. Aquatic microbes are highly specific, because the most efficient probiotics for aquaculture may be different from those of terrestrial species. The survival of land probiotics is uncertain in the GI tract of aquatic animals and thus reduces the desired beneficial effect [1]. Most of the aquaculture probiotics belong to Lactic acid bacteria, Genus *Vibrio*, Genus *Pseudomonas* and genus *Bacillus*.

Gatesoupe [24] reported that the application of probiotics is promising in aquaculture; and clearly pointed out the necessity of the suitable method of introduction of probiotics with effective dose and solution to keep the probiotic alive in dry pellets [24].

12.3.3 SCREENING, IDENTIFICATION AND CHARACTERIZATION OF AQUACULTURE PROBIOTICS

It is recommended to isolate the probiotic bacterial strains from fish (gills, fish body, GI tract) and applied to the same fish for better establishment/colonization results. Basic screening involves Gram's stain, colony morphology characteristics, shape, color, and motility (Figure 12.2). Among the dominant isolates, further screening is necessary to determine their inhibitory spectrums against the pathogenic bacteria. Biochemical identification followed by their molecular identification (16S rDNA gene sequence) confirms the phylogenetic relationship of probiotic bacteria. Bacteriocins are a proteinaceous compound produced by a microorganism that inhibit the growth of other bacteria; and BLIS assay of probiotic strains are also effective against controlling pathogens considered as efficient weapons to protect diseases.

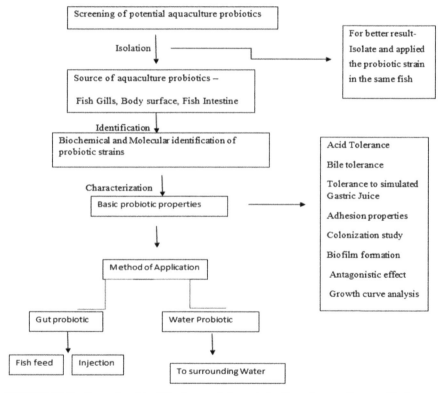

FIGURE 12.2 Screening, identification, and characterization of aquaculture probiotics.

12.3.4 QUALITIES OF GOOD PROBIOTIC [22]

- It should modulate the immune response.
- It should not be harmful to the host applied.
- Maintain viability for prolonged period and field conditions.
- Preferably a large number of viable probiotic exerts better results.

- Probiotic bacteria have the capacity to exist and metabolize in acidic conditions present in GI tract, bile salts, organic acids, and proteases.
- Probiotic should be from fish.
- Probiotic strain must be competent enough to exert a beneficial (e.g., Growth promoter, disease resistance) effect on the host applied.
- Produce antibacterial compounds like organic acids, bacteriocins, siderophores, and hydrogen peroxide.

12.3.5 INFLUENCE OF PROBIOTICS ON SURVIVAL, GROWTH, AND PRODUCTION OF FISHES

Several research studies have proven the influence of probiotics on the growth rate of fishes. The study was conducted on the influence of probiotics on the growth and gut microbial supplementation of *Carassiusauratus* [2]. Scientists also studied [2] the effects of different concentrations of *Bacillus* spp., and enrichment times on growth factors and survival rate in Persian sturgeon *Acipencerpersicus* larvae via feeding by bio-encapsulated Daphnia magna [19]. The influence of dietary administration of two closely linked probiotic species (*Shewanella putrefaciens* and *S. baltica*) on growth, biochemical composition, histology, and intestinal microbiota of farmed juvenile *Solea senegalensis* [23] was investigated. The growth performance, the innate immune response, and resistance to pathogens of a commercially obtained probiotic supplement diets were evaluated in cobia *Rachycendroncanadum* [25]. Study on *Bacillus* probiont enriched feed of live-bearing ornamental fishes also confirmed growth performance, survival, healthy gut environment and it was proved that the supplementation of higher concentration of the probiont will not lead to significantly improved health status of the host applied [26]. Table 12.1 indicates selected aquaculture probiotics.

12.3.6 FUNCTIONAL FOOD AND PROBIOTICS

Foods provide taste, aroma, and nutritive value, therefore all foods are considered as functional. New trend is that beyond nutrition the food may modulate for various physiological functions to reduce chronic diseases or optimize health. Globally these foods are referred to as functional foods [31]. Japan in the 1980s first conceptualized functional foods and is recognized as Foods for Specific Health Use (FOSHU). 'Nutrigenomics' studies the inter-relationships between diet and diseases [29]. Globally many research works

TABLE 12.1 List of Some Aquaculture Probiotics

Probiotics	Source	Application	Probiotic effects	Reference
Bacillus coagulans	Pond sediment of shrimp	Larvae of (*Penaeus vannamei*)	Water quality, survival rate, digestive enzymes	[66]
Bacillus spp.	Commercial product	Aquaculture	Produce bacteriocin, provide immunostimulation	[60]
Bacillus subtilis	*Macrobrachiumrosen-bergii*	Larvae of *Macrobrachium rosenbergii*	Growth, development, and survival of larvae	[35]
Bacillus subtilis		*Cyprinus carpio* fingerlings	Growth promoting	[8]
Bacillus subtilis, Bacillus licheniformis, Lactobacillus spp. and*Arthrobacter spp.*	Fish feed	Cobia (*Rachy- centron canadum*).	Non-specific immunity and protection against the fish pathogen *Vibrio harveyi*	[25]
Bacillus subtilis, Lactococcuslactis), Saccharomyces cerevisiae), (BF + B. subtilis, heat-killed bacteria of B. subtilis, L. lactis and S. cerevisiae)	Microbial Type Culture Collection (MTCC), Chandigarh, India	Fish feed Rohu, *Labeorohita* fingerlings	Growth, nutrient digestibility, intestinal enzyme functions	[40]
Carnobacterium sp	*Hepialusgonggaensis* larvae	Aquaculture	Growth performance and intestinal enzyme functions	[65]
Lactobacillus acidophilus	Vitagen, a yogurt drink commercial product	Fingerlings of African catfish (*Clarias gariepinus*)	Growth, hematology, and immunoglobulin concentration	[3]
Lactobacillus delbrueckiidelbrueckii	Adult sea bass gut	Live Feed additives	IGF–1, myostatin, cortisol gene expression	[10]
Streptococcus thermophilus	Turbot larvae	Feed additives	Growth promoter	[24]
Vibrio alginolyticus	Beach Sand	Feed	Pathogen inhibition	[6]

have identified the mode of action of functional foods to inhibit chronic diseases for the improvement of health. Most of the available functional foods are from dairy products, and it has limited or nil application in aquaculture. Applications of functional food for aquaculture are based on agro-industrial residue (especially cereals like wheat bran, rice bran, etc.) as a substrate for functional food because of high micro and macronutrients, rich dietary fiber. Biotechnology can greatly influence the functional foods and nutrigenomics. The biotechnology-derived products such as golden rice and iron-enriched rice have the potential to prevent iron deficiency anemia and Vitamin A deficiency [29]. Probiotic based functional foods are commonly used in health-promoting and therapeutic, prophylactic, and growth supplements aspects for humans and animals [56, 58].

12.3.6.1   PROBIOTICS AS GROWTH PROMOTER

Currently, the use of antibiotics as growth promoter (AGP) is declining, and the focus is concentrated on manipulating microbiota within the animal to desired flora through diet. Products like probiotics, prebiotics, enzymes, minerals, etc. are available in the market for the maximum production of the animal in a healthier way [11]. Probiotics are used for growth improvement in both edible and ornamental fishes. Probiotics when applied for a longer time, it colonizes, and multiplies within the GI tract of fishes and produce their multiple benefits. The use of probiotic Biogens® as a feed additive for Nile tilapia stimulates the productive growth performance and nutrient consumption [13].

12.3.6.2   PROBIOTICS INHIBIT PATHOGENS

Antibiotics are used to prevent diseases in humans, animals, and in aquaculture. However, it creates various health problems to the host, such as: generation of bacterial resistance, alteration in the gastrointestinal flora and the presence of antibiotic residue in host tissues. However, now its trend is for using natural products free of chemicals and antibiotics to prevent disease rather than for treatment. Toxic or inhibitory chemicals produced by the microorganisms against other microorganisms are called as antagonistic compounds. Probiotics can produce bacteriocins, which is a chemical substance with bacteriostatic or bactericidal effect on pathogens. Other than bacteriocins, the antibactericidal effect of the probiotics is also due to production, siderophores, antibiotics, and enzymes like proteases, lyso-zymes, production of H_2O_2 [60]. The release of a chemical substance by the microbes plays an important role in retaining a balance between useful normal microbiota and harmful microbiota. The presence of aquaculture probiotic in the intestine, or on its body, or in its water surrounding is to help for removing pathogenic bacteria or prevent its multiplication [46]. The exact mechanism of the antiviral capability of probiotic bacteria has not been studied properly. Bacterial antagonism of probiotics has effectively been proved *in vitro* and but has failed *in vivo* that is one of the major limitations of probiotic research. More studies are required to clarify the exact mode of action probiotic bacteria in aquaculture.

12.3.6.3 PROBIOTICS IMPROVES NUTRIENT DIGESTION

Probiotic bacteria can synthesize digestive enzymes amylases, proteases, and lipases and also growth factors like vitamins, fatty acids, and amino acids. Supplied probiotic feed fishes can absorb more efficiently nutrients [13]. The probiotic yeast *Dicentrarchus labrax* produces polyamines, spermine, and spermidine essential for differentiation and maturation of the GI tract in mammals and also secrets trypsin and amylase that can help in digestion [59]. *Bacillus* species contribute to improved nutrient digestion, and some positive results have been observed in a diet supplemented with 0.5 g of *Bacillus cereus* strain. Supplementation of *B. subtilis*, *B. licheniformis* and exo-enzymes secreted by *Bacillus* positively supports the activities of the fishes and increase enzymatic digestion by secreting chitinase, protease, cellulase, lipase, and trypsin.

12.3.6.4 PROBIOTICS IMPROVE IMMUNE ACTIVITY

Fishes play a vital role in defense against pathogens and have some simi-larity like that of mammals. It can be either specific or non-specific, and specific may be innate or acquired. Regardless of the pathogens, both humoral and cellular nonspecific immune system provides innate immunity to fishes. The supplementation of viable probionts on rainbow trout induces higher immune activity in the head kidney leucocyte phagocytosis, and serum complements [48]. Immunological activity of probiotics may be due to enhancing of pathogen phagocytosis [47], stimulating specific antibody secreting cell response and also by cytokine production [49]. Probiotics produce bioactive peptide by the hydrolization of milk proteins, which will activate gut immune responses.

Interaction of probiotics with the epithelial cells regulates the secretion of anti-inflammatory cytokines which results in decreased level of inflammation. In rainbow trout gut cells IL–1β, IL–8, TNF-α, and TGF-β expression was not induced by the application of the probiotic *Carnobacteriumdivergens*B33 and *C. maltaromaticum* B26. But significantly higher levels of IL–1β and TNF-α expression was detected in head kidney cells induce the anti-inflammatory effect.

12.3.6.5   PROBIOTICS IMPROVE WATER QUALITY

Gram-positive bacteria in the water can efficiently transform organic matter to carbon dioxide, than the Gram-negative bacteria. Research has proved that probiotic present in aquaculture ponds balances the production of phytoplankton and also helps to reduce the deposition of dissolved and particulate organic carbon during the growing time. *Bacillus, Pseudomonas, Cellulomonas, Enterobacter, Nitrobacter*, and *Rhodopseudomonas* are some of the species that can improve water quality [60]. Nitrification is the process of transformation of ammonia to nitrate by the ammonia-oxidizing bacteria and nitrite-oxidizing bacteria. Very limited published evidence is available for improving water quality by probiotics. During application, various probiotics like photosynthetic bacteria, *Bacillus, nitrifiers*, and *denitrifiers* are combined, and therefore probiotic application are labeled as multifunctional.

Nile tilapia supplemented with a combination of Bacillus *subtilis* and *B. licheniformis* assessed the water quality parameters such as dissolved oxygen concentration of 5.7–6.3mg/ L, ammonia level of 0.36–0.42mg/L, and pH between 6.3 and 8.2 [13]. Chinese application of probiotics in aquaculture mainly focused on photosynthetic bacteria. Addition of photosynthetic bacteria to the feed or to the pond has been observed to improve the growth of the fishes and water quality [51].

12.3.7   CHALLENGES FOR INCORPORATING PROBIOTICS IN FISH FEED

Once the probiotic has been isolated, identified, and characterized as aquaculture probiotic; next step is to find out the efficient method of application. Based on the mode of application, probiotics are classified as 'gut probiotics (one blended with feed and applied orally to enhance the microbiota of the gut)' and 'water probiotics (one that can multiply in a water medium and eliminate the pathogens by utilizing the available nutrients)' [55]. Studies have proved that the oral administration of probiotic incorporated feed would give better results for the establishment within GI tract and simultaneously will transform through the surrounding water also. Heating is another major problem, which affects the viability of probiotics during the manufacturing of pellets [4].

The basic requirement of probiotic feed is that it should remain viable and contain enough numbers of probiotic strains, up to the expiration date [20]. Marketing and technological challenges faced by probiotic based functional foods include the retention of viability of the probiotic bacteria [39]. During preparation, storage, and transportation, many active probiotic cultures can become dead, and some probiotic strains become dead on passing through the GI tract. Because of low temperature, the formation of crystals occurred on bacterial cells during refrigeration, which also reduces the lives of probiotic bacteria [50]. In general, 50 to 200% of probiotic cells are incorporated into products.

Plant-based fish feed formulation faces a major crisis of having anti-nutritional factors (ANFs), which will inhibit the growth of probiotic strains, affects the digestion and absorption of nutrients. Much of the ANFs present in the plants are non-labile except in soybeans which can be easily inactivated by normal soybean processing. Reports indicate that the fermentation of plant materials may considerably reduce the anti-nutritional factors [52]. Traditionally fish meal was used as the main protein source; but because of the increased rate, modern feed formulation promotes soybean meal as the principle protein source. Unprocessed soybean meal has the presence of trypsin inhibitors, which bind to trypsinogen and chymotrypsinogen and prevent its conversion to active form thereby limiting protein digestion [12].

12.3.8 BIOTECHNOLOGICAL APPLICATION OF PROBIOTICS IN AQUACULTURE

Environmental biotechnology is defined as the use and application of living cells (bacteria, fungi, etc.) and their enzymes for the sustainable production of biomaterials, biochemicals, and biofuels. 'Red' biotechnology is confined to pharmaceutical (health care) and 'Green' biotechnology is related to the agricultural and food sector. Industrial (white) biotechnology offers economically feasible eco-friendly opportunities to the society (Figure 12.3). Fermentation technology is one of the core areas in industrial biotechnology. Raw materials or substrates employed for fermentation are by-products from agriculture (wheat bran, rice bran, copra cake, dried tapioca chips) and household sectors. By employing microorganisms, these organic products can be converted into the desired products (vitamins, enzymes) and the remaining microbial mass after fermentation can be converted into animal feed or fertilizers. Commercial production of feed must be a farmer-friendly and economically feasible, there comes the role of solid-state fermentation.

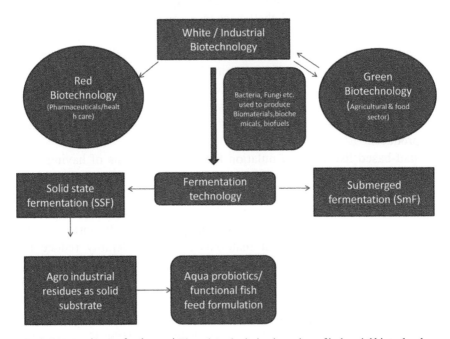

FIGURE 12.3 **(See color insert.)** Flowchart depicting branches of industrial biotechnology.

12.3.9 *ROLE OF SOLID STATE FERMENTATION IN FEED FORMULATION*

Fermentation is the process in which there is a biological conversion of complex organic substrates into simple compounds by various microorganisms like fungi and bacteria. In addition to these, secondary metabolites or otherwise bioactive compounds like antibiotics, peptides, enzymes, and growth factors are also produced. Solid-state fermentation (SSF) is the fermentation involving solids substrates in the absence (or near absence) of free water; though the substrate must have sufficient moisture to promote the growth and metabolism of micro-organism [45]. SSF has been used for producing chemicals, enzymes, and value-added products like protein supplemented cattle feed, which involves the utilization of cheap agricultural residues. Nature of solid substrate involved in SSF is the key factor, and its selection depends on economic feasibility, availability, and nutritional quality. An optimum substrate size is necessary in SSF: small ones provide increased surface area for microbial growth, and if it is too small it results in poor microbial growth and substrate agglomeration [44, 45].

Food, animal feed, pharmaceutical industries and biofuel use agro-industrial residues and are a good source of renewable carbon and energy; and using this for SSF is an added environmental friendly advantage by removing agricultural waste. The absence of free aqueous water is also environment protective (ecological) advantage of SSF. From the economic point of view, SSF requires only minimum sterility and instrument.

Our ancestors knew various health benefits by consuming fermented products and since 1970s scientists have focused on the effects of consumption of fermented probiotic products. Two major types of fermentation technology are SSF and submerged fermentation (SmF). In the 1940s without any logical reason, SSF was completely avoided in western countries by the discovery of penicillin through SmF. During the 1950s–1960s, steroid transformations from fungal cultures were reported by SSF; and another milestone occurred in 1960s–1970s when mycotoxin production was reported by SSF. Globally, researchers again got attention to SSF by the production of protein-enriched cattle feed involving utilization of agro-industrial residues. Since then, many research and many patents and scientific publications have been produced on basic features of SSF, fermenter designing, production of products like primary and secondary metabolites, food, feed and bioprocesses like bioremediation, bioleaching, biopulping, and biobeneficiation [44].

Feed formulation is the most expensive component of modern aquaculture. The best method of application of probiotics in aquaculture is incorporating it into the feed. For continuous usage of probiotics incorporated feed in aquaculture, the method of feed formulation must be economically feasible. Farm made feeds prepared with locally available cheap plant sources have much relevance. Organic aquaculture is a process of production of aquatic plants and animals with the use of only organic inputs in terms of seed and supply of nutrients and management of diseases following aquatic principles. Use of pre- or probiotic cultures in fish feed offers an alternative to antibiotic treatment for sustainable aquaculture. Probiotic based functional feed can be an alternative for vaccination, where no vaccines are developed or not possible.

Feed formulation and its preparation involve combining individual ingredients to form a complex mixture that will satisfy the nutritional requirements of fish. Sophisticated feed formulation technology improves the quality of feed. SSF with probiotic bacteria in optimum water level increases the number of bacteria, secrete digestive enzymes, converts complex substrate to simpler easily consumable form, and produce vitamins and organic acids. It is better to isolate and applied the presumed probiotics from the same host or its environment for easy acceptance and its better effect. Probiotic bacteria added in the feed will enter into GI tract, and it also colonizes in the fish skin and in the surrounding water. The impact of using SSF reflects on both the environment and economics.

12.3.10   FUNCTIONAL FISH FEED FORMULATION

Locally available cereal based solid state fermentation using aquaculture probiotics for feed formulation will increase the nutritional quality of feed with minimum price and less effort. Solid-state fermentation with probiotics can be done with lower manpower and non-skilled persons. Only minimum sterile conditions are required for SSF; therefore farm made fresh probiotic based functional feed for aquaculture has been developed by the farmer himself.

12.3.10.1   PROXIMATE ANALYSIS OF FERMENTED SUBSTRATE

Total moisture of the fermented substrate is determined by drying the samples at elevated temperature, and the percentage of moisture influences the viability of the probiotic organisms [5]. Crude ash is the residue remaining after the sample has been ignited until it is free of carbon, usually at a temperature of $550^{\circ}C$ and the fermented substrate. Total protein determination is by the Kjeldahl method, in which organic compound is oxidized by sulphuric acid to form CO_2 and H_2O and release of N_2 as ammonia. The ammonia exists in the H_2SO4 solution as ammonium sulphate, but the CO_2 and H_2O are driven off. Digestion mixture is used to accelerate the digestion. (Organic Compound: $H_2SO4+H_2O+ (NH4)_2SO_2$).

The ammonia is thus trapped in the boric acid by steam distillation. The acid is then back titrated to determine how much ammonia has been distilled. Standard and blank are also run under same conditions. Crude lipid in the dried form of the sample is extracted repeatedly using petroleum spirit as solvent at high temperature. The fat-soluble in hot petroleum ether, except phospholipid, is extracted from the sample and then evaporated. The amount of fat present in the sample is estimated gravimetrically. Crude fiber is determined by digesting moisture and fat-free sample (2 g) with 1.25% of H_2SO4 and boils for 30 min. Filter and wash the residues with distilled water. The residue is boiled with 1.25% NaOH. Filter and wash with alcohol and ether. The residue is then transferred to a previously weighed crucible and dried overnight at 30–100°C. Heat the crucible at 600°C for 2–3hours in a muffle furnace. Loss of weight represents the crude fiber.

12.4 PREBIOTIC

Prebiotic is a non-digestible food component that positively affects the host by selectively stimulating the growth and/or activity of one or a limited number of probiotic bacteria in the colon and results in improvement of host health [27]. Carbohydrates can be classified into digestible and non-digestible based on their biochemical and physiological properties. Non-digestible oligosaccharides (NDO) or short chain oligosaccharides (SCC) or Low digestible carbohydrates (LDCs) contain anomeric Carbon (C) atom of the monosaccharide, and this configuration makes its glycosidic bonds indigestible to the hydrolytic activity of the human/animal digestive enzymes [38].

Prebiotics are classified into monosaccharides, oligosaccharides or poly-saccharides based on the number of monosaccharide units and molecular size. According to the International Union of Pure and Applied Chemistry (IUPAC) nomenclature, sugar moieties between three and ten are termed as oligosaccharides. Prebiotics available in the market are oligosaccharides such as: fructooligosaccharides (FOS), soybean oligosaccharides (SOS) and inulin. For the development of food additives, the NDOs are used because these contain monosaccharide unit's fructose, galactose, glucose or xylose. Carbohydrates are a good source of dietary fiber, which may be either water-soluble (inulin, oligosaccharide), insoluble (cellulose) or mixed (bran).

The viability of live probiotic bacteria multiplies by solid-state fermenta-tion, and the application of prebiotic will enhance the selective growth of native fish gut microbiota. Fermentable carbohydrates are most promising in terms of the positive effect on the composition and activity of the indigenous microbiota of the GItract [27]. However compared to the application studies of prebiotics for terrestrial animals, research, and application of orally administered prebiotics is in its infancy in aquaculture.

The criteria required for using food component as prebiotic are: It must not be hydrolyzed or absorbed in the upper GI tract, it must undergo selective fermentation by potentially healthy bacteria, alteration of the composition of the colonic microbiota towards a healthier composition and also the food ingredient should exert the beneficial effects to the host's health. Indigestible carbohydrates, some peptides and proteins, and certain lipids are some of the dietary components that get into the colon, and therefore they are considered as prebiotic [21]. Table 12.2 gives of the list of some aquaculture prebiotics.

12.5   SYNBIOTICS

The combination of probiotics and prebiotics are referred to as synbiotics. By using symbiotic, we can manage the microbiota and results will be positive. For example, fructose oligosaccharide (FOS) and *Bifidobacterium* strain as a symbiotic and in combination, the effect will be better survival of the probiotic (here it is *Bifidobacterium*) in the specific and readily available substrate (here it is FOS) for fermentation; and additionally, possess individual advantages of pro- and prebiotic.

Much of the prebiotic work is done on humans, and terrestrial animals and very limited or less information are available on fishes [53]. The effect of synbiotics may be directed towards both the small and the large intestines of the gastrointestinal tract, i.e., the probiotic strain utilizes the prebiotic carbohydrate for the growth and proliferation in the gut [34].

12.6   COBIOTICS

The term cobiotics was coined in 2013, which is related to products that provide nutritional benefits for the host. It contains probiotics, prebiotics, and additionally contains digestive enzymes in the host gut to improve the growth of healthy bacteria and to suppress the pathogenic bacterial growth, and thus improve the overall health of the host. The cobiotic concept is more functional than symbiotic, and it is a combination of probiotics,

TABLE 12.2 List of Some Aquaculture Prebiotics

Prebiotic	Fish name	Results	Reference
Inulin	Grass carp (Ctenopharyngodon idellus), Tilapia (Tilapia aureus)	Higher survival rates against towards fish pathogens Aeromonas hydrophilia and Edwardsiellatarda.	[61]
Fructooligosaccharides (FOS)	Atlantic salmon	Increased feed intake, growth rate, and digestibility.	[30]
	Hybrid tilapia (O. niloticus♀ × · O. aureus♂); turbot larvae and soft-shell turtle (Triortyxsinensis).	Better growth rate increased survival rate and non-specific immunity.	[32]
Galactooligosaccharides (GOS)	Red drum	Increased protein and organic matter ADC values.	[9]
Mannanoligosaccharides (MOS)	Channel catfish (Ictalurus punctatus)	Increased survival rate against the fish pathogen Edwardsiellaictaluri	[63]
Short chain fructooligosaccharides (scFOS)	Hybrid tilapia (O. niloticus♀ ×· O. aureus♂)	Improved growth rate, feed intake, feed conversion & survival rate.	[66]
Xylooligosaccharides (XOS)	Crucian carp (Carassius auratusgibelio)	—	[64]

prebiotics, and digestive enzymes (Figure 12.4). By the incorporation of digestive enzymes and prebiotics, the cobiotic concept boosts up the nutritional value of synbiotics. In Belgium, a first cobiotic product was made with Registration number NUT/PL/AS 1164/22 by Federal Public Service, Health Food Chain Safety; and the main ingredients are probiotic, Prebiotic (Inulin, Dextrine), Rice Bran, Glutamine, Digestive enzymes (Amylase, Invertase, Lactase, Xylanase, Pectinase, Lipase), Vitamin A, Vitamin B5, Vitamin B6, Vitamin B9, Vitamin B12, Vitamin C, Vitamin D, Vitamin E, Zinc. Suggested use is 5–10 g daily (maximum 15 g/day), with meals or between the meals.

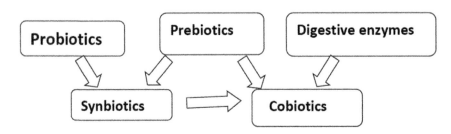

FIGURE 12.4 Portraying probiotic, prebiotic, symbiotic, and cobiotic concept.

12.7 CASE STUDY 1: PROBIOTIC (*BACILLUS CLAUSII*) APPLICATION IN GUPPY (*POECILIA RETICULATA*)

When *Bacillus* species are incorporated into the fish feed, the gut of the animals colonized with Bacillus species that will later compete with pathogenic Vibrio species and protect it from diseases. Pelletization and drying are essential steps of feed formulation, and *Bacillus* species has the advantage to survive the pelletization process [13]. Based on the results of the previous study, authors of this chapter also studied *Bacillus clausii* isolated from Enterogermia a probiotic product in the market, and it was cultivated in nutrient broth and incubated at 37°C. Solid state fermentation with wheat bran and *Bacillus clausii* and feed were formulated after fermentation, and a 60-days trial of probiotic-based feed study was designed with guppy fish. In parallel, a control feed without probiotic and a commercial feed was also fed to fishes, and the results indicated that *Bacillus clausii* treated fishes showed increased feed efficiency, growth rate and better survival than the control and commercial feed.

12.8 CASE STUDY 2: COBIOTIC APPLICATION IN HOLSTEIN-FRIESIAN COW

'YESTURE,' a cobiotic preparation for animal nutrition was formulated [57]. It directly influences the productive performances of dairy cows and udder health. The cobiotic 'YESTURE' was composed of live yeast cells *Saccharomyces cerevisiae* in blend with *Lactobacillus casei, Streptococcus faecium, Aspergillus oryzae, L. acidophilus* and enzymes (1,3-b and 1,6 D-Glucan, hemicellulase, protease, cellulase, alpha-amylase), which modified the fermentation in rumen stimulating the development of ruminal bacteria and increasing the fiber digestion. Positive effects of this preparation have been studied and were established for Holstein-Friesian cows. Research proved that Yeasture influenced the quantity and composition of the milk. Based on the positive results from a cow, the possibility of application of cobiotic in fisheries /aquaculture is encouraged.

12.9 SUMMARY

The study of basic biological process in the natural environment is more complex, but aquaculture ponds can be considered as a semi-natural environment in which the researcher has some control over it. Probiotic use in aquaculture is eco - friendly method of the betterment of aquatic environmental quality and is the key element for healthy aquaculture. By utilization of agro-industrial residue as a solid substrate, probiotics based fermentation can reduce the cost of fish feed, increases the number of probiotic strains and thereby increasing the nutrient as well as a preventive measure towards selected fish pathogens of the feed. The incorporation of probiotics into the substrate during solid-state fermentation increase the protein, ash, and lipid content of the organic substrates. For a sustainable eco- friendly aquaculture, the regular use of the probiotic based functional feed developed by solid state fermentation offers antibiotic free alternatives and prophylaxis for infection caused by the fish pathogens.

It is time to encourage the application of the organic concept in water bodies for the sustainable aquatic eco-system. Probiotic based functional fish feed supports organic aquaculture in a sense eliminating the use of antibiotic growth promoters for disease management. Recently science-based knowledge on probiotics, prebiotics, synbiotics, and cobiotics application has increased and has fabulous future and scope in aquaculture. Innovative application of genetic engineering and microbiology and its practicable implementation will solve many of the constraints in aquaculture.

KEYWORDS

- anti-microbial resistance
- cobiotics
- eco friendly
- feed additives
- functional fish feed
- normal microbiota
- organic aquaculture
- prebiotics
- probiotics
- solid state fermentation
- synbiotics

REFERENCES

1. Abraham, T. J., Mondal, S., & Babu, C. S., (2008). Effect of commercial aquaculture probiotic and fish gut antagonistic bacterial flora on the growth and disease resistance of ornamental fishes, *Carassius auratus* and *Xiphophorus helleri*. *Journal of Fisheries and Aquatic Sciences, 25*(1), 27–30.
2. Ahilan, B., Shine, G., & Santhanam, R., (2004). Influence of probiotics on the growth and gut microbial load of juvenile goldfish (*Carassius auratus*). *Asian Fisheries Science, 17,* 271–278.
3. Al-Dohail, M. A., Hashim, R., & Aliyu-Paiko, M., (2009). Effects of the probiotic, Lactobacillus acidophilus, on the growth performance, hematology parameters and immunoglobulin concentration in African Catfish (*Clarias gariepinus*, Burchell 1822) fingerling. *Aquaculture Research, 40*(14), 1642–1652.
4. Angelis, M., Siragusa, S., Berloco, M., Caputo, L., Settanni, L., Alfonsi, G., et al., (2006). Selection of potential probiotic lactobacilli from pig feces to be used as additives in pelleted feeding. *Research in Microbiology, 157,* 792–801.
5. AOAC (Association of Official Analytical Chemists), (2005). *Official Method of Analysis* (18ᵗʰ edn., p. 334). AOAC, Gaithersburg, MD, USA.
6. Austin, B., (2006). The bacterial microbiota of fish, revised. *The Scientific World Journal, 6,* 931–945.
7. Balcazar, J., Blas, I., Ruizzaezuela, I., Cunningham, D., Vendrell, D., & Muzquiz, J., (2006). The role of probiotics in aquaculture. *Veterinary Microbiology, 114*(3/4), 173–186.
8. Bisht, A., Singh, U. P., & Pandey, N. N., (2012). *Bacillus subtilis* as a potent probiotic for enhancing growth in fingerlings of common carp (*Cyprinus carpio* L.). *Indian J. Fish, 59*(3), 103–108.

9. Burr, G., Hume, M., Neill, W. H., & Gatlin, D. M., (2008). Effects of prebiotics on nutrient digestibility of a soybean-meal-based diet by red drum *Sciaenops ocellatus* (Linnaeus). *Aquaculture Research*, *39*(15), 1680–1686.

10. Carnevali, O., De Vivo, L., Sulpizio, R., Gioacchini, G., Olivotto, I., Silvi, S., & Cresci, A., (2006). Growth improvement by probiotic in European sea bass juveniles (*Dicentrarchus labrax*, L.), with particular attention to IGF–1, myostatin and cortisol gene expression. *Aquaculture*, *258*(1–4), 430–438.

11. Dibner, J. J., & Richards, J. D., (2005). Antibiotic growth promoters in agriculture: History and mode of action. *Poultry Science*, *84*(4), 634–643.

12. Dozier, W. A., & Hess, J. B., (2011). Soybean meal quality and analytical techniques. In: Hany El-Shemy, (ed.), *Soybean and Nutrition* (pp. 111–124). www.intechopen. com, ISBN: 978–953–307–536–5, http://www.intechopen.com/books/soybean-and-nutrition/soybean-meal-quality-and-analytical-techniques,Accessed on July 31, 2017.

13. EL-Haroun, E. R., Goda, A. M. A. S., & Chowdhury, M. A., (2006). Effect of dietary probiotic Biogen supplementation as a growth promoter on growth performance and feed utilization of Nile tilapia *Oreochromis niloticus* (L.). *Aquaculture Research*, *37*(14), 1473–1480.

14. FAO (Food and Agriculture Organization), (2006). *State of World Aquaculture* (p. 132). FAO, Rome.

15. FAO (Food and Agriculture Organization), (2007). *Study and Analysis of Feeds and Fertilizers for Sustainable Aquaculture Development* (p. 142). FAO, Rome.

16. FAO (Food and Agriculture Organization of the United Nations), (2012). *World Fisheries and Aquaculture* (p. 87). FAO, Rome.

17. FAO/WHO (Food and Agriculture Organization of United Nations & World Health Organization), (2002). *Joint FAO/WHO Working Report on Drafting Guidelines for the Evaluation of Probiotics in Food* (p. 68). FAO, Rome.

18. Farzanfar, A., (2006). The use of probiotics in shrimp aquaculture, *FEMS Immunology Medical Microbiology*, *48, 149–158*

19. Faramarzi, M., Jafaryan, H., Patimar, R., Iranshahi, F., Boloki, M. L., Farahi, A., et al., (2011). The effects of different concentrations of probiotic *Bacillus* spp. and different bioencapsulation times on growth performance and survival rate of Persian Sturgeon (*Acipencer persicus*) Larvae. *World Journal of Fish and Marine Sciences*, *3*(2), 145–150.

20. Fasoli, S., Marzotto, M., Rizzotti, L., Rossi, F., Dellaglio, F., & Torriani, S., (2003). Bacterial composition of commercial probiotic products as evaluated by PCR-DGGE analysis. *International Journal of Food Microbiology*, *82*(1), 59–70.

21. Fooks, L. J., & Gibson, G. R., (2002). Probiotics as modulators of the gut flora. *The British Journal of Nutrition*, *88*(1), 39–49.

22. Fuller, R., (1989). Probiotics in man and animals. *The Journal of Applied Bacteriology*, *66*(5), 365–378.

23. García de La Banda, I., Lobo, C., León-Rubio, J. M., Paniagua, S. T., Balebona, M. C., Moriñigo, M. A., et al., (2010). Influence of two closely related probiotics on juvenile Senegalese sole (*Solea senegalensis*, Kaup 1858) performance and protection against *Photobacterium damselae* subsp. piscicida. *Aquaculture*, *306*(1–4), 281–288.

24. Gatesoupe, F. J., (1999). The use of probiotics in aquaculture. *Aquaculture*, *180*, 147–165.

25. Geng, X., Dong, X. H., Tan, B. P., Yang, Q. H., Chi, S. Y., Liu, H. Y., & Liu, X. Q., (2012). Effects of dietary probiotic on the growth performance, non-specific immunity and disease resistance of cobia, *Rachycentron canadum*. *Aquaculture Nutrition*, *18*(1), 46–55.

26. Ghosh, S., Sinha., & Sahu, C., (2008). Dietary probiotic supplementation on growth and health of live-bearing ornamental fishes. *Aquaculture Nutrition, 14*, 289–299.

27. Gibson, G. R., & Roberfroid, M. B., (1995). Dietary modulation of the human colonic microbiota: Introducing the concept of prebiotics. *The Journal of Nutrition, 125*(6), 1401–1412.

28. Godfray, H. C. J., Beddington, J. R., Crute, I. R., Haddad, L., Lawrence, D., Muir, J. F., et al., (2012). Food security: The challenge of feeding 9 billion people. *Science, 327*, 812–818.

29. Grajek, W., Olejnik, A., & Sip, A., (2005). Probiotics, prebiotics, and antioxidants as functional foods. *Acta Biochimica Polonica, 52*(3), 665–671.

30. Grisdale-Helland, B., Helland, S. J., & Gatlin, D. M., (2008). The effects of dietary supplementation with mannan oligosaccharide, fructooligosaccharide or galactosaccharide on the growth and feed utilization of Atlantic salmon (*Salmo salar* L.). *Aquaculture, 283*, 163–167.

31. Hasler, C. M., (2002). Functional foods: Benefits, concerns, and challenges - A position paper from the American Council on Science and Health. *The Journal of Nutrition,* 3772–3781.

32. He, S., Xu, G., Wu, Y., Weng, H., & Xie, H., (2003). Effects of IMO and FOS on the growth performance and non-specific immunity in hybrid tilapia. *Chinese Feed., 23*, 14–15.

33. Holzapfel, W. H., Haberer, P., Snel, J., Schillinger, U., Jos, H. J., & Veld, H., (1998). Overview of gut flora and probiotics. *International Journal of Food Microbiology, 41*, 85–101.

34. Holzapfel, W. H., & Schillinger, U., (2002). Introduction to pre- and probiotics. *Food Research International, 35*(2–3), 109–116.

35. Keysami, M. A., Saad, C. R., Sijam, K., Daud, H. M., & Alimon, A. R., (2007). Effect of *Bacillus subtilis* on growth development and survival of larvae *Macrobrachium rosenbergii* (de Man). *Aquaculture Nutrition, 13*(2), 131–136.

36. Kozasa, M., (1989). Probiotics for animal use in Japan. *Rev. Sci. Tech. Off. Int. Epiz., 8*(2), 517–531

37. Lee, J. S., Damte, D., Hossain, M. A., Belew, S., Kim, J. Y., Rhee, M. H., Kim, J. C., & Park, S. C., (2015). Evaluation and characterization of a novel probiotic *Lactobacillus pentosus* PL11 isolated from Japanese eel (*Anguilla japonica*) for its use in aquaculture. *Aquaculture Nutrition, 21*(4), 444–456.

38. Marteau, P., & Flourié, B., (2007). Tolerance to low-digestible carbohydrates: Symptomatology and methods. *British Journal of Nutrition, 85,* 1–17.

39. Mattila-Sandholm, T., Myllärinen, P., Crittenden, R., Mogensen, G., Fondén, R., & Saarela, M., (2002). Technological challenges for future probiotic foods. *International Dairy Journal, 12*(2/3), 173–182.

40. Mohapatra, S., Chakraborty, T., Prusty, K., Das, P., Paniprasad, K., & Mohanta, K. N., (2012). Use of different microbial probiotics in the diet of rohu, *Labeo rohita* fingerlings: Effects on growth, nutrient digestibility and retention, digestive enzyme activities and intestinal microbiota. *Aquaculture Nutrition, 18*(1), 1–11.

41. Moriarty, D. J. W., (1998). Control of *luminous Vibrio* species in penaeid aquaculture ponds. *Aquaculture, 164*(1–4), 351–358.

42. Moriarty, D. J. W., (1997). The role of microorganisms in aquaculture ponds. *Aquaculture, 151*, 333–349.

43. Olafsen, J. A., (2001). Interactions between fish larvae and bacteria in marine aquaculture. *Aquaculture*, *200*(1/2), 223–247.
44. Pandey, A., (2003). Solid-state fermentation. *Biochemical Engineering Journal*, *13*(2/3), 81–84.
45. Pandey, A., Soccol, C. R., & Mitchell, D., (2000). New developments in solid state fermentation: I-bioprocesses and products. *Process Biochemistry*, *35*(10), 1153–1169.
46. Pandiyan, P., Balaraman, D., Thirunavukkarasu, R., George, E. G. J., Subaramaniyan, K., Manikkam, S., & Sadayappan, B., (2013). Probiotics in aquaculture. *Drug Invention Today*, *5*(1), 55–59.
47. Panigrahi, A., Kiron, V., Kobayashi, T., Puangkaew, J., Satoh, S., & Sugita, H., (2004). Immune responses in rainbow trout Oncorhynchus mykiss induced by a potential probiotic bacteria *Lactobacillus rhamnosus* JCM 1136. *Veterinary Immunology and Immunopathology*, *102*(4), 379–388.
48. Panigrahi, A., Kiron, V., Kobayashi, T., Puangkaew, J., Satoh, S., & Sugita, H., (2005). The viability of probiotic bacteria as a factor influencing the immune response in rainbow trout *Oncorhynchus mykiss*. *Aquaculture*, *243*(1–4), 241–254.
49. Panigrahi, A., & Azad, I. S., (2007). Microbial intervention for better fish health in aquaculture: The Indian scenario. *Fish Physiology and Biochemistry*, *33*(4), 429–440.
50. Porubcan, R. S., (1975). *Stabilized Dry Culture of Lactic Acid Producing Bacteria* (p. 23). United States Patent.
51. Qi, Z., Zhang, X., Boon, N., & Bossier, P., (2009). Probiotics in aquaculture of China: Current state, problems, and prospect. *Aquaculture*, *290*(1/2), 15–21.
52. Qazi, J. I., Nadir, S., & Shakir, H. A., (2007). Solid state fermentation of fish feed with amylase producing bacteria. *Punjab Univ. J. Zool.*, *27*(1), 1–7.
53. Ringø, E., Olsen, R. E., Gifstad, T., Dalmo, R. A., Amlund, H., Hemre, G. I., & Bakke, A. M., (2010). Prebiotics in aquaculture: A review. *Aquaculture Nutrition*, *16*(2), 117–136.
54. Robinson, S., Thomas, S., & Toulmin, C., (2012). The challenge of food security. *Science*, *327*, 812.
55. Sahu, M. K., Swarnakumar, N. S., Sivakumar, K., Thangaradjou, T., & Kannan, L., (2008). Probiotics in aquaculture: Importance and future perspectives. *Indian Journal of Microbiology*, *48*(3), 299–308.
56. Senok, A. C., Ismaeel, A. Y., & Botta, G. A., (2005). Probiotics: facts and myths. *Clinical Microbiology and Infection*, *11*(12), 958–966.
57. Stretonovick, L., Petrovic, M. P., Aleksic, S., Pantelic, V., & Katic V., (2008). Influence of yeast, probiotics, and enzymes in rations on dairy cows performances during the transition. *Biotechnology in Animal Husbandry*, *24*(5/6), 33–43.
58. Sullivan, Å., & Nord, C. E., (2002). The place of probiotics in human intestinal infections. *International Journal of Antimicrobial Agents*, *20*(5), 313–319.
59. Tovar, D., Zambonino, J., Cahu, C., Gatesoupe, F. J., Vázquez-Juárez, R., & Lésel, R., (2002). Effect of live yeast incorporation in the compound diet on digestive enzyme activity in sea bass (*Dicentrarchus labrax*) larvae. *Aquaculture*, *204*(1/2), 113–123.
60. Verschuere, L., Rombaut, G., Sorgeloos, P., & Verstraete, W., (2000). Probiotic bacteria as biological control agents in aquaculture. *Microbiology and Molecular Biology Reviews : MMBR*, *64*(4), 655–61.
61. Wang, W., & Wang, D. H., (1997). Enhancement of the resistance of tilapia hydrophila and *Edwardsiella tarda*. *Comparative Immunology, Microbiology, and Infectious Diseases*, *20*(3), 261–270.

62. Wang, Y. B., Li, J. R., & Lin, J., (2008). Probiotics in aquaculture: Challenges and outlook. *Aquaculture, 281*(1–4), 1–4.

63. Welker, T. L., Lim, C., Yildirim-Aksoy, M., Shelby, R., & Klesius, P. H., (2007). Immune response and resistance to stress and *Edwardsiella ictulari* challenge in channel catfish, *Ictalurus punctatus*, fed diets containing commercial whole-cell or yeast subcomponents. *Journal of the World Aquaculture Society, 38*(1), 24–35.

64. Xu, Z., Yan-bo, W., & Wei-fen, L. B., (2009). Effect of prebiotic xylooligosaccharides on growth performances and digestive enzyme activities of allogynogenetic crucian carp (*Carassius auratus gibelio*). *Fish Physiology and Biochemistry, 35*(3), 351–357.

65. Yin, Y., Mu, D., & Chen, S., (2011). Effects on growth and digestive enzyme activities of the *Hepialus gonggaensis* larvae caused by introducing probiotics. *World Journal of Microbiology and Biotechnology, 27*, 529–533.

66. Zhou, X., Wang, Y., & Li, W., (2009). Effect of probiotic on larvae shrimp (*Penaeus vannamei*) based on water quality, survival rate, and digestive enzyme activities. *Aquaculture, 287*(3/4), 349–353.

PART IV
Marine Microbiology

CHAPTER 13

MICROBIOLOGY OF MARINE FOOD PRODUCTS

RESHMA B. NAMBIAR, S. PERIYAR SELVAM, P. ANAND BABU,
M. MAHESH KUMAR, EMMANUEL ROTIMI SADIKU, and
SHANMUGAM KIRUBANANDAN

ABSTRACT

Marine food has become an important part of the healthful diet, which not only provides lean protein and omega-3 fatty acids but also reduces the risk of cardiovascular diseases and diabetes. Whereas, intake of seafood could also lead to the danger of eating spoiled marine food. The major cause of marine food linked to diseases is bacteria, viruses, and parasites. Although, most of the seafood are preserved by salting, smoking, drying, etc. These products are at risk of contamination due to stable heat biotoxins from fish and heat stable microorganisms. There might be a small number of infectious microbes in fish or shellfish at the time of harvest, and the contamination by remaining organism might have occurred during handling and processing or by unhygienic methods. In recent years, the isolation of beneficial probiotic bacteria such as *Lactobacillus, Lactococcus, Pediococcus*, etc. from the fish gut and traditional fermented fish product (Tungtap, Nagari, etc.) has been reported. These organisms confer a health benefit on the consumer like improving acute liver injury caused by D-galactosamine. In addition, using these organisms as a starter for fermented and salted fishes showed potent antimicrobial activity against infectious pathogens and suppressed biogenic amine accumulation in salt fermented marine-based foods. Therefore, this chapter gives an overview about the harmful and the beneficial effects of microorganism on seafoods.

13.1   INTRODUCTION

Marine foods are a nutrient-rich supplement of a healthy diet. It is a decent supply of protein, vitamins, minerals, omega-3 fatty acids, DHA, and EPA. Seafood consumption helps in preventing the common chronic disorders, and it is related to conceivable medical advantages that include: neurologic development during gestation and infancy [14], and it also lowers the danger of coronary illness [35, 36]. In 2013, the global per capita consumption of fish was estimated to be 19.7 kg, with fish accounting for about 17% of the worldwide population's consumption of animal proteins and 6.6% of all proteins consumed. Globally, in the average per capita intake of animal proteins, fish provides more than 3.1 billion people with almost 20% protein. The assessments for 2014 showed further progress in the per capita consumption, to around 20 kg, with the share of aquaculture production, in the total food supply, overtaking that of captured fisheries for the first time [20].

Marine foods are extremely perishable products, and they spoil easily due to enzymatic, chemical or microbial activities, while the microbial action is the primary reason for fresh fish spoilage. The factors [27] that render seafood sensitive to microbial attack include: water activity and comparatively increased pH. Biogenic amines are another major causative factor that renders seafood toxic. Biogenic amines, such as: histamine, cadaverine, dopamine, putrescine, serotonin, spermine, and tyramine, are bacterial spoilage products and their amount are commonly taken as an index to survey the keeping quality and shelf-life of fishery foods [41]. Okuzumi et al. [47] reported that the microbes, such as *Pseudomonas, Vibrio,* and *Photobacterium* are responsible for the formation of biogenic amines (putrescine, cadaverine, and histamine) by the decarboxylation of amino acids in horse mackerel. Numerous studies stated that the microorganisms present in seafood are directly related to the fishing ground, environmental factor, handling, and the methods of harvesting and storage [19].

A range of preservation techniques is applied to extend the keeping quality of seafood. The preservation techniques either inactivate or kill microorganisms. Some of the standard methods of seafood preservation are: refrigeration, or packing, acidification, and conventional methods may avoid marine food contamination and increase their shelf life [4]. Foods which are prepared by curing (for example salting, acidification, fermentation, smoking, additives, combinations, etc.) are categorized as mildly, semi or heavily salted. Mildly stored seafood contains less amount of salt. Sodium chloride, vinegar, is used to preserve the semi-preserved seafood products. Soaked and salted fish are natural foodstuffs of the group.

However, microbial growth during storage will depend on the preservation conditions. Gram and Huss [28] reported that the microorganisms responsible for spoilage of chilled fish were psychro-tolerant Gram-negative bacteria (*Pseudomonas* spp. and *Shewanella* spp.). Recently, the occurrence of beneficial microorganisms in fish products was reported. Lactic acid bacteria (*Lactobacillus plantarum* and *Lactobacillus casei*) exhibiting enhanced antimicrobial properties towards bacteria, were isolated from budu (fish sauce) and funasuzhi [54].

This chapter describes the different types of microorganisms present in marine food products and their causes on the quality deterioration of fish and shellfish.

13.2 MICROORGANISMS AFFECTING MARINE FOOD AND FOOD PRODUCTS

13.2.1 HARMFUL MICROORGANISM (BACTERIA)

13.2.1.1 VIBRIO

Vibrio organism is a facultative anaerobic, Gram-negative, and halophilic bacterium that is usually found in the aquatic habitat. *Vibrio* infections are caused mainly by consuming spoil seafood or by the exposure of an open injury to seawater [44]. The two-major infection-causing strains, are *Vibrio parahaemolyticus* and *Vibrio vulnificus*. Among these, *V. vulnificus* is virulent, especially among patients who are highly susceptible to invasive diseases, such as liver disease and iron storage disorders [30]. Clinical symptoms, most often caused by *V. parahaemolyticus* illness, are: abdominal cramps, diarrhea, nausea, and vomiting. The symptoms that occur less commonly are wound infections and septicemia. Prevention methods include: thorough cooking of shellfish and avoiding the cross-contaminating of foods by raw seafood [12, 22]. Postharvest treatments, such as: irradiation, quick freezing, high-pressure treatment, and pasteurization, are available to make seafood safer.

13.2.1.2 LISTERIA MONOCYTOGENS

Genus *Listeria* consists of Gram-positive, short rod-shaped bacteria that have a flagellum, which gives distinctive tumbling motility at room temperature. Among all the species, only *L. monocytogenes* is currently related to human disease, and on rare occasions, *Listeria* infections lead to mortality. The *L. monocytogenes* commonly causes encephalitis, septicemia, and abortion in both human and animals. *L. monocytogenes* may lead to infectious diseases, and this omnipresent bacterium occurs in both terrestrial and aquatic environments. The microorganism is also found in fish, shrimps, clams, etc. [46]. Thorough cooking, cleaning, and proper handling can eliminate the pathogens. *L. monocytogenes* can survive in refrigerated conditions (4°C to 10°C).

13.2.1.3   SALMONELLA

Salmonella is Gram-negative, facultatively anaerobic, mesophilic, non-spore forming bacilli that have an optimum growth temperature of between 35°C and 37°C. *Salmonella* is an intracellular pathogen that commonly causes stomach infection with symptoms, such as: abdominal cramps, diarrhea, fever, and typhoid fever, bacteremia, and urinary tract infections. The above infections are caused by the adaptation to the host and their resistance to oxidative stress. The main *Salmonella* serotype *Typhi* reservoir is human, whereas birds, reptiles, and mammals also act as the repository for non-typhoidal serotypes. Control measures to counter *Salmonella* infection, include the monitoring of fecal coliform counts in water; this was found to be useful in decreasing the pre-harvest *Salmonella* contamination of seafood. The *Salmonella* contamination also occurs during the processing and storage of the seafood [6, 25]. Seafood spoilage by *Salmonella* is controlled by proper and adequate cooking, post-harvest storage, processing, and the reduction of cross-contamination of seafood during handling [21].

13.2.1.4   ESCHERICHIA COLI

Escherichia coli are Gram-negative, facultative anaerobic bacilli. It occurs, most commonly, in the lower intestine of the host. Fecal contamination leads to the occurrence of this bacterium in seafood. Contamination is chiefly caused by water contamination and unhygienic conditions, during the handling process. In addition, the quality of the ice used for conservation and the food processing plants are frequently cited as the likely cause for *E. coli* contamination. Clinical symptoms include: abdominal cramping, diarrhea, fatigue, fever, etc. Preventive measures include: adequate cooking and avoiding cross infectivity at the time of cleaning of the fish [13].

13.2.1.5   SHIGELLA

Shigella species are Gram-negative, facultative anaerobic and rod-shaped bacteria. Clinical symptoms of *Shigella* are: diarrhea, fever, stomachache, and diarrhea. In small children, *Shigella* causes lethal dilation of colon, bacteremia, Reiter's syndrome and anemia and kidney failure disorder. The major reservoir of *Shigella* is humans, and it spreads mainly, by contact with feces of deceased persons, directly or indirectly. The preparation of food by an infected person, improper handling and harvesting seafood from water contaminated with sewage, can cause seafood contamination. Even though *Shigella* can live outside the host, cooking can, readily, kill the organism. The strategies to avoid *Shigella* infection, include: checking the fecal coliform count in harvested water, preventing the harvest from areas contaminated with sewage, regulating of sewage dumping overboard and observing the seafood handling guidelines in eateries.

13.2.1.6   CLOSTRIDIUM BOTULINUM

It is widely distributed in nature. A neurotoxin botulinum is produced by the organism, which can cause flaccid paralytic disease in humans and animals. Home-canned foods are most often associated with foodborne botulism infection. The occurrence of *C. botulinum* type *E* spores is generally associated with anaerobically fermented seafood, which favors the *C. botulinum* germination [42, 68]. Symptoms of *Clostridium* infection include: diarrhea, colitis, sepsis, etc. Preventive measures to avoid contamination by seafood originated botulism include: educational awareness to encourage appropriate methods for food fermentation and to adopt proper cooking of fermented foods before eating, particularly if the food were tightly stored in sealed plastic or glass containers [8].

13.2.1.7   OTHER TOXIN FORMING BACTERIA

Tetragenococcus halophilus is the main organism that causes histamine accumulation in fish products. Fish and soy sauces are typical forms of seasoning of fish in South East Asia, which are the main reasons for several instances of histamine poisoning. Histamine can likewise occur in salted foods, e.g., sardines, herring [70], fish sauce [60] and salted anchovies. Elevated amounts of histamine in foods can have significant vasoactive effects in humans [38]. Histamine poisoning almost mimics food allergy; hence incidents of histamine poisoning after the consumption of these salted and fermented foods might have gone unnoticed.

Staphylococcus aureus, Clostridium perfringens, and *Bacillus cereus* may produce toxic substances, which can lead to heightened gastrointestinal problems. Vomiting, nausea, and minor diarrhea, after the consumption of the infected food, are common symptoms. However, symptoms caused by *C. perfringens* starts within few hours after consumption of contaminated food and the time of the symptoms is generally less than 24 h; hence, many illnesses are not diagnosed. Only some of the strains of *S. aureus* produce the toxin, which leads to gastrointestinal disorders. Humans are the major reservoirs, and they carry microbes in the skin, nasal passages or in their injuries. *S. aureus*-infected seafood arises majorly because of improper handling during the preparation of food [6, 63]. Toxin-mediated illness can be prevented by refrigeration. Epidemics are commonly linked with foods stored at unsuitable temperatures, for a prolonged time, thereby allowing the growth of microbes and the production of enterotoxin [48].

13.2.2   VIRUSES

13.2.2.1   NOROVIRUS

Norovirus is an RNA virus (winter vomiting virus); it commonly causes acute gastrointestinal disorders in human. Norovirus is responsible for the shellfish-borne outbreak, and it is transmitted *via* contaminated food, water or fecal-oral route. Common symptoms of the diseases are diarrhea, vomiting, nausea, abdominal cramps, headache, myalgias, and mild fever. Humans are the only identified reservoir. The dose for infection is low; however, attack rates are high. Seafood harvested from water that has been contaminated with sewage can cause epidemics of Norovirus gastrointestinal disorders. Major epidemics have been linked to the intake of shellfish that are poorly cooked, such as: oysters and clams. It is a widely known fact that, shellfish, such as bivalve mollusks normally gather huge numbers of viruses. The major reasons of the outbreak include contamination of harvest areas with human sewage, sewage run-off into harvest areas after heavy rains and inadequate cooking practices, e.g., cooking mollusks only till it opens thus not killing the Noroviruses by cooking them at high temperatures [45].

13.2.2.2   HEPATITIS A VIRUS (HAV)

It is a non-enveloped RNA virus, which spreads one of the most severe viral infections related to seafood consumption. Most common symptoms are: fever, anorexia, jaundice, and vomiting and the symptoms can persist from a few days to months. The primary reservoir of hepatitis A virus (HAV) is the human and transmission usually occurs through the fecal-oral route. The major sources of food-borne epidemics of HAV outbreaks are related to the consumption of oysters and clams [16]. HAV is mostly resistant to heat, and it also withstands steaming. Adequate cooking of marine foods decreases the possibility of consuming this virus live. Countries in which high standard of hygiene is maintained, hepatitis A related epidemic, are less prevalent.

13.2.3   PARASITES

13.2.3.1   HELMINTHS

It is worm-like endoparasite, which is visible through naked eye when matured. Infections are mainly caused by nematodes, trematodes, cestodes, and species of *Diphyllobothrium* [58] and *Anisakidae* [69]. Helminthic infections are related to the consumption of uncooked seafood. Clinical symptoms of infection are mild, severe gastrointestinal infection, allergic responses, intestinal perforation, and invasive diseases. Numerous worms usually occur in the sea, and fresh waters and these parasites infect most of the marine animals. However, helminths are unable to multiply in food. Hence, the outbreaks caused by helminths in seafood are seldom reported. Extensive incubation time makes it very challenging for the identification of the infection source, and due to a lack of timely diagnostic analyses, infections are likely to be under-diagnosed and under-reported, [61]. Helminths are preventable by suitable chilling and cooking and other strategies that can be put in place in order to control the risk of helminthic infections, are visual inspection of fish and candling since the parasites are big enough to be identified visually.

13.2.3.2 PROTOZOA

Giardiasis is an infection caused by protozoa (*Giardia*), and it is the most common seafood-associated illnesses in the United States [50]. Abdominal pain, bloating, nausea, a prolonged course of watery, smelly diarrhea and asymptomatic infections are some of the clinical symptoms associated with giardiasis. Protozoa are heat-sensitive, and they are also sensitive to prolonged freezing. Methods of preventing protozoa transmission include sufficient cooking and the prevention of the contamination of food during preparation.

13.2.3.3 WHITE SPOT SYNDROME VIRUS (WSSV)

The WSSV is a large rod or elliptical shaped, dsDNA virus of around 300 kbp. The WSSV has an extremely wide host range among crustaceans, thereby killing the host; however, humans are not susceptible. It infects several tissues and multiplies in the nucleus of the target cell. Prevention techniques include the stocking of healthy shrimp seeds and raising them separately in a healthy environment and site selection.

13.2.4 BENEFICIAL MICROORGANISM

Conventional techniques of processing of fish, such as: fermentation, salting, drying, and smoking are the primary methods of fish preservation. The application of the above methods may result in producing traditional foods with unique sensory attributes and prolonged shelf life. The main microflora present in the fermented food is lactic acid bacteria and yeasts.

13.2.4.1 LACTIC ACID BACTERIA (LAB)

Probiotics are live microorganisms; they are generally lactic acid bacteria, which, when consumed in sufficient quantities, exerts a positive impact on health. The health benefits, include: anticholesterol, immunomodulatory, and anti-inflammatory activities. The major source of probiotics for the human is dairy products [17]. Some of the common examples of probiotic microorganisms present in fermented milk products are: *Bifidobacterium lactis*, *B. animalis*, *Lactobacillus casei*, *L. acidofilus*, *L. rhamnosus*, *L. johnsonii*, *L. reuteri*, and *Enterococcus faecium*. *Saccharomyces boulardii* yeast is another probiotic that is commercially available in the probiotic market. Recently, it was demonstrated that the LAB occurs in the gut microflora of fish from the first few days [53, 54]. In addition, lactic acid bacteria are known to inhibit another seafood-contaminating microorganism, e.g., *Listeria.*

13.3 MICROBIOLOGY OF MARINE FOODS

13.3.1 HARMFUL BACTERIA IN SEAFOOD PRODUCTS

13.3.1.1 FISH

It is estimated that one-fourth of the global food source [31] and about 30% of landed fish [3] are wasted, chiefly due to the microbial spoilage. Between 4–5 million tons of shrimp, fish are wasted because of the decay caused by bacteria and fungi each year, resulting from improper onsite storage. Fresh fish spoilage occurs rapidly once it is caught, starting (Rigor mortis) within the initial 12 h in tropics. Most fish species spoil because of oxidation, enzymatic, bacterial, and fungal decay [1, 49]. The new compounds formed during fish spoilage, due to the breakdown of many compounds are the reasons for the variations in sensory properties of the fish meat. Fish microflora depends mainly on the composition of the microorganism present in the water in which the fish live. Most common microorganism present in fish, are bacteria such as: *Pseudomonas, Alcaligenes, Vibrio, Serratia, and Micrococcus* [27].

It was demonstrated that the organic enrichment of sediment might impact on the composition of bacterial groups in fish [33]. The impact of frozen storage (60 days) on the microbiological, chemical, and sensory profiles of tilapia fish, obtained from a research pond of Agricultural Development Project in Nigeria, was determined every 10 days. It was observed that the total coliform count increased with the period of storage and also the pH, fat, and protein contents and it was found that the sensory quality (texture, odor, and color) decreased with increasing storage time.

The aquaculture industry is a growing industry. More than 90% of the world's aquaculture comes from Asia. The increasing production has also led to an increase in the spread of the virus. Spring Viremia of Carp Virus (SVCV) causes contagious disease in various groups of wild and cultured fishes, including the common carp, cyprinids, pike (escocids) and wels catfish (*Silurus glanis*). Zhang et al. [71] screened large numbers of ornamental fish in China and reported the existence of this virus. Common signs of infection in fish are: exophthalmia, swelling of the stomach, skin darkening, petechial hemorrhages on the skin and gill, swollen, and jutted vent and equilibrium loss. Other viruses that can infect [11] fish include: hematopoietic necrosis virus (IHNV), SVCV, etc.

Recently, food-borne parasitic contaminations were recognized as a major public health crisis that has economic impacts in regard of: disease, and medical costs. Improper handling techniques, hygiene, and conventional approaches to the preparation of food have enhanced the food-related trematode outbreaks [51]. Khalil et al. [34] surveyed the variety of fish parasites in Saudi Arabia. Approximately 163 fish were surveyed, and it was reported that parasites, such as *Heterophyes heterophyes* and *Haplorchispumilio*, occurred in muscle and *Capillaria* sp. in the gastrointestinal tract. Even though, the number of recorded instances is high; the general effect on human health is minimal. Decreasing the cooking time when preparing seafood product increases the risk of fish getting contaminated by the parasites. In addition, they concluded that specific fish manufacture sectors may offer seafood parasites and there is a need to control infections in the reservoir host.

13.3.1.2   MOLLUSKS

Phylum Mollusk is a major and most diverse group in the animal kingdom. Phylum Mollusk constitutes more than 50, 000 defined species and nearly 30, 000 of the species are present in the ocean. They are soft-bodied animals; however, a vast majority of them are protected by a hard-protecting shell and inside this shell, is tissue, known as the mantle, which encloses the internal organs of the animal. A great number of diverse species of mollusks are consumed worldwide, either raw or cooked. Certain mollusk species have been commercially exploited and exported as part of the international trade in shellfish, while other species are harvested, sold, and eaten locally. Mollusk shellfish are filter feeders and accumulates bacteria and virus particles in their flesh, from polluted waters. Raw and partially cooked shellfish are vectors of disease transmission.

13.3.1.3   OYSTERS

Oysters are the most harvested shellfish globally; they are extremely perishable, and they have a short shelf life. The major reason for oyster spoilage is through microbial contamination. The microflora of shellfish depends on various factors, which include: aquatic habitat, salinity, environmental factors, bacterial counts, the temperature of water, diet, mode of catch and the freezing conditions. Cao et al. [7] obtained Pacific oysters from the Yellow sea and performed the microbiological analysis. The raw oysters were cleaned and stored separately in plastic bags at different temperature (0°C, 5°C and 10°C) and after 48 h, the microbiological analysis of oysters was carried out. After determining the genus of the organisms, by employing biochemical tests, it was revealed that a wide variety of microorganisms existed in oysters. Adequate and appropriate cooking and responsible sanitization are highly recommended to eliminate these pathogens.

Many marine bivalve mollusks are parasitized by protozoan parasites of *Perkinsus* spp., *P. marinus* and *P. olseni*, which are the most destructive species and the disease caused by these organisms, are collectively known as perkinsosis [66]. Substantial disease-associated mortalities, have led to heavy losses to the oyster industry [29] and has also caused the degradation of the oyster reefs, which has adversely affected the overall health of the ecosystem, loss of natural habitats and reduced water purity.

13.3.1.4   MUSSELS

Mussels flesh is rich in vitamins (A, B1, B2, B6, B12, and C) and minerals (selenium, calcium, iron, magnesium, and phosphorous), which makes it ideal for human consumption. Mussel also contains a high amount of PUFA (polyunsaturated fatty acids, 37–48%), majorly ω-3 PUFA [23], which are important in reducing the risk of cardiovascular diseases [15, 65]. Recently, there has been increasing popularity of fresh mussels rather than chilled mussels; this is because of their greater nutritional value. However, high water activity ($a_w > 0.5$), pH (6.7–7.1), glycogen, and amino acid contents, make them suitable substrates for the growth of microorganisms. Levesque et al. [40] conducted a pilot study in order to document the existence of fecal and intestinal disease-causing organisms in *Mytilus edulis* in population in Nunavik, Quebec. The researchers collected the mussels from low tidal region in each community and examined for the presence of fecal indicators, such as: *Enterococci, Escherichia coli*, F-specific coliphages and *Clostridium perfringens*, by enumeration methods and pathogens, such as: Norovirus, *Salmonella* spp., *Campylobacter jejuni, Campylobacter coli* and *Campylobacter lari*, verocytotoxin secreting *E. coli, Shigella* spp. and *Yersinia enterocolitica* that were determined by molecular identification method. Microscopy and molecular methods were performed to conclude the existence of *Giardia duodenalis* and *Cryptosporidium* spp. in 5 communities and molecular methods were employed to detect *Toxoplasma gondii*. *Clostridium perfringens* were detected in 2 samples and none of the other bacterial or viral pathogens, including *Toxoplasma gondii*, were identified in the mussels. Although, *G. duodenalis* (18%) and *Cryptosporidium* spp. (73%) were present in the sample investigated for the pathogens, the mussels tested had a low microbial count. However, they had a high number of zoonotic protozoa. Hence, it was suggested that the mollusks must be cooked thoroughly, before consumption.

The presence of the virus in wild mussels has also been reported. Mussels were obtained from the Baltic Sea near the Polish coast and were tested for the presence of human Noroviruses of genogroups I and II and HAV [5]. Out of 120 shellfish, NoV GI virus was identified in 22, GII in 28 and HAV in 9.

Recently, parasitism has generally been perceived as a variable that impacts on the overall well-being of animal populations [10]. Parasites can reduce the development, existence or propagative output of their hosts and adjust their behavior; however, it often does not cause the death of the host. Hence, it has a direct effect on the establishment of natural properties and facilities. Mussels are known to get parasitized by platyhelminths. Duck mussels are the initial transitional host of the Bucephalidae), which then spreads by the subsequent in-between host, the common roach before infecting its specific hosts, either the perch or the zander. Swan mussels are the transitional host of the Gorgoderidae, while digenean trematode and zebra mussels are the intermediate hosts of *Phyllodistomum folium* [37].

13.3.1.5   CLAMS

Clams are bivalve mollusks that inhabit in mud and bottom rock layer in estuaries and hence, they indicate the quality of the water from where they were collected. They are filter feeders, which filters a significant amount of water and retain particulate matter, including pathogenic microorganisms [52]. Amoah et al. [2] isolated about twenty bacterial species by enumerating fecal coliforms and total aerobic bacteria from Volta clam (*Galatea paradoxa*). The 16S and 23S rRNA analyses revealed that the microorganisms were: *P. aeruginosa*, *K. ornithinolytica*, *F. meningosepticum*, *E. aerogenes*, *E. agglomerans*, *E. cloacae*, *A. sobria*, *Acinetobacter sp*, *V. cholerae*, *M. radiodurans*, *S. faecalis*, *S. aureus*, *E. coli*, *C. freundii*, *C. violaceum*, *M. morganii*, *P. mirabilis*, *Salmonella sp.*, *S. marcescens* and *Y. intermedia*. In the PCR studies, it was also observed that the bacterial count was higher in the mantle of clam and not the abdomen.

There have been some reports of viral contamination of clams. In December 2001, a food poisoning outbreak was reported among people who consumed purple Washington clam (*Saxidomuspurpuratus*) at a restaurant in Hamamatsu City [24]. Acute gastroenteritis, such as: diarrhea, vomiting, and fever were observed amongst 22 out of the 57 diners. ELISA test, RT-PCR, and real-time PCR analyses of the fecal specimens of 4 individuals, showed the presence of Norovirus (Norwalk-like virus, NV) genogroup I (GI) and genogroup II (GII). After one month, these individuals also developed hepatitis A (HAV). RT-PCR and real-time PCR analyses of the fecal specimens demonstrated the presence of HAV gene. The above study demonstrated the importance of adequate cooking of bivalve mollusks.

Parasite infection of Manila clams found in Europe and Asia and the carpet shell *R. decussatus* in Europe and North Africa (Tunisia) was reported by El Bour et al. [18]. *Perkinsus olseni,* a protozoan, previously named *P. atlanticus*, was the causative agent that similarly infects various clam spp. like: *P. rhomboids*, *P. aurea, and V. pullastra*, but to a lower amount, about incidence, and infection severity. Mortalities in clam communities in Southern Portugal and North-West Spain [66] and the coastal regions of Korea, Japan, and China [9], were caused by *Perkinsus olseni*. It also caused between 50 to 80% of *R. decussatus* mortality in Algarve (S. Portugal) [57].

13.3.1.6 CRUSTACEANS

Crustaceans belong to a large group of arthropods that comprises of: crabs, lobsters, shrimp, etc. Most of the crustaceans are free-living aquatic animals. However, few are terrestrial. Worldwide, fish, and crustacean consumption represents an important source of animal protein for humans. Due to the living environment of crustaceans (i.e., brackish lagoon waters and shallow sea waters), there is a high risk of pollution with ecological pollutants and transmissible agents, such as: bacteria, virus, and parasites. White spot syndrome virus (WSSV) is known to infect [39] crustaceans, and they have a wide range of hosts among decapod crustaceans, such as: sea and freshwater shrimp, crabs, lobsters, etc. It is a DNA virus, which is the only member of the family Nimaviridae [67]. Other viruses that affect the crustaceans are Taura syndrome virus [64], yellow head virus, etc. Certain parasites are also known to parasitize crustaceans, e.g., Cirripedia (*Anelasmasqualicola*) and Rhizocephala.

13.3.1.7 LOBSTER

Total seafood consumption has significantly increased in the past few years, and there is a great increase in the consumption of lobster. However, lobsters are infected by many pathogens, mainly bacteria. Tall et al. [62] characterized and determined the infection mechanisms associated with the infection in the American lobster, triggered by a *Vibrio fluvialis,* e.g., microorganism. Among the19 strains and each of the isolates resembled *V. fluvialis.* The lobster was extremely vulnerable to a diverse of tested antibiotics and required 1% NaCl for growth. The study supported Koch's postulates that these *V. fluvialis*-like organisms were the causative agent for the systemic infection.

13.3.1.8 CRAB

The crab industry is one of the important seafood industries. Microorganisms, like *Vibrio, Aeromonas* and Rhodobacteriales-like organism, chitinoclastic bacteria, and *Spiroplasma,* cause bacterial diseases of crabs. Several bacterial pathogens, for example, *V. cholerae* and *V. vulnificus*, are normally present in blue crab hemolymph [32] and they can cause possible diseases to human beings. Therefore, a lot of attention should be paid to these agents, since these organisms can cause serious diseases, if consumed raw. In the past few years, due to the advance in aquaculture science, new diseases related to novel pathogens, e.g., spiroplasmas, and Rhodobacteriales-like organisms were identified in commercially used crab species. The adoption of numerous promising methods for the elimination of the bacterial disease in crab, might be useful in the aquaculture industry.

13.3.1.9 SHRIMP

Seafood is highly perishable due to the action of enzymes and microorganism that speed up the spoilage mechanism. Also, since prawns are filter feeders, they accumulate several bacteria and viruses in their bodies. The microbiological, sensory, and K value of the Indian white prawn was analyzed by Ginson et al. [26]. Headless Indian white prawns were packed under vacuum in alcohol film, and they were exposed to various pressure treatments and kept in ice (2±1°C). The samples were then sporadically analyzed for variation in their total *Enterobacteriaceae* count, K value and acceptability. It was observed that the *Enterobacteriaceae* count and K value decreased rapidly under higher pressure, whereas the total viable count and K value increased during chilling storage and the *Enterobacteriaceae* count decreased. The study demonstrated that high pressure of ~270 MPa has a positive effect on prawn quality.

Shrimps quality, often deteriorates because of inappropriate handling and further processing can, by no means, retain its freshness. Poor hygienic conditions, such as: improper processing, preservation, and storage conditions, can often, lead to the contamination of shrimp [64]. Shrimps available in local markets can be contaminated by the microorganism, intrinsically or extrinsically. Samia et al. [59] conducted a study that tried to determine the amount of these microbes in shrimps obtained from Dhaka, Bangladesh. Out of the 7 groups of shrimp evaluated, each was noted to be infected with *Staphylococcus* spp., *Aeromonas spp., Klebsiella* spp., *Pseudomonas* spp. and *Shigella* spp. and comparatively, high frequency of contamination was caused by *Klebsiella* spp., *Staphylococcus* spp. and *Aeromonas* spp.

13.3.2 BENEFICIAL BACTERIA IN SEAFOOD

Most of the lactic acid bacteria are considered as safe for consumption, and it is also known to exert benefits, such as: protection from liver injury, antibacterial activity against pathogens by producing bacteriocins and the inhibition of biogenic amines on the host. *Lactobacilli*, notably *Lactobacillus plantarum*, were isolated from Atlantic salmon Pollock, Arctic char and cod. Salinity and stress of the water are some of the factors that affect the quantity of LAB. The frequency of *Lactobacilli* present in the digestive tract of *Salvelinus alpinus*, was lesser in the fish raised in seawater when compared to freshwater; even though the frequency of *Leuconostoc* and *enterococci* were similar [55]. The effectiveness of certain strains isolated from marine foods can be examined in order to produce a new dietary supplement.

Carno bacterium strain from the digestive tract of Atlantic salmon was effectively inserted in salmonids, which was able to reduce the infection by *A. salmonicida*, *V. ordalli* and *Yersinia ruckeri* [56]. In another study, *Lactobacillus fructivorans* isolated from seabream gut, considerably enhanced the viability of seabream larvae and enhanced the immunity. The application of the potentials of the LAB for salmonids, as pre- and probiotic was reviewed by Merrifield et al. [43]. The knowledge of probiotic in aquaculture is in its early stages and hence, there is an actual concern in producing feed with advantageous influence on fish.

13.4 SUMMARY

In recent years, seafood consumption has increased considerably. However, contamination by microorganism leads to health risk for the consumer. Seafood spoilage can be reduced by appropriate storage, handling, and by following good manufacturing practices. Time, temperature, and hygiene are other factors that play a significant role in the processing of seafoods. The spoilage of seafood can be reduced by the quick process, appropriate freezing and cleanliness (good hygiene) in processing operations.

KEYWORDS

- abdominal cramps
- aquaculture
- biotoxin
- clam
- crab
- histamine
- immunomodulatory
- lactic acid bacteria
- lobster
- marine foods
- mussel
- omega-3 fatty acids
- oyster
- probiotic
- seafood
- shrimp
- spoilage
- Vibrio spp.
- virus

REFERENCES

1. AMEC (2003). *Management of Wastes From Atlantic Seafood Processing Operations*. AMEC Earth and Environmental Limited, Dartmouth, Nova Scotia, Canada., http://aczisc.dal.ca/nparpt.pdf.

2. Amoah, C., Brown, C., Ofori-Danson, P. K., & Odamtten, G. T., (2011). Microbiological quality of the Ghanaian volta clam (*Galatea paradoxa*). *Journal of Food Safety, 31*(2), 154–159.

3. Amos, B., (2007). *Analysis of Quality Deterioration at Critical Steps/Points in Fish Handling in Uganda and Iceland and Suggestions for Improvement*. United Nations University, Uganda. http://www.unuftp.is/static/fellows/document/amos 06prf.pdf

4. Ashie, I. N. A., Smith, J. P., & Simpson, B. K., (1996). Spoilage and shelf life extension of fresh fish and shellfish. *Critical Reviews in Food Science and Nutrition, 36*(1/2), 87–21.

5. Bigoraj, E., Kwit, E., Chrobocińska, M., & Rzeżutka, A., (2014). Occurrence of norovirus and hepatitis A virus in wild mussels collected from the Baltic Sea. *Food and Environmental Virology, 6*(3), 207–212.

6. Bryan, F. L., (1980). Epidemiology of foodborne diseases transmitted by fish, shellfish and marine crustaceans in the United States, 1970–1978. *Journal of Food Protection, 43*(11), 859–876.

7. Cao, R., Xue, C-H., Liu, Q., & Xue, Y., (2009). Microbiological, chemical, and sensory assessment of Pacific oysters (*Crassostrea gigas*) stored at different temperatures. *Czech Journal of Food Sciences, 27*(2), 102–108.

8. CDC, (2017). *Home Canning and Botulism - Keeping Your Family Safe From Botulism*. https://www.cdc.gov/features/homecanning/index.html accessed on Jun 19.

9. Choi, K. S., & Park, K. I., (1997). Report on the occurrence of *Perkinsus* sp. in the Manila clam, Ruditapesphilippinarum, in Korea. *Korean Journal of Aquaculture, 10*, 227–237.

10. Combes, C., (1996). Parasites, biodiversity and ecosystem stability. *Biodiversity & Conservation, 5*(8), 953–962.

11. Crane, M., & Hyatt, A., (2011). Viruses of fish: An overview of significant pathogens. *Viruses, 3*(11), 2025–2046.

12. CSTE, (2006). *CSTE Position Statement: National Reporting for Non- Cholera Vibrio Infections*. CSTE, Atlanta, GA. http://c.ymcdn.com/sites/www.cste.org/resource/resmgr/PS/06-ID-05FINAL.pdf, Accessed on, September.

13. Dang, S. T., & Dalsgaard, A., (2012). *Escherichia coli* contamination of fish raised in integrated pig-fish aquaculture systems in Vietnam. *Journal of Food Protection, 75*(7), 1184–1358.

14. Daniels, J. L., Longnecker, M. P., Rowland, A. S., & Golding, J., (2004). Fish intake during pregnancy and early cognitive development of offspring. *Epidemiology, 15*(4), 394–402.

15. Daviglus, M. L., Stamler, J., Orencia, A. J., Dyer, A. R., Liu, K., Greenland, P., Walsh, M. K., Morris, D., & Shekelle, R. B., (1997). Fish consumption and the 30-year risk of fatal myocardial infarction. *The New England Journal of Medicine, 336*(15), 1046–1053.

16. Desenclos, J. C., Klontz, K. C., Wilder, M. H., Nainan, O. V., Margolis, H. S., & Gunn, R. A., (1991). A multistate outbreak of hepatitis A caused by the consumption of raw oysters. *American Journal of Public Health, 81*(10), 1268–1272.

17. Ebringer, L., Ferencik, M., & Krajcovica, J., (2008). Beneficial health effects of milk and fermented dairy products. *Food Microbiology, 53*(5), 378–394.

18. EL Bour, M., Dellali, M., Boukef, I., Lakhal, F., Mraouna, R., El Hili, H. A., et al., (2012). First assessment of Perkinsosis and brown ring disease co-infection in Ruditapes *decussatus* in the North Lake of Tunis (southern Mediterranean Sea). *Journal of the Marine Biological Association of the United Kingdom, 92*(7), 1579–1584.

19. Eze, E. I., Echezona, B. C., & Uzodinma, E. C., (2011). Isolation and identification of pathogenic bacteria associated with frozen mackerel fish (*Scomberscombrus*) in a humid tropical environment. *African Journal of Agricultural Research, 6*(7), 1918–1922.

20. FAO, (2014). *The State of World Fisheries and Aquaculture 2014.*

21. FDA, (2001). *Fish and Fisheries Products Hazards and Controls Guide* (3ʳᵈ edn.). FDA, Rockville, MD.

22. FDA, (2006). Posting date. *Fresh and Frozen Seafood: Selecting and Serving it Safely.* FDA, Rockville, MD. http://www.fda.gov/Food/ResourcesForYou/Consumers/ucm077331.htm.

23. Fewtrell, M. S., Aboott, R. A., Kennedey, K., Singhal, A., Morley, R., Caine, E., et al., (2004). Randomized, double-blind trial of long-chain polyunsaturated fatty acid supplementation with fish oil and borage oil in preterm infants. *Journal of Pediatrics, 144*(4), 471–479.

24. Furuta, T., Akiyama, M., & Nishiya, O., (2003). A food poisoning outbreak caused by purple Washington clam contaminated with norovirus (Norwalk–like virus) and hepatitis A virus. *KansenshogakuZasshi, 77*(2), 89–94.

25. Gangarosa, E. J., Bisno, A. L., Eichner, E. R., Treger, M. D., Goldfield, M., DeWitt, W. E., et al., (1968). Epidemic of febrile gastroenteritis due to *Salmonella java* traced to smoked whitefish. *American Journal of Public Health, 58*(1), 114–121.

26. Ginson, J., Kamalakanth, C. K., Bindu, J., Venkateswarlu, R., Das, S., Chauhan, O. P., et al., (2013). Changes in K value, microbiological and sensory acceptability of high pressure processed Indian white prawn (*Fenneropenaeusindicus*). *Food and Bioprocess Technology, 6*(5), 1175–1180.

27. Gram, L., & Dalgaard, P., (2002). Fish spoilage bacteria– problems and solutions. *Current Opinion in Biotechnology, 13*(3), 262–266.

28. Gram, L., & Huss, H. H., (2000). Fresh and processed fish and shellfish. In: Lund, B. M., Baird-Parker, A. C., & Gould, G. W., (eds.), *The Microbiological Safety and Quality of Food* (pp. 472–506). Chapman and Hall, London, U. K.

29. Gutierrez, J. L., Jones, C. G., Strayer, D. L., & Iribarne, O. O., (2003). Mollusks as ecosystem engineers: Their functional roles as shell producers in aquatic habitats. *Oikos, 101*(1), 79–90.

30. Hlady, W. G., & Klontz, K. C., (1995). The epidemiology of *Vibrio* infections in Florida, 1981–1993. *Journal of Infectious Diseases, 173*(5), 1176–1183.

31. Huisint, V. J. H. J., (1996). Microbial and biochemical spoilage of foods: An overview. *International Journal of Food Microbiology, 33*(1), 1–18.

32. Huq, A., Huq, S. A., Grimes, D. J., O'brien, M., Chu, K. H., & Capuzzo, J. M., (1986). Colonization of the gut of the blue crab (*Callinectessapidus*) by *Vibrio cholerae*. *Applied and Environmental Microbiology, 52*(3), 586–588.

33. Kawahara, N., Shigematsu, K., Miyadai, T., & Kondo, R., (2009). Comparison of bacterial communities in fish farm sediments along an organic enrichment gradient. *Aquaculture, 287*(1/2), 107–113.

34. Khalil, M. I., El-Shahawy, I. S., & Abdelkader, H. S., (2014). Studies on some fish parasites of public health importance in the southern area of Saudi Arabia. *Brazilian Journal of Veterinary Parasitology, 23*(4), 435–442.

35. Kris-Etherton, P. M., Harris, W. S., & Appel, L. J., (2002). Fish consumption, fish oil, omega-3 fatty acids, and cardiovascular disease: *AHA scientific statement. Circulation, 106*(21), 2747–2757.

36. Kromhout, D., Keys, A., Aravanis, C., Buzina, R., Fidanza, F., Giampaoli, S., et al., (1989). Food consumption patterns in the nineteen sixties in Seven Countries. *The American Journal Clinical Nutrition, 49*(5), 889894.

37. Laruelle, F., Molloy, D. P., & Roitman, V. A., (2002). Histological analysis of trematodes in *Dreissenapolymorpha:* Their location, pathogenicity, and distinguishing morphological characteristics. *Journal of Parasitology, 88*(5), 856–863.

38. Lehane, L., & Olley, J., (2000). Histamine fish poisoning revisited. *International Journal of Food Microbiology, 58*(1/2), 1–37.

39. Leu, J. H., Yang, F., Zhang, X., Xu, X., Kou, G. H., & Lo, C. F., (2009). Whispovirus. *Current Topics in Microbiology and Immunology, 328*, 197–277.

40. Lévesque, B., Barthe, C., Dixon, B. R., Parrington, L. J., Martin, D., Doidge, B., Proulx, J. F., & Murphy, D., (2010). Microbiological quality of blue mussels (*Mytilus edulis*) in Nunavik, Quebec: A pilot study. *Canadian Journal of Microbiology, 56*(11), 968–977.

41. Masniyom, P., (2011). Deterioration and shelf-life extension of fish and fishery products by modified atmosphere packaging. *Journal of Science and Technology, 33*(2), 181–192.

42. McLaughlin, J. B., Sobel, J., Lynn, T., Funk, E., & Middaugh, J. P., (2004). Botulism type E outbreak associated with eating a beached whale, Alaska. *Emerging Infectious Diseases, 10*(9), 1685–1687.

43. Merrifield, D. L., Dimitroglou, A., Foey, A., Davies, S. J., Baker, R. T. M., Bogwald, J., Castex, M., & Ringo, E., (2010). The current status and future focus of probiotic and prebiotic applications for salmonids. *Aquaculture, 302*(1/2), 1–18.

44. Morris, J. G., & Black, R. E., (1985). Cholera and other vibrioses in the United States. *The New England Journal of Medicine, 312*(6), 343–350.

45. Morse, D. L., Guzewich, J. J., Hanrahan, J. P., Stricof, R., Shayegani, M., Deibel, R., et al., (1986). Widespread outbreaks of clam- and oyster-associated gastroenteritis. Role of Norwalk virus. *The New England Journal of Medicine, 314*(11), 678–681.

46. Nakamura, H., Hatanaka, M., Ochi, K., Nagao, M., Ogasawara, J., Hase, A., Kitase, T., Haruki, K., & Nishikawa, Y., (2004). *Listeria monocytogenes* isolated from cold smoked fish products in Osaka city, Japan. *International Journal of Food Microbiology, 94*(3), 323–328.

47. Okuzumi, M., Fukumoto, I., & Fuji, T., (1990). Changes in bacteria flora and polyamine contents during storage of horse mackerel meat. *Nippon Suisan Gakkaishi, 56*(8), 1307–1312.

48. Olsen, S. J., Aucott, J. N., & Swerdlow, D. L., (2002). Food poisoning. In: Blaser, M., Smith, P., Ravdin, J., Greenberg, H., & Guerrant, R., (eds.), *Infections of the Gastrointestinal Tract* (2nd edn., pp. 199–214). Lippincott Williams and Wilkins, Philadelphia, PA.

49. Orban, E., Di Lena, G., Nevigato, T., Casini, I., Santaroni, G., Marzetti, A., & Caproni, R., (2002). Quality characteristics of sea bass intensively reared and from lagoon as affected by growth conditions and the aquatic environment. *Journal of Food Science, 67*(2), 542–546.

50. Osterholm, M. T., Forfang, J. C., Ristinen, T. L., Dean, A. G., Washburn, J. W., Godes, J. R., Rude, R. A., & McCullough, J. G., (1981). An outbreak of foodborne giardiasis. *The New England Journal of Medicine, 304*(1), 24–28.

51. Phan, V. T., Ersboll, K. A., Nguyen, V. K., Madsen, H., & Dalsgaard, A., (2010). Farm-level risk factors for fish-borne zoonotic trematode infection in integrated small-scale fish farms in North Vietnam. *PLOS Neglected Tropical Diseases*, *4*(7), 742.

52. Rehnstam-Holm, A. S., & Hernroth, B., (2005). Shellfish and public health: A Swedish perspective. *Ambio*, *34*(2), 139–144.

53. Ringo, E., (2008). The ability of carnobacteria isolated from fish intestine to inhibit growth of fish pathogenic bacteria: A screening study. *Aquaculture Research*, *39*(2), 171–180.

54. Ringo, E., & Gatesoupe, F. J., (1998). Lactic acid bacteria in fish: A review. *Aquaculture*, *160*(3/4), 177–203.

55. Ringo, E., & Strom, E., (1994). Microflora of arctic char, *Salvelinusalpinus* (L.), gastrointestinal microflora of free-living fish, and effect of diet and salinity on the intestinal microflora. *Aquaculture Research*, *25*(6), 623–629.

56. Robertson, P. A. W., O'Dowd, C., Burrels, C., Williams, P., & Austin, B., (2000). Use of *Carnobacterium* sp. as a probiotic for Atlantic salmon (*Salmo salar* L.) and rainbow trout (*Oncorhynchus mykiss*, Walbaum). *Aquaculture, 185*(3/4), 235–243.

57. Ruano, F., & Cachola, R., (1986). Outbreak of a severe epizootic of *Perkinsus marinus* (Levin-78) at Ria de Faro clam's culture beds. *Proc 2nd Int. Colloq. Pathol. Mar. Aquac. (PAMAQ II)* (pp. 41, 42), Oporto, Portugal.

58. Ruttenbur, A. J., Weniger, B. G., Sorvillo, F., Murray, R. A., & Ford, S. A., (1984). Diphyllobothriasis associated with salmon consumption in Pacific Coast states. *American Journal of Tropical Medicine and Hygiene*, *33*(3), 455–459.

59. Samia, S., Galib, H. T., Tanvir, A. S., Basudeb, C. S., Md. Walliullah, T. A., et al., (2014). Microbiological quality analysis of shrimps collected from local market around Dhaka city. *International Food Research Journal*, *21*(1), 33–38.

60. Sato, Y., Wada, H., Horita, H., Suzuki, N., Shibuya, A., Adachi, H., Kato, R., Tsukamoto, T., & Kumamoto, Y., (1995). Dopamine release in the medial preoptic area during male copulatory behavior in rats. *Brain Research*, *692*(1/2), 66–70.

61. Suleria, H. A. R., Masci, P. P., Addepalli, R., Chen, W., Gobe, G. C., & Osborne, S. A., (2017). *In vitro* anti-thrombotic and anti-coagulant properties of blacklip abalone (Haliotisrubra) viscera hydrolysate. *Anal. Bioanal. Chem., 409*(17), 4195–4205.

62. Tall, B. D., Fall, S., Pereira, M. R., Ramos-Valle, M., Curtis, S. K., Kothary, M. H., et al., (2003). Characterization of *Vibrio fluvialis*-like strains implicated in limp lobster disease. *Applied and Environmental Microbiology*, *69*(12), 7435–7446.

63. Tranter, H. S., (1990). Foodborne staphylococcal illness. *Lancet*, *336*(8722), 1044–1046.

64. Tu, C., Huang, H. T., Chuang, S. H., Hsu, J. P., Kuo, S. T., Li, N. J., Hsu, T. L., Li, M. C., & Lin, S. Y., (1999). Taura syndrome in Pacific white shrimp Penaeus vannamei cultured in Taiwan, *Diseases of Aquatic Organisms, 38*, 159–161.

65. Vareltzis, K., (1996). Mussels as food. *Fishing News*, *11*, 38–47.

66. Villalba, A., Reece, K. S., Ordas, M. C., Casas, S. M., & Figueras, A., (2004). Perkinsosis in mollusks: A 1681 review. *Aquatic Living Resources*, *17*, 411–432.

67. Vlak, J. M., Bonami, J. R., Flegel, T. W., Kou, G. H., Lightner, D. V., Lo, C. F., et al., (2005). Family nimaviridae, In: Fauquet, C. M., Mayo, M. A., Maniloff, J., Desselberger, U., & Ball, L. A., (eds.), *Virus Taxonomy: VIIIth Report of the International Committee on Taxonomy of Viruses* (pp. 187–192). Elsevier, Academic Press, London.

68. Wainwright, R. B., Heyward, W. L., Middaugh, J. P., Hatheway, C. L., Harpster, A. P., & Bender, T. R., (1988). Food-borne botulism in Alaska, 1947–1985: Epidemiology and clinical findings. *Journal of Infectious Diseases*, *157*(6), 1158–1162.

69. Wittner, M., Turner, J. W., Jacquette, G., Ash, L. R., Salgo, M. P., & Tanowitz, H. B., (1989). Eustongylidiasis—a parasitic infection acquired by eating sushi. *The New England Journal of Medicine, 320*(17), 1124–1126.

70. Yatsunami, K., & Echigo, T., (1992). Occurrence of halotolerant and halophilic histamine-forming bacteria in red meat fish products. *Bulletin of the Japanese Society of Scientific Fisheries, 58,* 515–520.

71. Zhang, N. Z., Zhang, L. F., Jiang, Y. N., Zhang, T., & Xia, C., (2009). Molecular analysis of spring viraemia of carp virus in China: A fatal aquatic viral disease that might spread in East Asian. *PLoS ONE, 4*(7), 6337.

CHAPTER 14

MARINE GENOMICS: AN EMERGING NOVEL TECHNOLOGY

KARUNA SINGH and MONIKA THAKUR

ABSTRACT

Genomic approach in marine science helps to create an understanding among the relationships between genome structure, function, and its biotic and abiotic environments, which get revolutionized due to numerous advances at different biological levels. It provides comprehensive information and rapid investigation from ecosystems down to genes. Various functional marine genomics approaches are: metagenomics, metatranscriptomics, and proteomics. Environmental monitoring uses various methods like mass-sequencing of seawater samples, monitoring the abundance of microplankton species, etc. Several compounds extracted from marine have been used by pharmaceutical and food industries as thickeners, proteins or bioactive molecules. Marine organism synthesizes chemicals with bioactive components such as: proteins, enzymes, polysaccharides, and lipid and can lead to new industrial processes. Marine genomics has an application to produce a valuable commercial product having various structural and functional properties. Various marine resources have been exploited by food, health, and pharmaceutical industries to produce various products and components of their use. Therefore, this chapter summarizes the importance, scope, and potential use of marine genomics.

14.1 INTRODUCTION

Genomics deals with the study of genomes, chromosomes, genes and polymer information of an organism [3]. It is a branch of biology that deals with the structure, the function of genomes of organisms, which keep on changing continuously through events like genome rearrangements, duplications until their evolution. These evolutions increase the usefulness and functioning of genetic components [26].

The ability to analyze and understand the relationship of the genome struc-ture, function, and its biotic and abiotic environment get revolutionized due to numerous advances at different biological levels [34]. At the DNA level, the development of sequencing of DNA and using advanced technologies and techniques cause a great enhancement [23]. Using digital gene expression technology and microarray at the gene expression level ("transcriptomics") and tandem mass spectrometry for the analysis of protein ("proteonomics") help to understand genome structure and functions; and advanced techniques like pyrolysis gas chromatography and infrared spectroscopy have been used at the metabolic level to study the detailed cellular constituents ("metabolo-mics"). Three other technological advances, i.e., microtechnology, computing, and communication are associated with the genomics revolution [30]. The application of genomics has three salient points:

1) Initially, genomics field was initiated and developed based on studies of model organisms like *Saccharomyces cerevisiae*, *Drosophila melanogaster*, etc. Today the use of improved ecologically-well characterized species has been allowed. Laboratory-based model clarified the basic processes of growth and development and their response to ecosystem and environmental changes. Marine DNA sequence data along with sequence libraries have increased the understanding and illustration of marine texa [34].

2) "Metagenomics" has been used for the isolation and sequencing of DNA directly from the environment [20].

3) The enhanced awareness has shifted its focus of environmental management to a functional relationship between biotic diversity and ecosystem processes. This helps in application and use of genomics in various fields like pharmaceuticals, food, and nutrient supple-ment industry. This allows discoveries and development of various commercial products. It is very difficult to exploit marine microorgan-isms to produce new products as it is difficult to keep them in cultures. Metagenomics libraries give insight into physiological mechanisms and act as a powerful tool for their use as biocatalytic enzymes. As a result, increased knowledge of their genome can aid in gene discovery. This knowledge has an extensive versatility for industrial application.

Marine environment provides a good percentage of oxygen and seques-tering about 50% of the CO_2. The consumption of aquatic food has globally increased in the recent decade because of the increased understanding of their health benefits and functional properties. Marine resources with nutraceutical value have been utilized as a functional ingredient. Food industry and health sectors are utilizing various products and by-products derived from marine organisms. Marine organism synthesizes chemicals with bioactive components such as: proteins, enzymes, polysaccharides, and lipid and can lead to new industrial processes.

14.2   APPLICATIONS OF GENOMICS IN HEALTHCARE AND DISEASE CONTROL

Genomics has a significant role in the creation of new drugs. This field provides a better understanding of the genes and its association and relation with various diseases like diabetes, cancer, heart diseases, etc. Genomics provides the application of genetic information for the identification of genes, which are pre-disposed and susceptible to the disease and information of genes that are defective and the process of their effect on the human body.

Most of our medicines come from natural resources and scientists are still exploring the organisms. Various commercially available products from marine origin include: antibiotic cephalosporin from marine fungi, cytostatic cytarabine from the sponge, anthelmintic insecticide kanic acid from red algae, and analgesic zincototide from mollusk [27]. Some products like blood clotting compounds from cone snail, inflammatory ointment from a sea sponge, anticancer substances and disinfectant from shark are under development. Various enzyme inhibitors have been utilized for the study of enzyme structure and reaction mechanism and are being used in the field of pharmaceutical and in agriculture and food industry [9]. Along with this, marine microbe having immense genetic and biochemical diversity are used as a naval and effective drug, e.g. actinomycetes is used to produce antibiotic as well as some bioactive compounds. Research studies during 1998–2006 have reported about 592 marine compounds having anti-tumor and cytotoxic activity, 666 additional chemicals depicting various functional properties like anti-bacterial, anti-coagulant, anti-inflammatory, anti-fungal, anti-viral that have various health benefits [13].

Genomics has brought a new era in controlling infectious diseases. New drugs and agents like anti-viral, anti-bacterial, anti-inflammatory, antibiotics were formulated utilizing the information generated from pathogen genes, their expression and their interaction in the body. Similarly, the genome of marine plants and animals will provide additional benefits like various immunization products to replace the vaccine. [28].

The oceans are now becoming the focus of research. The microbial flora of the deep sea has barophiles and bacteria, which thrive under high pressures (approximately 600 atmospheres). These marine-based microorganisms generate various bioactive compounds that inhibit colon tumor cell growth and prevent replication of viruses [36].

14.3 MARINE FUNCTIONAL INGREDIENTS AND APPLICATIONS IN FOOD SECTOR

Marine resources like a marine plant, sponges, and microorganisms are very rich in nutraceutical value and are used as functional ingredients. They contain many unique biomolecules [15]. Another growing source of marine-based food ingredients has been fish and seafood by-products, which include types of oils like fish oil, algal oil, and shark liver oil. Various products include: shark cartilage, chitin, various enzymes, vitamins, protein hydrolysates and isolates that have been were used in pharmaceutical and food industries.

Various marine functional ingredients, its health benefits, and potential food applications have been investigated. Proteins like collagen, gelatin, albumin, crustaceans, bioactive peptides derived from crustaceans have various functional properties and were utilized by industry for their unique properties like film formation, foaming property, antioxidant, anticoagulant, and antimicrobial activity. Lipids and pigments derived from microalgae and fungi are utilized for the prevention of various diseases. Food industry uses protamine as a natural antibacterial preservative. Marine polysaccharides such as alginis, carrageenans, agar, fucans, hydrocolloids have been utilized by the beverage industry. These molecules have varied biological functions and properties like anti-inflammatory activity, anti-bacterial, etc.

Production of various functional ingredients and bioactive compounds like enzymes using advances in biotechnological tools provides various useful materials and products to the food industry. Enzymes isolated from marine biomass like lipase, polyphenol oxidase, trans-glutamase, etc. were involved in various types of processes like starch degradation by the various food industry. While other enzymes like alkaline phosphate, hyaluronidase, acetylglucosaminidase, chitinase, and protease isolated from fish have been used by food sector for manufacturing of various products and like cheese, fish sauce, and fish protein hydrolysates and concentrate.

Various new and innovative use of underutilized marine organism especially fish biomass is utilized and converted to fish protein hydrolysates. These protein hydrolates have rich nutritional and pharmaceutical applications. Fish protein hydrolates show various functional properties, which have been utilized by the food sector for providing holding, gelling, whipping, and emulsifying of the food products.

Seaweeds are a rich source of iodine, glutathione, and carbohydrates such as alginates. Various algal polysaccharides are of industrial importance. Glucosamine is present in connective and cartilage tissues as a component of glycosaminoglycans. Along with this, various natural amino polysaccharides like chitin and chitosan polymers have unique structure and properties, which have been utilized by the food industry.

14.4 APPLICATIONS IN NUTRIENT AND NUTRITIONAL SUPPLEMENTS

Marine ecosystems comprising of the diversity of living organisms are a source of many biocomponents required for health, prevention of diseases and maintaining the human body in a good state of nutrition. They are dietary supplements, functional food, and nutraceuticals. Microalgae are better sources of various micronutrients like Vitamin A, ascorbic acid, Vitamin E, Thiamine, Riboflavin, Pyridoxine and B_{12}, and various macronutrients like carbohydrate (polysaccharide) and lipids (polyunsaturated fatty acids) [6, 12, 29, 35]. These microalgae are used by the food industry as food additives. These are incorporated into food products like infant milk formulations and dietary supplements to fulfill various nutrient needs of different age groups [32]. Likewise, macroalgae comprise of various components like proteins, lipids like polyunsaturated fatty acids, pigments, minerals, insoluble, and soluble dietary fiber [21]. Various components like peptides, ether, steroids, phenolic compounds terpenoids, and alkaloids can be derived from invertebrates like sponges, mollusks, and crustaceans [7].

Marine fish is consumed because of various nutritional benefits. They are rich in various nutrients like proteins, PUFA, omega-3 fatty acids. They provide a good amount of minerals and vitamins. Research has shown the therapeutic usage of fish for treating many diseases [14, 33].

Marine oils extracted from various types of fish, fungi, microalgae, macroalgae, and krill are used for the treatment of various neuro related and vision-related diseases and help in reducing the prevalence of risk of various lifestyle-related diseases like hypertension and cardiovascular problems. The non-protein nitrogenous compound, such as taurine, is also derived from fish cut fractions along with protein, peptides, and lipid extraction.

Marine Phytoplankton is a very rich source of various nutrients (Table 14.1). For example, it contains 9 amino acids that the body cannot make. Along with this, they are a rich source of various essential fatty acids (ω-3 and ω-6), water-soluble and fat-soluble vitamins and major and trace minerals. Hence Marine Phytoplanktons are utilized for the preparation of

TABLE 14.1 Sources, Applications, and Health Perspectives of Marine-Derived Nutrients

Nutrient	Bioactive molecules	Major marine sources	Health Perspectives	Reference
Protein and Peptides	Collagen	Fish like tuna, grunt, yellow sea bream, and toadfish.	Various functional properties like anti-oxidant, anti-aging antihypertensive.	[11, 18]
	Gelatin	Cold water fish, like Pollock, cod, and cusk	Prevent and treat chronic gastritis and gastrointestinal problems	[5]
	Albumin	Mollusks, crustaceans, fish with low fat	Anti-coagulant and anti-oxidant properties	[17]
Poly-Saccharides	Carrageenan	Macroalgae	Anti-coagulant properties and anti-HIV property	[31]
	Agar agar	Red Alga like *Gelidium*, Gigartina	Antimicrobial, anti- inflamatory, anticancer	[4]
	Fucans and fucanoids	Brown algae cell wall, and cell wall of sea cucumbers	Shows various functional properties like anti-coagulant, anti-viral, anti-thrombotic, and anti-inflammatory	[1]
	Chitin, chitosan derivatives	Shrimp, crab, lobster, prawn, krill, crab, lobster	Good source of dietary fiber, reduce lipid absorption and depict properties like anti-tumor, anti-bacterial, and antifungal activity	[24]
Fatty acids	ω-3 and ω-6 fatty acids	Almost all marine sources	Health benefits like reduction in Cardiovascular diseases, diseases of the eye, nerve-related diseases arthritis, etc.	[25]
Vitamins and Minerals	Fat soluble and water-solublevit, macro and trace minerals	Almost all marine sources, Seaweeds	They are regulatory nutrients and help in regulating various functions in the body like transport of material across the cell and as coenzymes in metabolic processes	[19]

dietary supplements [2]. Certain marine microalgae (blue-green algae) have many vitamins in higher concentrations also used as food supplements. Ingestion of relatively small quantities of these marine resources can fulfill many macro and micronutrients and can be utilized to combat various nutrition-related diseases.

14.5 SUMMARY

This chapter begins with the discussion of Genomics and Genomic approaches that insight into the biodiversity of the marine environment. Various functional marine genomics approaches and its applications in health care and disease control were discussed. Marine functional ingredients with various bioactive components such as proteins, enzymes, polysaccharides, and lipid and its application in the food sector have been discussed. The marine genomics in the production of valuable products has been emphasized in this chapter. The utilization of various nutrients and nutritional supplements from the available knowledge is the next point discussed in the chapter. The chapter finally concluded with the discussion on the importance and potential use of marine genomics in health care and nutrient supplements.

KEYWORDS

- **functional food**
- **genomics**
- **glycosaminoglycan**
- **immunization**
- **marine organism**
- **metabolomics**
- **metagenomics**
- **microalgae**
- **omega 3 fatty acids**
- **polysaccharides**
- **protein hydrolytes**
- **proteomics**
- **transcriptomics**

REFERENCES

1. Berteau, O., & Mulloy, B., (2003). Sulfated fucans, fresh perspectives: Structures, functions, and biological properties of sulfated fucans and an overview of enzymes active toward this class of polysaccharide. *Glycobiology*, *13*, 29R–40R.
2. Castle, D., Cline, C., Daar, A. S., Tsamis, C., & Singer, P. A., (2007). *Science, Society, and the Supermarket: The Opportunities and Challenges of Nutrigenomics* (p. 163), John Wiley & Sons Inc. Hoboken, NJ.
3. Chandonia, J. M., & Brenner, S. E., (2006). The impact of structural genomics: Expectations and outcomes. *Science*, *311*(5759), 347–351.
4. Freile- Pelegrín, Y., & Murano, E., (2005). Agars from three species of *Gracilaria* (Rhodophyta) from the Yucatán Peninsula. *Bioresour. Technol.*, *96*, 295–302.
5. Gomez-Guillen, M. C., Turnay, J., Fernandez-Dıaz, M. D., Olmo, N., Lizarbe, M. A., & Montero, P., (2002). Structural and physical properties of gelatin extracted from different marine species: A comparative study. *Food Hydrocoll.*, *16*, 25–34.
6. Grobbelaar, J. U., (2004). Algal biotechnology: Real opportunities for Africa. *South Afr. J. Bot.*, *70*, 140–144.
7. Hu, G. P., Yuan, J., Sun, L., She, Z. G., Wu, J. H., Lan, X. J., et al., (2011). Statistical research on marine natural products based on data obtained between 1985 and 2008. *Mar. Drugs*, *9*, 514–525.
8. Hurst, D., (2006). Marine functional foods and functional ingredients. *Marine Foresight Series*, *5*, 17–20.
9. Imada, C., (2004). Enzyme inhibitor of marine origin with pharmaceutical importance. *Mar. Biotechnol.*, *6*(1), 193–198.
10. Joshua, L., & Alexa, T., (2001). Omics - A genealogical treasury of words. *The Scientist*, *15*, 7–9.
11. Lai, G., Yang L., & Guoying, L., (2008). Effect of concentration and temperature on the rheological behavior of collagen solution. *Int. J. Biol. Macromol.*, *42*, 285–291.
12. Luiten, E. E., Akkerman, I., Koulman, A., Kamermans, P., Reith H., Barbosa, M. J., et al., (2003). Realizing the promises of marine biotechnology. *Biomol. Eng.*, *20*, 429–439.
13. Mayer, A., Glaser, K. B., Cuevas, C., Jacob, R. S., Kern, W., Little, R. D., et al., (2010). The odyssey of marine pharmaceuticals: A current pipeline perspective. *Trends Pharmacol Sci.*, *31*, 255–265.
14. Mayer, A. M. S., Rodriguez, A. D., Berlinck, R. G. S., & Hamann, M. T., (2009). Marine pharmacology in 2005–2006: Marine compounds with anthelmintic, antibacterial, anticoagulant, antifungal, anti-inflammatory, antimalarial, antiprotozoal, antituberculosis, and antiviral activities, affecting the cardiovascular, immune and nervous systems, and other miscellaneous mechanisms of action. *Biochem. Biophys. Acta*, *1790*, 283–308.
15. Miyashita, K., (2009). Function of marine carotenoids. *Forum Nutr.*, *61*, 136–146.
16. Ngo, D., Wijesekara, I., Vo, T., Ta, Q. V., & Kim, S., (2011). Marine food-derived functional ingredients as potential antioxidants in the food industry: An overview. *Food Res. Int.*, *44*, 523–529.
17. Nicholson, J. P., Wolmarans, M. R., & Park, G. R., (2000). The role of albumin in critical illness. *Br. J. Anaesth.*, *85*, 599–610.
18. Noitup, P., Garnjanagoonchorn W., & Morrissey, M. T., (2005). Fish skin type I collagen. *J. Aquat. Food Prod. Technol.*, *14*, 17–28.

19. Parr, R. M., Aras, N. K., & Iyengar, G. V., (2006). Dietary intakes of essential trace elements: Results from total diet studies supported by the IAEA. *J. Radioanal. Nucl. Chem., 270*, 155–161.

20. Peers, G., & Price, N. M., (2006). Copper-containing plastocyanin used for electron transport by an oceanic diatom, *Nature, 441*, 341–344

21. Ramirez, J. C., & Morrissey, M. T., (2003). *Marine Biotechnology* (p. 214). First joint trans-Atlantic fisheries technology conference (TAFT), Reykjavik, Iceland.

22. Rasmussen, R. S., & Morrissey, M., (2007). Marine biotechnology for production of food ingredients. *Adv. Food Nutr. Res., 52*, 237–292

23. Ruthberg, J. M., & Leamon, J. H., (2008). The development and impact of 454 sequencing. *Nat Biotechnol., 26*(10), 117–124

24. Shahidi F., & Abuzaytoun, R., (2005). Chitin, chitosan, and co-products: Chemistry, production, applications, and health effects. *Adv. Food Nutr. Res., 49*, 93–135.

25. Sijtsma, L., & De Swaaf, M. E., (2004). Biotechnological production and applications of the omega-3 polyunsaturated fatty acid docosahexaenoic acid. *Appl. Microbiol. Biotechnol., 64*, 146–153.

26. Tang, B., & Lyons, E., (2012). Unleashing the genome of *Brassica rapa. Front. Plant Sci., 3*, 172.

27. Thakur, N. L., & Muller, W. E. G., (2004). Biotechnological potential of marine sponges. *Curr. Sci., 86*, 1506–1512

28. Titilade, P. R., & Olalekan, E. I., (2015). The importance of marine genomics to life. *Journal of Ocean Research, 3*(1), 1–13.

29. Tseng, C. K., (2001). Algal biotechnology industries and research activities in China. *J. Appl. Phycol., 13*, 375–380.

30. Van Straalen, N. M., (2006). Roelof, Dick. *An Introduction of Ecological Genomics* (p. 307). Oxford University Press, Oxford.

31. Vlieghe, P., Clerc, T., Pannecouque, C., Witvrouw, M., De Clercq, E., Salles, J. P., & Kraus, J. L., (2002). Synthesis of new covalently bound kappa-carrageenan-AZT conjugates with improved anti-HIV activities. *J. Med. Chem., 45*, 1275–1283.

32. Volkman, J. K., (2003). Sterols in microorganisms. *Appl. Microbiol. Biotechnol., 60*, 495–506.

33. Wijesekara, I., & Kim, S. K., (2010). Angiotensin-i-converting enzyme (ACE) inhibitors from marine resources: Prospects in the pharmaceutical industry. *Mar. Drugs, 8*, 1080–1093.

34. Wilson, W. H., Schroeder, D. C., & Allen, M. J., (2005). Complete genome sequence and lytic phase transcription profile of a coccolithovirus. *Science, 309*, 1090–1092.

35. Yap, C. Y., & Chen, F., (2001). Polyunsaturated fatty acids: Biological significance, biosynthesis, and production by microalgae and microalgae-like organisms. In: Chen, F., & Jiang, Y., (eds.), *Algae and Their Biotechnological Potential* (pp. 1–32). Kluwer Academic Publishers: Dordrecht, The Netherlands.

36. Yayanos, A., (1995). Microbiology to 10, 500 meters in the deep sea. *Annual. Rev. Microbiol., 49*, 777–805.

GLOSSARY OF TECHNICAL TERMS

Algae refer to a large, diverse group of photosynthetic organisms, which are not necessarily closely related and are thus polyphyletic.

Anaerobic digestion is a collection of processes, where microorganisms break down biodegradable material in the absence of oxygen.

Antibiotics, also called antibacterials, are a type of antimicrobial drug used in the treatment and prevention of bacterial infections.

Antibody (Ab), also known as an immunoglobulin (Ig), is a large, Y-shaped protein produced mainly by plasma cells that are used by the immune system to identify and neutralize pathogens such as bacteria and viruses.

Antimicrobial refers to destroying or inhibiting the growth of microorganisms and especially pathogenic microorganisms

Antioxidant is a substance that inhibits oxidation, especially one used to counteract the deterioration of stored food products.

Aquaculture, also known as aquafarming, is the farming of fish, crustaceans, mollusks, aquatic plants, algae, and other aquatic organisms.

Aroma refers to a characteristic, usually pleasant odor, as of a plant, spice, or food.

Ash content refers to the inorganic residue remaining after the water, and organic matter has been removed by heating in the presence of oxidizing agents, which provides a measure of the total amount of minerals within the food.

Autolysis refers to the destruction of cells or tissues by their own enzymes, especially those released by lysosomes

Bacteriocin is a protein produced by bacteria of one strain and active against those of a closely related strain.

Bioactive component is a compound that has a health effect on a living organism, tissue or cell.

Biochar is a charcoal made from biomass via pyrolysis.

Biogenic amines are biogenic substances with one or more amine groups. They are basic nitrogenous compounds formed mainly by decarboxylation of amino acids or by amination and transamination of aldehydes and ketones.

Biological oxygen demand (BOD) is the amount of dissolved oxygen needed by aerobic biological organisms to break down organic material present in a given water sample at a certain temperature over a specific time period.

Biomagnification, also known as bioamplification or biological magnification, is the increasing concentration of a substance, such as a toxic chemical, in the tissues of organisms at successively higher levels in a food chain.

Biot number detects the nature of major resistance offered to heat transfer when a body is immersed in the fluid. The body could face two resistances: One is convective resistance in the fluid surrounding the body and other is conductive resistance within the body. Biot number is the ratio of the internal resistance to heat transfer in the solid body to external transfer in the fluid.

Biotechnology is the use of living systems and organisms to develop or make products, or "any technological application that uses biological systems, living organisms or derivatives thereof, to make or modify products or processes for specific use."

Brine is a solution of salt and water. It is mainly used as a preservative for vegetables, fish, fruit, and meat through a process known as brining. The high salt content in Brine prevents the growth of bacteria and thus helps to preserve the food for a long time without creating any difference in taste. Brine is widely used all over the world for various food preparations and is best suited for delicacies made from meat and fish

Byproducts are incidental or secondary products made in the manufacture or synthesis of some other products.

Cardiovascular disease (CVD) is a class of diseases that involve the heart or blood vessels.

Cellulase is any of several enzymes produced chiefly by fungi, bacteria, and protozoans that catalyze cellulolysis, the decomposition of cellulose of some related polysaccharides.

Chelation is a chemical process in which a synthetic solution-EDTA (ethylenediaminetetraacetic acid) to remove heavy metals and/or minerals from the fluid.

Chemical oxygen demand (COD) is commonly used to measure the amount of organic compounds in water indirectly.

Chitosanases is an enzyme which catalyzes the hydrolytic degradation of chitosan.

Coagulation (also known as clotting) is the process by which sample changes from a liquid to a gel, forming a precipitate.

Contamination is the presence of an unwanted constituent, contaminant or impurity in a material, physical body, natural environment, or workplace.

Critical size refers to the size below which ice crystal formation is inhibited.

Crustaceans are taxons with the very large number and diverse types of arthropods (organisms) which includes crabs, lobsters, crayfish, shrimp, krill, woodlice, and barnacles.

Cured fish refers to fish which has been cured by subjecting it to fermentation, pickling, smoking, or some combination of these before it is eaten. These food preservation processes can include adding salt, nitrates, nitrite or sugar, can involve smoking and flavoring the fish, and may include cooking it.

Degree of deacetylation (DD) is one of the main parameters to determine whether the biopolymer is chitin or chitosan.

Demersal are marine organisms living on or near the bottom of ocean or lake waters.

Derivatization is a technique which transforms a chemical compound into a product of similar chemical structure, called a derivative.

Diabetes mellitus is a disease characterized by the raised level of blood glucose.

Diadromous fish migrate between freshwater and saltwater. The migration patterns differ for each species and have seasonal and lifecycle variations.

Dietary fiber is the indigestible portion of food adds bulk to the diet. It has many health benefits. It can reduce your risk of heart disease, diabetes, and some cancers, and also help weight control.

Dietary supplement is a product intended for ingestion that contains a "dietary ingredient" intended to add further nutritional value to (supplement) the diet.

Drug delivery refers to approaches, formulations, technologies, and systems for transporting a pharmaceutical compound in the body as needed to achieve its desired therapeutic effect safely.

Empirical Modeling refers to the modeling based on empirical observations rather than the mathematically descriptive relationships of the system modeled.

Enthalpy is the heat content per unit mass of food material. The absolute value of enthalpy cannot be defined easily, and a zero value is arbitrarily defined at -40°C to 0°C or any other convenient temperature. Since the measurement of latent and sensible heats separately in frozen food is difficult, enthalpy gives a good measure of energy in frozen food.

Entrails are animal's intestines or internal organs, especially when removed or exposed.

Enzymes are macromolecular biological catalysts which accelerate, or catalyze, chemical reactions.

Eutectic point is the temperature when the whole system (solute + solvent) starts to freeze together.

Evisceration is the removal of viscera (internal organs, especially those in the abdominal cavity). This can refer to disembowelment, removal of the internal organs of an animal.

Fat oxidation is a process by which the stored, giant lipid molecules are broken back down into their smaller parts, triglycerides, and fatty acids. The larger molecules can't be used as energy by the body, so, they just stack up.

Fermentation is a metabolic process that converts sugar to acids, gases, or alcohol.

Fermented fish is a traditional preparation of fish. Before refrigeration, canning and other modern preservation techniques became available, fermenting was an important preservation method. Fish rapidly spoils or goes rotten, unless some method is applied to stop the bacteria that produce the spoilage.

Fish paste is fish, which has been chemically broken down by a fermentation process until it reaches the consistency of a soft, creamy purée or paste.

Fish sauce is an amber-colored liquid extracted from the fermentation of fish with sea salt. It is used as a condiment in various cuisines.

Flash frying refers to the frying of food for a brief period of time at very high temperatures.

Flavor is the sensory impression of a food or other substance and is determined primarily by the chemical senses of taste and smell.

Flocculation is a process wherein colloids come out of suspension in the form of floc or flake; either spontaneously or due to the addition of a clarifying agent.

Food additives are added to food to preserve flavor or enhance its taste and appearance.

Food spoilage is the process by which food deteriorates to the point in which it is not edible to humans, or its quality of edibility becomes reduced.

Fortification is the process of adding micronutrients(essential trace elements and vitamins) to food.

Freeze drying is a dehydration process to preserve a perishable material or make the material more convenient for transport

Freezing or solidification is a phase transition in which a liquid turns into a solid when its temperature is lowered below its freezing point.

Functional food is a food given an additional function (often one related to health promotion or disease prevention) by adding new ingredients or more of existing ingredients.

Genomics refers to branch of biotechnology concerned with applying the techniques of genetics and molecular biology to the genetic mapping and DNA sequencing of sets of genes or the complete genomes of selected organisms, with organizing the results in databases, and with applications of the data (as in medicine or biology)

Glazing refers to the application of a layer of ice to the surface of a frozen product by spraying; brushing, dipping that confers protection to the product against dehydration and oxidation during cold storage.

Graft polymers are segmented copolymers with a linear backbone of one composite and randomly distributed branches of another composite.

Hakarl is a national dish of Iceland consisting of a Greenland shark (Somniosus microcephalus) or another sleeper shark which has been cured with a particular fermentation process and hung to dry for four to five months. hakarl has a strong ammonia-rich smell and fishy taste.

Halophiles are organisms that thrive in high salt concentrations. They are unique because they require high levels of salt that would be lethal to most organisms.

Handling is the coordination and integration of operations, such as: packaging, unpacking, re-packing, and movement of materials or goods over certain distances.

Harvest is the process of gathering fish/crop from the field or sea.

Immersion freezing refers to the process of immersing in liquid nitrogen (at −320°F) to apply an instant crust-freeze, or to fully freeze diced products or other small items in a flash.

Immune response refers to the reaction of the cells and fluids of the body to the presence of a substance, which is not recognized as a constituent of the body itself.

Latent heat of fusion is the amount of heat released or absorbed at a specific temperature when a unit mass of food material is transformed from a liquid phase to a solid phase.

Lipid is generally considered to be any molecule that is insoluble in water and soluble in organic solvents. Lipids are a group of naturally occurring molecules that include fats, waxes, sterols, fat-soluble vitamins (such as vitamins A, D, E, and K), monoglycerides, diglycerides, triglycerides, phospholipids, and others.

Marine biodiversity means the variability among living organisms from all marine sources and the ecological complexes of which they are a part; this includes diversity within species, between species and of ecosystems.

Marine refers to objects or things relating to the sea or ocean, such as marine biology, marine ecology, and marine geology. It also refers t life forms living in sea or ocean.

Marine toxins are chemicals and microorganisms that can contaminate certain types of seafood.

Marine waste refers to marine debris that has deliberately or accidentally been released in a lake, sea, ocean or waterway.

Melanosis is a condition of excessive production of melanin in the skin or other tissue.

Metabolomics is the scientific study of chemical processes involving metabolites. Specifically, metabolomics is the "systematic study of the unique chemical fingerprints that specific cellular processes leave behind," the study of their small-molecule metabolite profiles. The metabolome represents the collection of all metabolites in a biological cell, tissue, organ or organism, which are the end products of cellular processes.

Microorganism is a single-celled or multicellular microscopic organism, which is not visible to the naked eye.

Mollusks refer to a group of soft-bodied animals that includes snails, clams, and sea slugs which forms the large phylum of invertebrate animals.

Mucoadhesion refers to the adhesion between two materials, at least one of which is a mucosal surface.

Myofibrils are composed of long proteins including actin, myosin, and titin, and other proteins that hold them together and are organized into thick and thin filaments called myofilaments.

Nanoparticle is defined as a small object that acts as a whole unit with respect to its transport and properties.

Nucleation refers to the formation of the cluster by agglomeration of molecules of water, which is strong enough to support the growth of ice crystals.

Nutraceuticals is a broad umbrella term that is used to describe any product derived from food sources with extra health benefits in addition to the basic nutritional value found in foods.

Nutrigenomics is a branch of nutritional genomics and is the study of the effects of foods and food constituents on gene expression.

Omega-3 fatty acids: refers to a class of essential fatty acids found in fish oils, especially from salmon and other cold-water fish, that acts to lower the levels of cholesterol and LDL (low-density lipoproteins) in the blood. (LDL cholesterol is the "bad" cholesterol.)EPA (eicosapentaenoic acid) and DHA (docosahexaenoic acid) are the two principal omega-3 fatty acids.

Opportunistic pathogen is an organism that is capable of causing disease only when the host's resistance is lowered, for example, by other diseases or drugs.

Organic aquaculture is a holistic method for farming marine species in line with organic principles.

Osmosis is the spontaneous net movement of solvent molecules through a semi-permeable membrane into a region of higher solute concentration, in the direction that tends to equalize the solute concentrations on the two sides.

Oxidation is the loss of electrons or an increase in oxidation state by a molecule, atom, or ion.

Pelagic are marine organisms that survive in the pelagic zone of ocean or lake, i.e., neither close to the bottom nor near the shore.

Peptides are biologically occurring short chains of amino acid monomers linked by peptide (amide) bonds.

Pigments are chemical compounds which reflect only certain wavelengths of visible light. This makes them appear "colorful." More important than their reflection of light is the ability of pigments to absorb certain wavelengths.

Polyelectrolytes are water-soluble polymers carrying ionic charge along the polymer chain. Depending upon the charge, these polymers are anionic or cationic.

Polysaccharides are polymeric carbohydrate molecules composed of long chains of monosaccharide units bound together by glycosidic linkages and on hydrolysis give the constituent monosaccharides or oligosaccharides. Three important polysaccharides, starch, glycogen, and cellulose, are composed of glucose.

Prebiotics are substances that induce the growth or activity of microorganisms (e.g., bacteria and fungi) that contribute to the well-being of their host

Preservation refers to the set of activities that aims to prolong the life of a record and relevant metadata, or enhance its value, or improve access to it through non-interventive means.

Probiotic denotes a substance which stimulates the growth of microorganisms, especially those with beneficial properties (such as those of the intestinal flora).

Protease (also called a peptidase or proteinase) is an enzyme that performs proteolysis, that is, begins protein catabolism by hydrolysis of the peptide bonds that link amino acids together in a polypeptide chain.

Protein denaturation refers to the unfolding of the quaternary and tertiary structure of proteins. Due to unfolding, protein loses its primary function.

Proteomics is a branch of biotechnology concerned with applying the techniques of molecular biology, biochemistry, and genetics to analyzing the structure, function, and interactions of the proteins produced by the genes of a particular cell, tissue, or organism, with organizing the information in databases, and with applications of the data.

Pulverization is a process in which a solid substance is reduced to tiny loose particles.

Rakfisk is a Norwegian fish dish made from trout or sometimes char, salted and fermented for two to three months, or even up to a year, then eaten without cooking.

Rancidity refers to the auto-oxidation of double bonds present in fatty acids and lipids, which results in deterioration of flavor, aroma, and appearance of food items. Rancidity could be the result of action or oxygen exposure of enzymes.

Reducing sugar is any sugar possessing a free aldehyde group or a free ketone group and capable of acting as a reducing agent, for example, all monosaccharides.

Rigor mortis is the state of muscle hardening after the death of a fish. This process has three stages: delay phase, onset phase, and finally completion phase.

Seafood is any form of sea life regarded as food by humans and prominently includes fish and shellfish viz., mollusks, crustaceans, and echinoderms.

Seaweed refers to several species of macroscopic, multicellular, marine algae. The term includes some types of red, brown, and green algae.

Sewage is a water-carried waste, in solution or suspension that is intended to be removed from a community.

Smoked fish is a fish that has been cured by smoking. Foods have been smoked by humans throughout history. Originally, this was done as a preservative.

Solid state fermentation (SSF) is the cultivation of microorganisms under controlled conditions in the absence of free water.

Specific heat is the quantity of heat that is gained or lost by a unit of mass of product to accomplish a unit change in temperature, without a change in state.

Spoilage is the mechanism in which food quality deteriorates to a point at which, it is no longer edible by humans.

Stefan Number refers to the ratio of sensible heat to latent heat.

Sterols are a subgroup of the steroids and an important class of organic molecules. They occur naturally in plants, animals, and fungi, with the most familiar type of animal sterol being cholesterol.

Storage is the action or method of storing something for future use.

Supercooling is the cooling of liquid below its freezing point without solidification or crystallization.

Supercritical fluid extraction (SFE) is the process of separating one component from another using supercritical fluid as the extracting solvent.

Surströmming is fermented Baltic Sea herring that has been a staple of traditional northern Swedish cuisine since at least the 16th century. Just enough salt is used to prevent the raw fish from rotting (chemical decomposition).

Synbiotics are referred to as food ingredients or dietary supplements combining probiotics and prebiotics in the form of synergism.

Thaw rigor is a process in which frozen fish shows contraction upon thawing.

Therapeutic effect is a consequence of medical treatment of any kind, the results of which are judged to be desirable and beneficial.

Thermal arrest refers to the cooling curve formed when the material is solidified due to the latent heat of fusion.

Thermal conductivity is the amount of heat that will be conducted per unit time through a unit thickness of the material if a unit temperature gradient exists across that thickness.

Thermal diffusivity measures the ability to store thermal energy. Products of large values will respond quickly to changes in their thermal environment, while materials of small values will respond more slowly. It also determines how rapidly heat front moves or diffuses through a material.

Total suspended solids (TSS) are solids in water that can be trapped by a filter which includes a wide variety of material, such as silt, decaying plant and animal matter, industrial wastes, and sewage.

Ultra-centrifugation refers to the process that utilizes a high-speed centrifuge to sediment colloidal and other small particles so as to determine sizes of such particles and molecular weights of large molecules.

INDEX

Printed and bound by CPI Group (UK) Ltd, Croydon, CR0 4YY

23/10/2024

01777703-0014